SAFETY MANAGEMENT

A
HUMAN APPROACH
SECOND EDITION

SAFETY MANAGEMENT

A
HUMAN APPROACH

SECOND EDITION

DAN PETERSEN

MANAGEMENT CONSULTANT
SAFETY/ORGANIZATIONAL BEHAVIOR

PROFESSIONAL & ACADEMIC PUBLISHER

GOSHEN, NEW YORK

Library of Congress Cataloging-in-Publication Data

Petersen, Dan.
    Safety management.

    Bibliography: p.
    Includes index.
    1. Industrial safety—Management.   2. Psychology,
Industrial. I. Title.
HD7262.P56 1988   658.3'82   88-10544
ISBN 0-913690-12-0

To Nadine

Table of Contents

i

Preface

In the world as I perceive it, safety is fundamentally a people problem. OSHA is here, MSHA is here, Consumerism is here, and all the beliefs usually accompanied by legislation that say things cause accidents—let's fix it with new laws and we will all be safe.

But, we will not, in my opinion, achieve substantial breakthroughs in our safety records until we deal with the fundamental cause of accidents—the error-making human being.

For that reason I started to collect my thoughts and, more importantly, the thoughts of the behavioral scientists about what is known and believed about the human causes of accidents, about what is known in behavioral science theory, in management theory, and in other applicable fields to the field of safety management.

I believe most of it is directly applicable. And that is what this book is all about—the behavioral approach to safety management.

In late 1967 when I started to write *Techniques of Safety Management*, I had been in the safety field for some thirteen years, eight of those in learning the fundamentals and five in wondering if there wasn't a better conceptual framework than what we were using at the time. Working with D. A. Weaver in safety management training from 1962 to 1966 convinced me there was and *Techniques* attempted to bring together that framework with some techniques to make the principles real. From initial writing in 1967 to publishing in 1971 a lot happened in safety management, including the Occupational Safety and Health Act (OSHA).

It seems to me the techniques described in *Techniques of Safety Management* are real and meaningful regardless of OSHA. Furthermore, the principles espoused are real—more real than OSHA. In fact, I am not sure what principles undergrid OSHA; in actuality it seems the fundamental principle is "accidents are caused by things, not by people."

I do not believe in that principle. I believe that accidents are caused by people.

My training as an industrial engineer tells me that it is extremely difficult and costly to design the foolproof anything. My experience as a safety engineer tells me that accidents result from errors—human errors. My training as an industrial psychologist tells me that it is normal and natural for human beings to exhibit behavior that can result in accidents. And my education in organizational behavior and management, and my consulting with a number of organizations, both large and small, tell me that there are a number of specific actions that management can take to lessen the probability of those behaviors, and those errors.

There are a number of changes in this second edition. In the early sections the historical perspective is brought up to date with the inclusion of the new eras in which we are now. Five new principles have been added to the five previously presented. There is a greater emphasis on role definition within the organization. The new Human Error Reduction model is presented. A new chapter has been added incorporating the newer theories on attitude development and change. There are updates on recent experiences in behavior modification, in Safety by Objectives (SBO) programs, and in participative safety programs. New research is presented on climate, communications, and token systems. And finally, the future is looked at again with new insights from Japan, Germany, Europe, and this country.

Some chapter updates are due to new thinking in the area—new concepts that have been presented or researched. Some updates are because today we can cite practical applications that we did not have in 1975. For instance, SBO had been tried only once at the time of the first edition; today it is almost commonplace. We can show results today where we could not before. Behavior Modification was only an idea in safety in the first edition. Today we can show the results of numerous applications in organizations. Climate and culture were recognized by only a few researchers; today they are management's buzz words.

New chapters in this edition reflect the changes in the field. A new Part VII includes chapters on Subjective Injuries, Stress, Wellness Programs, and Substance Abuse—all of major importance today. They were little discussed in 1975.

This revision just happened to coincide with the clear need to revise another book, *Techniques of Safety Management*. As a result, I took the opprotunity to reexamine both in depth, to remove as many duplications as possible in subject matter, and to build a clear interface between the two. Conceptually I have attempted to build this second edition as a book that examines the people approach to the safety problem and to build the third edition of *Techniques of Safety Management* to examine the management or systems side of safety.

These books aim at:

- The Safety Professional, who needs to know more about the people causes of accidents and their control. Most of us in the profession began with an engineering background and a technical orientation. Over the years we learned the management side of safety. We only now are beginning to focus on the human side.
- Executives and Managers. While safety may be only a small part of the executive responsibility, the approach described here is usable to achieve not only safety, but also every other management goal.
- The Student of Safety. Tomorrow's safety professionals, line managers, and executives need to know how tomorrow's safety programs will run.

My hope is that safety professionals consider both approaches as they examine their own personal philosophy and as they perform their professional function.

The people paramount in the development of this book are the behavioral scientists, the people whose names are so familiar to managers today, and the people in safety management who are leading the way. To name a few:

From the behavioral sciences: Argyris, Bennis, Blake and Mouton, Berne, Bass and Vaughan, Gellerman, Guion, Herzberg, Hovland, Harris, Lawler and Porter, Lewin, Likert, Mager, Mayfield, McClelland, McGregor, McGehee and Thayer, Myers, Skinner, and Sherif.

From the business theorists: Drucker, Humble, Kelly, Kollatt, Luthans, Odiorne, Reddin, and Sayles.

From safety management or allied fields: Bird, Burns, Brody, Currie, Health, Heinrich, Haner, Hughes, Kerr, McFarland, Planek, Pollina, Pope, Sinaiko, Surry, Tarrants, Vilardo, Weaver, and Woodson.

And for this Second Edition particularly: Brown, English, Festinger, Hammer, Hannaford, Ketchum, Norem, Sanders et al., Vogel, Wallace and Zohar.

Finally, a special thanks to a number of individuals in industrial organizations in this country and others who "bought" the concepts expressed in the first edition and tried them in their organization, often with some super results:

Nick Andrews, then with the ICG Railroad, Chicago
Chuck Bailey, then with the DM&IR Railway, Duluth
John Bash, Mobay Chemical, Kansas City
Hugh Braly, Amoco Production, Denver
Jerry Burns, then with Union Pacific, Omaha
Steve Clark, Caterpillar Tractor, Peoria
Chip Dawson and Bill Golden, Eastman Kodak, Rochester
Tracy East, then at Amoco's Texas City Refinery
Dick Erickson, Hallmark Cards, Kansas City
Lou Gergely, Ohio Industrial Commission, Columbus
E. W. Harrington, B. F. Goodrich Chemical, Cleveland
Bob Huskey, Duke Power, Charlotte, NC
Ted Ingalls, Battelle Institute, Columbus
The late Frank Kaylor, then with Southern Railway, Atlanta
Rollie Kuhlmann, Burlington Northern Railway, Overland Park
Don Lambert, Esso Production, Toronto
Al Lepine, Esso Resources, Calgary
Ken Lewis, then with General Motors (Delco), Anderson, IN
Jerry Mithen, American Institute of Baking, Manhattan, KS
Paul Mueller, Pillsbury, Minneapolis
Orville Pilcher, Southern Pacific, San Francisco
Guy Pollard, then with the Frisco Railroad, Springfield, MO
Jafar Rasheed, Province of Alberta, Edmonton
Marilyn Scott, State Accident Insurance Fund, Salem, OR
Ken Sullivan, then with EBI Insurance, San Jose
Terry Swain, British Columbia Safety Council, Vancouver
Bob Tranter and Dave Klaiber, Armco, Inc., Middletown, OH
Art Ulbrich, DuPont, Orange, TX
Phil Ulmer, ARCO Alaska, Anchorage
Bill Van DerMohlen, IBM, San Jose
John Watson, Johnson & Higgins and American Builders and Contractors
Dick Wilburn, National Park Service, Washington, DC
Dick Williams (and many others) State Insurance Fund, San Francisco
Mike Williamsen, Lyle Mueller, John Lautenschlager, Steve Milewski (and others), Frito-Lay, Inc., Dallas, TX
Mark Wright, Adolph Coors, Golden, CO
Mike Zabetakis, then of the MSHA Academy, Beckley, WV

Dan Petersen

Introduction

The intent of the following pages is to relate what is known and what is theorized in the behavioral sciences to the practical, everyday world of safety management. The role of the safety manager is to identify how and why accidents happen and to install or suggest the installation of controls to limit their happening.

If we subscribe to the thesis (as we will discuss later) that to exhibit behavior that could lead to accidents is normal, logical human behavior in the kinds of situations in which the average worker operates, then we, as safety managers, are in the position of having to understand and cope with (or change) normal human behavior. If we believe that the normal person works unsafely because it is normal and logical to work unsafely, then, as safety managers, we must either change normal human behavior or change the situation.

I believe the above is exactly the situation as it exists; it is reality. I believe that the modern safety manager must change all types of situations if there is hope of success: physical condition situations, group pressure situations, psychological situations, boredom situations, no involvement situations, etc. Those kinds of situations make an accident the logical and normal result. And that is what the following pages discuss: situations that exist that make accidents logical and normal; situations that can and must be changed for success to be possible.

This book is intended to be:

1. A book for the practicing safety professional, who, I believe will find this approach different, and, I hope, usable.
2. A textbook for the kind of course that is not being taught today on the human causes of accidents and the removal of these causes. Courses such as this will be a part of the future.

THE SCOPE OF THIS BOOK

This book is divided into eight parts.

Part I This part is concerned with where we are and how we got there. It attempts to briefly summarize our history, our old axioms and our new principles that will replace them. It summarizes my book *Techniques of Safety Management*[127] very quickly and then moves on to the field of behavioral science, describing what it is, who is in it, and how it is being used in both safety and nonsafety fields. This part is only to set the stage.

Part II This part deals with the front-line supervisor, Heinrich's key person. It looks at what inspires activity in behavioral theory. It looks at what kinds of things determine performance in safety. And finally, it looks at how the behavioral scientist would analyze performance to assess how we can get improvement in safe work habits.

Part III This part looks at the "key-chain holders," the people who decide through their thoughts, words, deeds, and systems whether or not the key persons will in fact be performers in safety. It looks at top management, at the middle managers, and at the safety staff.

Part IV This part is about climate and style. Ultimately the success of safety programming depends on what happens between worker and supervisor, between supervisor and management, between worker and management. And what happens in each of these situations depends on the relationships that have been established, the climate that exists, and the style of the supervisor.

Part V This part looks at safety programming. To do this we first of all look at why programs don't fly, why workers don't buy, why workers are "turned off" on safety programs. We then look at some ways to turn them on: through enrichment, through participation, through group norms, and through communications.

Part VI Here we look at the individual worker, first at whether or not we can predict safe working performance, then at how we can change behavior through our traditional means (training) and through some newer approaches (behavior modification). Maintaining safe behavior is looked at by presenting some techniques of improving interpersonal skills. Finally, we look at how we build a safe work environment for the individual worker with what we know about human factors. The final chapter of this part is on Safety by Objectives (SBO).

Part VII This part looks at some current trends and at the future directions in safety management.

Part VIII (Appendixes) Here we look at some workable safety management techniques from *Techniques for Safety Management*[127] and other references that relate to the earlier sections. This, then, is the scope of the book. It intends to cover most of the behavioral science theory that seems to have an application to safety management.

In *Techniques of Safety Management*[127] I suggest any consultant ought to determine the operating level (inspector, engineer, or consultant). I would like to suggest that any safety manager ought to examine his or her operating level by answering the following questions. Choose one answer for each that best describes your safety program.

1. How am I measured?
   a. I'm not
   b. By frequency rate
   c. By concrete achievements
2. How are supervisors measured in safety?
   a. They're not
   b. By frequency rate
   c. By concrete achievements and results
3. How are supervisors rewarded for safety performance?
   a. They're not
   b. They get zapped when it's bad
   c. The results influence their income
4. Which supervisors receive safety training?
   a. None of them
   b. All of them
   c. Those with an identified need
5. What is top management's role in safety?
   a. None
   b. They support me
   c. It's their program

6. What is middle management's role?
   a. None
   b. They'll zap supervisors any time I tell them to
   c. It's their program
7. What is the employee's role?
   a. None
   b. To follow the rules
   c. To make the rules
8. What is your role as the safety manager?
   a. It's my program
   b. To help the supervisors
   c. To advise top management
9. What is the supervisor's role?
   a. None
   b. The supervisor's responsible
   c. It's the supervisor's program
10. When top management speaks on safety
    a. Workers yawn
    b. Workers laugh
    c. Workers listen
11. How often is discipline used to enforce safety rules?
    a. Sometimes
    b. Always
    c. Never
12. Who gets continuing safety training on the job?
    a. Nobody
    b. Everybody
    c. Those who need it
13. When is attention paid to the worker?
    a. When the worker performs an unsafe act
    b. Never
    c. When the worker performs properly
14. When things go wrong who finds it?
    a. Me
    b. The supervisor
    c. The worker
15. What does the big boss expect from you?
    a. Nothing
    b. Trouble
    c. Advice

Each of the above questions has three choices. In each case answer *a* is the least desirable, answer *b* so-so, and answer *c* the most desirable. Scoring one, two, or three points, your safety program is behind the times by twenty-five years at 15 points, by five years at 30 points and you are ahead of the pack of 45 points.

These questions give some insight into the content of this book.

Part I

Safety Management Today

Chapter One

The Eras of Safety Management

Progress slowed in the 1960s indicating that something was wrong. We reexamined ourselves and found some things wrong. Where do we go from here?

Safety management is still a young profession that is only now beginning to mature. The term *safety management* and the title *safety manager* are relatively new. To be sure there have been people in industry involved in safety for perhaps seventy years or more, but it is only in the last twenty years that we have connected the management half of the title onto the safety half.

About as far back as the early 1900s we can begin to see progress being made in safety, however, progress in industrial safety prior to then was practically nonexistent. With no workers' compensation laws, all states handled industrial injuries under a common law, which gave defenses to the management of industry that almost ensured that they would not have to pay for accidents. Without this financial incentive, little was achieved in safety. Workers' compensation legislation provided the financial atmosphere for industrial safety. Workers' compensation in effect states that regardless of fault, the injured employee will be compensated for injuries if they occur on the job. The passage of these laws, starting in 1911, marks the beginning of the first era in industrial safety management.

AN INSPECTION ERA

When management found itself in the position, through legislation, of having to pay for injuries on the job, it began to decide that it would be financially better to stop the injury from happening. This decision by industry gave birth to the industrial safety movement.

In the early years of the safety movement, management concentrated heavily, if not entirely, on cleaning up the impossible physical conditions that existed. Remarkable results were achieved during the period 1911 to 1931. In deaths alone (the least accurate indicator) the reduction was from an estimate of 18,000 to 21,000 lives lost in 1912 compared to about 14,500 in 1933. The frequency rate would show an even greater reduction. This reduction came from merely cleaning up the working areas. Cleaning up physical conditions came first—possibly because they were so obviously poor, possibly because people believed that these conditions were actually the cause of injuries.

It was in this era that W. H. Heinrich[54] published his text *Industrial Accident Prevention.* This volume was monumental in industrial safety—more so than most of us in safety realize to this day. It set the stage, in effect, for practically all organized safety work from that time on. The principles espoused by Heinrich in 1931 are actually the foundation for most elements of our current safety programs. Heinrich's text also ushered in the second stage of safety management.

THE UNSAFE ACT AND CONDITION ERA

Heinrich suggested that more accidents are caused by people than are caused by conditions. He suggested that *unsafe acts* are the cause of a high percentage of accidents (he said 88 percent) and *unsafe conditions* are the cause of the rest (except for some acts of God). This thought and other thoughts he had were a departure from the safety thinking at the time, and the safety professionals of the 1930s and 1940s started a two-pronged approach: cleaning up conditions and trying to teach and train workers in the "safe" way of working. Thus, 1931 ended the Inspection Only era and heralded in an era marked by a split concentration between removing unsafe conditions and stopping unsafe acts.

About this time another era was getting underway due to the various state legislatures. Historically workers who became sick as a result of job exposure to some occupational disease had not been covered by Workers' Compensation since no date of injury could be determined for the disease. When several states decided the date of injury could be the last day worked, suddenly diseases were covered and hence had to be controlled. This began the Industrial Hygiene era.

THE INDUSTRIAL HYGIENE ERA

Occupational diseases have been recognized since the beginning of civilization. Hippocrates wrote in 500 B.C. that many miners had difficulty in breathing, and in 100 B.C. respirators were in use by miners to prevent the inhalation of dust. Ramazzini in 1700 wrote a comprehensive book on occupational medicine in which he identified specific diseases related to certain occupations.

Until the twentieth century, physicians were the primary group interested in occupational diseases. An interest in occupational diseases was thrust on the safety professional when they became compensable in the early 1930s. Thus, the safety manager of the 1930s and 1940s split concentration three ways: looking at the physical conditions, looking at the behavior of workers, and looking at the environmental conditions.

THE NOISE ERA

In 1951 in Green Bay, Wisconsin, a worker in a drop forge plant put in a claim for the hearing that he felt he had lost while on the job and the fourth era of safety management was ushered in. Prior to this claim, loss of hearing had not been considered compensable, for deafness does not impair earning power and a fundamental concept of workers' compensation had been that its purpose was to compensate for loss of earning power as well as medical bills.

After years of litigation, scores of articles, and much talk, it became law in most states to reimburse employees on some basis for hearing loss. The result to the safety manager was that a company's efforts then had to concentrate in yet another direction: in protecting the workers from hearing loss and in protecting the company from paying for hearing lost elsewhere. A look at the professional literature since 1951 reflects the amount of interest in noise and the high percentage of time safety managers concentrate on this problem.

THE SAFETY MANAGEMENT ERA

During the 1950s and 1960s an era evolved which we might call the era of real safety management, although during that period and perhaps even now, the term is ill-defined, and the concepts are fuzzy. During the 1950s safety professionals started thinking in management terms for perhaps the first time. Safety engineers found that setting policy, defining responsibilities and clarifying authorities served their purpose. In this period the safety professional began to find tools from other disciplines that might be adapted. Statistical techniques used by quality control counterparts were useful in making control charts and in safety sampling. Management terms and concepts could be utilized, as could tools and techniques of the personnel people in selecting and placing people, as well as techniques used by the training manager. During this era systems safety evolved and the safety manager looked at human factors engineering. In short, in this era the scope of safety engineering broadened a great deal, and we built much more into the profession than what was there in the 1940s. The scope of safety management also widened to include more than injuries to employees on the job. Safety managers began to look at fleet safety, property damage control, off-the-job safety, etc. The field broadened so rapidly that most safety engineers became users of many techniques but masters of none and inventors of none of their own.

In the 1960s the safety professional began to think about professionalism: by attempting to better define the scope and functions of the safety professional, by developing curriculums for formal education to prepare a potential professional, and by evolving a professional certification program. In the late 1960s there was also considerable reexamination of safety's guiding principles and some disenchantment with the status quo in safety. Many in the field felt something was distinctly wrong and they referred to the record for proof of this.

Safety progressed markedly after 1931. Frequency rates, as measured by the measures of that time according to the National Safety Council, had dropped from 15.12 to 5.99 in 1961, severity rates from 1,590 to 666.[117] These indicators told a story that was a success. Safety people had indeed been proud of their accomplishments and these achievements had been made by doing what Heinrich had said to do in 1931. He seemed to have a formula that worked. However, in the 1960s things began to happen that made people reconsider their situation. In 1961 the record suddenly changed. From 1961 through 1981 the frequency rates consistently got worse; the severity rates still improved, but at a far lesser rate than they had in earlier years (and worsened in some years).

There may be other factors influencing the trend, but it does not take a statistician to see that the trend is there—the record which was worsening has triggered considerable reexamination of our approaches. Some of that reexamination will be discussed when we look at our fundamental principles.

This then is where we stood at the beginning of the 1980s. We had evolved to the point where we knew we must control physical conditions, environmental conditions, and the behavior of workers. We had found a variety of techniques from other disciplines and other systems that might help us to do this. While experimenting with these, we also had come to the realization that what we have been doing is no longer as effective as it once was. This realization forced us to examine critically not only what we have been doing but also the fundamental concepts on which our techniques were based.

Midway in this era we were on the threshold of a new era in safety management, the Psychology of Safety Management era. This era has as its foundation Heinrich's original thought in 1931 that accidents are caused by people, not things. This era has in it the evolving of safety programs which consist of new and different components, each based on things that have proven effective in influencing the behavior of people. We were on the threshold of an era which showed promise of reversing the trend of the 1960s and then came the Occupational Safety and Health Act (OSHA).

THE OSHA ERA

In 1970, the world of safety management changed, at least temporarily, and perhaps permanently. The Occupational Safety and Health Act was passed. Suddenly (it seemed) businesses were in a totally new ballgame without quite knowing the ground rules. Much has been written and said about OSHA's impact on safety, both pro and con, for there seems to be a considerable amount that can be said on each side.

In this historical overview of the eras of safety management, several things seem certain: OSHA delayed the Psychology of Safety Management era for some period of time and OSHA seemed, at least temporarily, to have placed safety management back into an earlier era where physical conditions received the primary (or only) emphasis. The OSHA era appeared to emphasize inspection with federal and state control and deemphasize the human approach. This is not necessarily to say it was bad, or that it should not have been; only the perspective of time will tell that. Most safety professionals agree that something needed changing because the record showed this need. Few safety professionals, however, would have selected a replay of the Inspection Era (with teeth) as the answer, especially in the light of the insights attained in the 1960s.

However, selected or not by the safety professional, OSHA became the era of the 1970s and required the safety professional to concentrate on two primary things: (1) removing those physical conditions that are mentioned in the standards and (2) documenting everything done. The more competent safety professionals and the more successful safety departments found they had two separate and distinct duties, complying with the law (the standards) and controlling losses, instead of only the one of controlling losses. In some cases the two are related and in some cases they are not. There was no question, however, as to priorities: complying came first. Discussion with hundreds of corporate safety people bore this out. The OSHA era was marked by changing physical conditions to federal standards and by documentation to protect.

THE ACCOUNTABILITY ERA

Sometime after the first edition of this book, it seemed safety professionals began to put OSHA into a better perspective. They seemed to learn what to fear and what not fear under OSHA. They seemed to get a handle on which standards were being enforced and which were not, which should be adhered to and which should be ignored. At the same time OSHA (for various reasons) seemed to shift directions from safety to health. Possibly because of the backgrounds for OSHA Assistant Secretaries, possibly because of pressures from the business community, possibly for other reasons, the emphasis came off adherence to safety regulations and was placed on the health standards. Compliance Officers were cross trained in Industrial Hygiene and compliance emphasis was shifted to health standards.

In the late 1970s it seems a new era was ushered in. The new era looked at different ways to measure performance, at new definitions of managerial roles, and at better definitions of what is acceptable safety performance at all levels of an organization. It seems that the key word in performance since that time has been accountability. In earlier eras we talked of management's responsibility; today the emphasis is on accountability. The shift in thinking is major, and as will be discussed in some detail later, is crucial to performance.

In the accountability era safety professionals are beginning to understand and utilize auditing systems, objective setting approaches, and the building of safety performance into performance appraisal systems. In this era we began to shift our emphasis from hazard finding to hazard correcting, looked less at how to find things wrong (inspecting, developing checklists, job safety analysis approaches) and began to look more at how to get things corrected and done (prioritizing, systems evaluations, organization).

It should be noted that this seems to be the direction of safety management at present. In my opinion, it is an extremely healthy direction and should result in considerably improved results. It should also be noted that this is not the direction of some in safety. OSHA is notable in that it is certainly not going in this direction; if anything, it goes further and further away from the mainstream of safety management thinking. It becomes more and more irrelevant as safety management thinking becomes more in tune with the thinking of the rest of management.

THE HUMAN ERA

Sometime after the start of the Accountability Era (perhaps in the early 1980s) another direction emerged—a direction that started to again look at the people side of the safety problem. The two directions, the accountability direction and the human direction, are not mutually exclusive; as a matter of fact they very much fit together, they complement each other. In reality they are to a large degree mutually inclusive as you cannot achieve a human approach in safety without a management system (and that requires accountability) to make it work.

In the Human Era, we have begun to use the principles of human behavior in our safety programs. In this era we have begun for the first time to structure safety programs out of things that make some psychological sense. We are now talking about safety programs that involve the hourly workers and seek their participation. Finally, we have some approaches being used, not just talked about, that utilize positiveness intead of only zapping people. Finally, we are concerned about the hourly employee's perceptions of the organization and the climate that we have created.

This book (as its previous edition) is concerned with the approaches to safety that make psychological sense—that utilize the managerial and behavioral principles that ensure results.

These are the eras safety management has gone through. We will now look at some of the re-examination that took place in the 1960s and then look at the new era, the human era.

Chapter Two

Safety Management's Principles

It you are using a 1931 road map it is hard to even find a freeway.

In 1931, Heinrich[54] spelled out a foundation for industrial safety programs for the first time. His foundation was a set of principles or axioms shown in Exhibit 2.1. Since that time we have relied heavily on the axioms. Consider for instance that:

1. Most accident investigation forms and procedures are built on the "domino theory" espoused in axiom 1.
2. Most record keeping we do is so organized around axioms 3 and 6 that we are constantly looking for trends in the belief that the trends will predict the future and that the severe injury will occur in the same way the minor one occurred.
3. We inspect, based on axiom 3, to find the hazard before the accident.
4. Based on axiom 8, we are constantly pleading for management support.
5. Most of our effort is directed toward the supervisor because we believe in axiom 9.
6. We preach hidden costs up and down the organization based on axiom 10.

It is only recently that we started examining these axioms. When we do we find some real questions about them. For the most part we have not changed many of our approaches to safety, but these traditional approaches will also have to be examined in the future more closely than they have been to date. Let us look again at some of the axioms and summarize the reexamination of them done in the 1960s.

THE DOMINO THEORY (AXIOM 1) VERSUS MULTIPLE CAUSATION

Most safety people have preached this theory many times. Many of us have actually used dominoes to demonstrate it. As the first domino tips, it knocks down the other four dominoes unless at some point a domino has been removed to stop the sequence. Obviously the easiest and most effective domino to remove is the center one—the one labeled "unsafe act or condition." This theory is quite clear; it is also quite practical and pragmatic as an approach to loss control. Simply stated, "If you are to prevent loss, remove the unsafe act or the unsafe condition."

We use this theory in two fundamental areas today: in accident investigation and in inspection. In accident investigation, almost invariably the forms that we use, or that we give to our supervisors to use, ask that one unsafe act and/or unsafe condition be identified and removed. This, of course, seems very logical, considering the statements and principles expressed by the domino theory. It is in fact a very practical and pragmatic approach. Perhaps, however, our interpretation of this domino theory has been too narrow. For instance, when we identify a

9

THE AXIOMS OF INDUSTRIAL SAFETY

1. The occurrence of an injury invariably results from a completed sequence of factors, the last one of these being the accident itself. The accident in turn is invariably caused or permitted directly by the unsafe act of a person and/or a mechanical or physical hazard.

X 2. The unsafe acts of persons are responsible for a majority of accidents.

3. The person who suffers a disabling injury caused by an unsafe act, in the average case has had over 300 narrow escapes from serious injury as a result of committing the very same unsafe act. Likewise, persons are exposed to mechanical hazards hundreds of times before they suffer injury.

X 4. The four basic motives or reasons for the occurrence of unsafe acts provide a guide to the selection of appropriate corrective measures. They are:

Improper attitude ✓   Physical unsuitability — COLOR-BLIND
Lack of knowledge or skill   Improper environment —
       PENCIL WHIP

5. Four basic methods are available for preventing accidents:

Engineering revision   Personnel adjustment
Persuasion and appeal   Discipline

X 6. The severity of an injury is largely fortuitous — the occurrence of the accident that results in injury is largely preventable.

7. Methods of most value in accident prevention are analogous with the methods for the control of quality, cost and quantity of production.

X 8. Management has the best opportunity and ability to initiate the work of prevention, therefore, it should assume the responsibility.

? 9. The supervisor or foreman is the key man in accident prevention. His application of the art of supervision to the control of worker performance is the factor of greatest influence in successful accident prevention. It can be expressed as a simple four-step formula:

Identify the problem
Find and verify the reason for the existence of the problem
Select the appropriate remedy
Apply the remedy

X 10. The humanitarian incentive for preventing accidental injury is supplemented by two powerful economic factors:

MONEY ! The safe establishment is efficient productively and the unsafe establishment is inefficient...

The direct employer of occupational injuries for compensation claims and for medical treatment is but one-fifth of the total cost which the employer must pay.

Exhibit 2.1 Heinrich's original principles.

single act and/or a single condition that caused the accident in the investigation procedures of today, how many other causes are we leaving unmentioned? When we remove the unsafe condition that we identify in our inspection, have we really dealt with the cause of the potential accident? Today we know that behind every accident there lie many contributing factors, causes, and subcauses. The theory of multiple causation is that these factors combine together in random fashion causing accidents.

Let us briefly look at the contrast between the multiple causation theory and our too narrow interpretation of the domino theory. We shall look at a common accident: a man falls off a stepladder. If we investigate this accident using our present investigation forms, we are asked to identify one act and/or one condition:

The unsafe act: climbing a defective ladder

The unsafe condition: a defective ladder

The correction: getting rid of the defective ladder

This would be typical of a supervisor's investigation of this accident under the domino theory.

Let us look at the same accident in terms of multiple causation. Multiple causation asks what are some of the contributing factors surrounding this incident? We might ask:

1. Why was the defective ladder not found in normal inspections?
2. Why did the supervisor allow its use?
3. Did the injured employee know not to use it?
4. Was the employee properly trained?
5. Was the employee reminded?
6. Did supervision examine the job first?

The answers to these and other questions would lead to the following kinds of corrections:

1. An improved inspection procedure
2. Improved training
3. A better definition of responsibilities
4. Prejob planning by supervisors

With this accident, as with any accident, we must find some fundamental root causes and remove them if we hope to prevent a recurrence. Defining an unsafe act of "climbing a defective ladder" and an unsafe condition of "defective ladder" has not led us very far toward any meaningful safety accomplishments. When we are looking at the act and the condition, we are looking only at symptoms, not at causes. Too often our narrow interpretation of the domino theory has led us only to accident symptoms. If we deal only at the symptomatic level, we end up removing symptoms but allowing root causes to remain to cause another accident or some other type of operational error.

Root causes often relate to the management system. They may be due to management's policies and procedures, supervision and its effectiveness, or training. Root causes are those which would effect permanent results when corrected. They are those weaknesses which not only affect the single accident being investigated, but also might affect many other future accidents and operational problems.

THE CAUSES OF FREQUENCY AND OF SEVERITY

Axiom 3 suggests that the causes of frequency are the same as the causes of the severe injury. Axiom 6 suggests this even more strongly. Many in safety work have believed in a predictable relationship between the frequency of accidents and their severity. Many studies have been made over the years to determine this relationship, with varying results. Common sense dictates totally different relationships in different types of work. For instance, the steel erector would no doubt have a different ratio of severity to frequency than the office worker. This very difference might lead us to a new conclusion. Perhaps circumstances which produce the severe accident are different from those that produce the minor accident.

Safety directors for years have been attacking frequency in the belief that severity would be reduced as a by-product. As a result, our frequency rates generally have been reduced much more than have our severity rates.[117]

If we study any mass data, we can readily see that the types of accidents that result in temporary total disabilities are different from the types of accidents that result in permanent partial disabilities or in permanent total disabilities or fatalities. The causes are different. There are different sets of circumstances surrounding severity. Thus, if we want to control serious injuries, we should try to predict where they will happen. Today we can often do just that.

Studies in recent years suggest that severe injuries are fairly predictable in certain situations. Some of these situations involve:

1. *Unusual, Nonroutine Work.* This is the job that pops up only occasionally or the one-of-a-kind type of situation. These situations may arise in production or in nonproduction departments. The normal controls that apply to routine work have little effect in the nonroutine situation.
2. *Nonproduction Activities.* Much of our safety effort has been directed to production work. But there is tremendous potential exposure to loss in nonproduction activities such as maintenance and research and development facilities. In these types of activities most work tends to be nonroutine. As it is nonproduction, it often does not get the attention from safety management. Severity is predictable here.
3. *Sources of High Energy.* We can usually associate high energy with severity. Electricity, steam, compressed gases, and flammable liquids are examples.
4. *Certain Construction Situations.* High-rise erection, tunneling, and working over water have high serious accident rates. (Actually, construction severity is an amalgam of the previously described high-severity situations.)

These are just a beginning point. A long list could be made which would more extensively specify the areas where severity is predictable.

MANAGEMENT BLESSING OR DIRECTION (AXIOM 8)

Consider the difference in how we handle safety compared to quality, cost, and quantity of production. How does management get other things done? When management wants a cerain level of production, it first tells somebody what is wanted or a policy and definite goals are set. Then management says to someone, "You do it." Management defines responsibility saying, "You have my permission to do whatever is necessary to get this job done." It grants authority, and finally, it says, "I'll measure you to see if you are doing it." Management fixes accountability. This is the way management motivates its employees to do what it wants in production, in cost control, in quality control, and in all other things except in safety.

In safety, industry has seemed to take a quite different track. Management officials have not effectively used the tools of communication, responsibility, authority, accountability; rather, they have chosen committees, safety posters, literature, contests, gimmicks, and a raft of other things that they would not consider using in quality, cost, or production control. In those other areas management has not worried too much about motivating people, but has decided what it wants and then made sure that it gets exactly that. In safety, we have gotten into the ludicrous position of pleading for management support instead of advising how management can better direct the safety effort to attain its specified goals.

WHO IS THE KEY MAN?

Axiom 9 states that the supervisor or foreman is the key in accident prevention. It is that person between management and the worker who translates management's policy into action. It is the person who has eyeball contact with the workers. Is that the key person? In a way, yes it is. However, although the supervisor is the key to safety, management has a firm hold on the key chain. It is only when management takes the key in hand and does something with it that the key becomes useful.

THE ROLE OF THE SAFETY FUNCTION

The role the safety function should play used to be that they would make periodic inspections, take part in training, participate in meetings to coordinate safety work and act as a liason with higher executives. It has been only in recent years that most safety professionals have been able to define their role in the safety work to be accomplished. What they do has changed and will continue to change as our concepts and principles continue to evolve. If permanent results can be effected by dealing with root causes, the safety professional must learn to go deeper than the symptomatic level.

If accidents are caused by management system weaknesses, the safety professional must learn to locate and define these weaknesses and unfold a method of doing this. This may or may not lead to doing the things done in the past. Inspection may remain one tool or it may not. Investigation may be a tool or it may not. Certainly new tools must be used and old tools modernized, for the direction is different today and the duties are also different.

BETTER ROLE DEFINITION

The key to an effective safety program today is the crisp, clear definition of roles at each level of the organization: who is to do what when and what precisely will be considered to be acceptable safety performance. The role of the safety staff is discussed further in Chapter 9. The role of upper, middle, and lower management is discussed in Chapters 5, 7, and 8. The roles of each of these can be summarized rather simply as follows:

The role of the supervisor in the safety program is to:

1. Carry out some previously agreed to tasks to an acceptable level of performance.

The role of middle and upper levels of management is to:

1. Ensure subordinate manager performance through a system of accountability (measurement).
2. Ensure the quality of that performance so that performance does get results.
3. Carry out some predetermined tasks personally that visibly shows all subordinates that safety performance is important and that even big bosses are involved.

The role of the supervisor is singular. The role of middle and upper managers is threefold. The role of safety staff is fourfold (see Chapter 9).

THE MOTIVES FOR UNSAFE ACTS (AXIOMS 4 AND 5)

Perhaps the greatest change in our thinking since the first writing of the ten axioms is in these two axioms. In the original axioms unsafe behavior was said to be caused by one of only four causes: improper attitude, lack of knowledge or skill, physical unsuitability, or improper environment. This then lead to the thinking of axiom 5 that there are only four areas of control that correspond: engineering revision, persuasion and appeal, personnel adjustment, and discipline.

As a result of these two axioms written in 1931, safety programs have been constructed ever since around these four control areas that were "allowable." These four are the basis for the famous "three Es of safety": engineering, education and enforcement. Since 1931, safety programs have consisted of various mixes of the three Es.

Our thinking about the causes of unsafe acts today rejects the thinking of axiom 4, and thus of axiom 5. We do not believe that the whole of human behavior can be pigeonholed into four simplistic causes. Today's thinking on accident causation might best be illustrated by the causation model shown in Exhibit 2.2, developed by the author and explained in the book, *Human Error Reduction,*[125] published in 1982. The model suggests that human error (unsafe behavior) is involved in every accident (incident) and that there are many reasons behind this behavior, not just four. It further suggests that these reasons can be identified, can be classified, and are caused by specific things. The model suggests that there are many things that managers can do to reduce the likelihood of the unsafe act beyond the two (or three) suggested originally by Heinrich. We simply are not limited to education and enforcement. Actually, we know today that education and enforcement are the two *least* effective things that we can do to improve behavior of the worker.

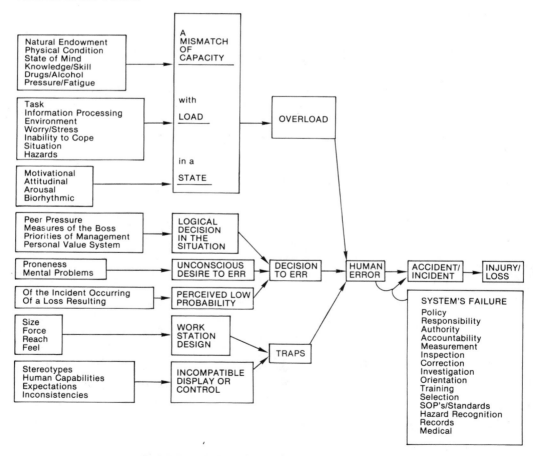

Exhibit 2.2 The Causation Model.[125]

NEW PRINCIPLES OF SAFETY MANAGEMENT

1. An unsafe act, an unsafe condition, an accident: all these are symptoms of something wrong in the management system.
2. Certain sets of circumstances can be predicted to produce severe injuries. These circumstances can be identified and controlled:

   Unusual, nonroutine   High energy sources
   Nonproductive activities *BUDGETED*   Certain construction situations

3. Safety should be managed like any other company function. Management should direct the safety effect by setting achievable goals, by planning, organizing, and controlling to achieve them.
4. The key to effective line safety performance is management procedures that fix accountability. *SOMEBODY WILL BE RESPONSIBLE*
5. The function of safety is to locate and define the operational errors that allow accidents to occur. This function can be carried out in two ways: (1) by asking why — searching for root causes of accidents, and (2) by asking whether or not certain known effective controls are being utilized.
6. The causes of unsafe behavior can be identified and classified. Some of the classifications are overload (improper matching of a person's capacity with the load), traps, and the worker's decision to err. Each cause is one which can be controlled. *PEOPLE DOING WRONG ON PURPOSE. HABIT—QUICKER?*
7. In most cases, unsafe behavior is normal human behavior; it is the result of normal people reacting to their environment. Management's job is to change the environment that leads to the unsafe behavior.
8. There are three major subsystems that must be dealt with in building an effective safety system; the physical, the managerial, and the behavioral.
9. The safety system should fit the culture of the organization.
10. There is no one right way to achieve safety in an organization; however, for a safety system to be effective, it must meet certain criteria. The system must:

    KNOW
    1. Force supervisory performance.
    2. Involve middle management.
    3. Have top management visibly showing their commitment.
    4. Have employee participation.
    5. Be flexible.
    6. Be perceived as positive.

Exhibit 2.3 New Principles.[127]

NEW PRINCIPLES

Exhibit 2.3 spells out the newer principles of safety management. Principle 1 embodies the concept of multiple causation, of symptoms versus causes, and of management system weaknesses. This principle suggests not that we boil down our findings to a single factor, but rather that we widen our findings to as many factors as seem applicable. Hence, every accident opens a window through which we can observe the system, the procedures, etc. Different accidents would unearth similar factors that might be wrong in the same management system. Also, the theory suggests that besides accidents, other kinds of operational problems result from the same causes. Production tie-ups, problems in quality control, excessive costs, customer complaints, and products failures are caused by the same things as accidents. Eliminating the causes of one organizational problem will eliminate the causes of others.

This applies to any accident. If we view that accident and its unsafe act and condition as only a symptom of what is wrong, not a cause, we must then look behind that act and condition to determine why. When we diagnose the causes and treat them, we effect permanent control.

The function of the safety professional then is similar to that of a physician: diagnose symptoms to determine causes, and then treat those causes or suggest appropriate treatment to management.

Principle 2 states that we can attack severity directly instead of merely hoping our attack on frequency will also affect severity.

Principle 3 restates the thought that safety is analogous with quality, cost, and quantity of production. It also goes further to bring the management function into safety (or rather safety into the management function). The management function by definition should include safety, but in practice it has not done so. Management has too often shirked its responsibility; it has not led the way, but at best it has given "support." Inherent in this principle is the thought that safety is and must be a line function. As management directs the effort by goal setting, planning, organizing, and controlling, it assigns responsibility to line managers and grants them authority to accomplish results. The word *line* here means not only the first-level supervisors, but also all management-level supervisors above them, up to the top.

Principle 4 is based on the belief that line managers will achieve results in those areas in which management is measuring them. The concept of *accountability* is important for this measurement. The lack of procedures for fixing accountability is safety's greatest failing. We have preached line responsibility for many years, and we had spend this time devising measurements for fixing accountability on line management, we would still be achieving a reduction in our accident record.

When someone is held accountable, that person, will accept the given responsibility. If not held accountable, in most cases responsibility is not taken. Efforts will be placed on those things that management is measuring: production, quality, cost, or wherever the current management pressure is.

Principle 5 defines safety's function as locating and defining operational errors involving (1) incomplete decision making, (2) faulty judgments, (3) administrative miscalculations, and (4) just plain poor management practices (borrowed from Pope and Cresswell[134]). They suggest that to accomplish our purposes, we in safety would do well to search out not what is wrong with people but what is wrong with the management system that allows accidents to occur. This new concept directs the safety professional to look at the management system, not at acts and conditions.

The second part of principle 5 suggests that a two-pronged attack is open to us: (1) tracing the sympton (the act, the condition, the accident) back to see why it was allowed to occur, or (2) looking at the system (the procedures) that our company has and asking whether or not certain things are being done in a predetermined manner that is known to be successful.

Principle 6 summarizes the causation model in Exhibit 2.3. It suggests that management's task with respect to safety is to identify and deal with the causes of unsafe behavior, not the behavior itself. Principle 7 reemphasizes that our task is to change the physical and psychological environment that leads people to unsafe behavior.

The role of safety will be defined in four elements: analysis, developing systems of control, communicating those systems to the line organization implementing them, and monitoring the results achieved. Principle 8 suggests these four must include systems in all three areas: physical condition control, the management system, and the behavioral environment.

Traditionally "safety programs" have dealt with the physical environment. Safety started here. Later we looked at management and attempted to build management principles into our safety programs. Today we recognize the need also to look at the behavioral environment—the climate and culture in which the safety system must live.

As indicated earlier, times have changed. The way we manage has changed markedly. And the way we manage safety must also change to be consistent with other functions (principle 9). To strive for an open and participative culture in an organization and then use a safety program that is directive and authoritarian simply does not work.

And finally principle 10 defines the criteria for success of a safety program.

These then are the new principles that our reexamination has developed. Many of these are as yet only principles without tested techniques available for implementation. Others are ready to go with well-tested methods of implementation available.

Many people in safety management today believe that we are no longer progressing because those principles which Heinrich enumerated and which we have relied on are suspect. Many people believe that some of the other principles which Heinrich stated (such as his belief in unsafe acts as a primary problem) are very true, but that we have ignored them. And even though we may believe that accidents are caused by people, we still concentrate a great deal of our time and energy on the environment—the physical surroundings and conditions. What little time is spent in working with the worker to eliminate the act is in large measure misdirected activity as we will see later. OSHA has ensured that even less of our available time is directed to people.

SAFETY IS A PEOPLE PROBLEM

We have made marked progress over a sixty-year span in industrial safety, but practically no progress or negative progress in the most recent twenty year period. Our initial progress was due to our control of the environment, and today, in most cases, we are controlling the environment in which we work relatively well. Safety today, then, is not as much an environmental problem as it was. Our primary job is not to control "things." Safety today is a people problem. We are in a situation where we must learn to understand people and where we must learn to control people's behavior.

We have not solved our problems in safety, either on the job or off. We have made fantastic progress since 1911, even so, we are still killing some 14,000 workers on the job each year, and many more off the job (or rather they are killing themselves). The problem will remain with us and continue to worsen until we learn to better understand and control the behavior of people. We live in a technically sophisticated society, but still know little about ourselves. We usually do not know why we act as we do.

We do know enough about ourselves, however, to know that we do not achieve anything (in safety or any other field) unless we as individuals want to achieve. Human beings act and perform well only when they want to. People act and perform in a safe manner only when they want to. Obviously, there are certain times and circumstances when this is not true. We all, at times, do things we do not like to do. We all, at times, operate in a manner in which we do not believe: when under direct orders, when the boss is closely supervising, when we are afraid to operate as we wish. In the large majority (perhaps 99 percent) of acts and duties we perform, we do them as we want to do them. Hence, many times we do them differently (sometimes less safely) than our bosses would wish because for some reason (usually unknown) we want to do them that way.

When we talk about the management control of accidents, we really mean to talk about how management can control the behavior of people. When we talk of how management can control the behavior of people, we really mean, for all practical purposes, how management can get those people to *want* to operate in the manner that management wants. This, of course, is motivation. In the final analysis all safety depends on motivation. The American workers today must be motivated to work and motivated to work safely. This does not mean they innately do not want to work or do not normally work safely. It does mean, however, that if management in any kind of work wishes to achieve results in accident control today, it must learn the secret of motivation of its people, of getting its people to want what it wants.

When safety professionals enter this area of motivation, they enter a whole new world—a unfamiliar world—a world they should begin to get to know. For here is the world of future safety progress. Progress in our accident records will not come through better, more sophisticated environments; it will come only from better understanding of people and better motivational techniques by management.

Here, then, is where we are in safety management. We have questioned our fundamental beliefs. We have developed new principles. We (regardless of OSHA) know that our future success depends on motivating (on controlling the behavior of) our people. And we really do not yet know how to do this. There is, however, knowledge available to us about people and how they behave and why they behave as they do. We should look at this knowledge and utilize whatever applies to safety management.

Chapter Three

A Human Approach

Our success depends on our control of the behavior of people. Behavioral science is the systematic study of the behavior of people. How can we achieve success without that body of knowledge?

The National Safety Management Society has indicated that "approaches to accident prevention should be directed toward improving the largest possible management system or subsystem as contrasted with the limitation of correcting localized hazards and practices," and that "a prime method of accomplishing safety objectives is the establishment and operation of systems, methods, and programs for identifying the causes and costs of operating errors."

It would seem then that the safety managers of today should turn their attention to the largest single source of operational error, the human being. Students of human factors will quickly point out that people are the least predictable, and often the least reliable, variable in any system. Or as Woodson[172] states: "At best, human variability is such that performance is subject to error from time to time for completely unexplained reasons." As Sinaiko[149] puts it: "The human component has many vulnerabilities. Many environmental conditions influence man's behavior: his health, his age, and his physical stamina are all examples of man's non-foolproof nature. He must be utilized with care." As we all know there are other personal and organizational conditions that influence behavior: "attitude," "feelings," pressure from the boss or the informal group, the organizational climate. Behavioral scientists attempt to look at some of these influences and to learn more about how people's behavior is changed by them.

WHO ARE BEHAVIORAL SCIENTISTS?

The academic disciplines that are included in the behavioral sciences are many and varied. They include the work of psychologists (social, organizational, experimental, industrial), sociologists, anthropologists, psychotherapists, socioeconomists, political scientists, educators, linguists, and others. The only unifying purpose to the staggering amount of special interest research done by these people is that all are concerned with human behavior in organizations.

In 1969, the National Industrial Conference Board surveyed 302 firms on their interest in and use of behavioral scientists. This survey indicated the following people as some of those who had the most influence on the thinking of modern managers:

Douglas McGregor (Theory X-Theory Y)
Frederick Herzberg (Motivation-Hygiene)
Rensis Likert (Systems I-Systems IV)
Chris Argyris (Incongruency Theory)
Robert Blake and Jane Mouton (Managerial Grid)

Each of these people has presented, proposed, or researched an approach to better explain human behavior. Some speak directly to human vulnerability and most have a real message for the modern safety manager. While we will look in some detail at these people and their theories later, we might briefly indicate some of the areas particularly applicable to safety management. In addition, we will briefly show how companies are utilizing these people at present.

MANAGEMENT STYLES

McGregor and Blake and Mouton (and others) spoke to the broad area of leadership styles. McGregor[98] identified the X manager as assuming that people dislike work and that they must be forced, controlled, or directed. He identified the Y manager as assuming that people view work as natural, that there are other means than control and punishment by which to manage, and that people seek responsibility and like to achieve. Obviously, the management style of X and Y would be totally different.

Most industrial safety programs are Theory X oriented, consisting of policies, rules and rulebooks devised by management, taught by management, and enforced by management. McGregor's Theory Y manager was the forerunner of the participative management approach. Participative safety programs are a logical outgrowth of this and only recently are beginning to appear on the scene.

Blake and Mouton[17] classify management styles similarly, drawing what they call a Managerial Grid. The two axes of the grid are Concern for People, or Relationships Orientation, and Concern for Production, or Task Orientation. With the grid we can further refine the management style identification and recognize that a single manager can be high in both the concern for people and the concern for production or job achievement. Conversely, the manager can be low in both; or high in one, and low in the other.

Safety managers need an understanding of Theory X versus Theory Y and of the types of managers that can be plotted on the Managerial Grid because their success depend on line managers whom they must influence and motivate into action. How do we as staff safety specialists motivate a manager with a high concern for people? Certainly differently than a manager with a high concern only for production. How does each of these types attempt to control the behavior of their people? Certainly quite differently. Their respective results will be quite different and their results are our results.

When the production-oriented manager faces a safety problem it is seen as a threat to production. Probably this very threat is the best motivation. It will be difficult to get this type of manager to train employees in safety. Safety training must be an integral part of job training or it will not get done, because this manager views job training as important to getting the job done and may perceive safety training as a waste of important production time.

When the people-oriented manager faces a safety problem it is seen as a threat to the welfare of the people. This manager probably will react readily in providing any needed safety training and in maintaining proper physical conditions. A weakness may be in the enforcement of safety rules, so it would seem our best sales point for this manager would be to appeal on the basis that employees could be injured, not on the basis of improved efficiency or lowered costs.

The managers who are both people- and production-oriented are sharp enough to see the relationship between safety and production. With them it is usually sufficient to merely identify the problem.

The managers who are neither people- nor production-oriented are a problem to us just as they are to top management. We may not accomplish much in safety with them, but it is likely they will not last long as managers.

The Managerial Grid and the Theory X-Theory Y approaches to management style typology could help safety managers by helping them to understand line managers and what motivates them. The theories infer that to motivate line supervisors successfully, there is probably no single "right" approach. Individual supervisors, depending on task and relationship orientation, will react differently to each motivational approach used.

INCONGRUENCY THEORY

Chris Argyris[5] has proposed and tested a theory which provides safety managers some insight into why people commit errors. He looks at the nature of people as a starting point, as did McGregor. Argyris analyzes people as they mature. As children human beings are passive and dependent on their parents. They exhibit little behavior; their interests are shallow and short-term. They are at all times subordinates in relationships with parents and are relatively lacking in self-awareness. As they mature this changes. Mature adults are active and independent creatures who like to stand on their own. They exhibit much behavior, and interests are deep and long-term in nature. They view themselves as equals in most relationships, not as subordinates, and they have self-awareness. These changes, this evolution, is what maturation is all about.

Argyris then looks at the characteristics of organizations. All organizations, whether they are industrial, governmental, mercantile, religious, or educational, are structured under certain principles:

1. They have a *chain of command*. This creates a superior-subordinate relationship. It creates dependency on the boss and passivity on the part of the worker, and creates shorter and shallower interests on the part of the worker.
2. The *span of control* must be small. This creates dependency and reduces the freedom and independence of the worker.
3. There is *unity of command*. There is only one boss, creating dependence and heightening the subordinate role of the worker.
4. *Specialization*. The work should be broken down into small simple tasks and each task assigned. This creates shorter, shallower interests, a lack of self-fulfillment and self-importance, dependency and passivity.

There is a basic incongruency between the characteristics of the mature worker and the characteristics of the organization. They are pulling at counter purposes. Worse yet, this seems to be an inevitable pull. We cannot select children to do the work.

Argyris suggests this incongruency causes people to quit (turnover), to quit mentally (apathy), to lose motivation and interest in the company and its goals, to form informal groups, to cling to the group norms instead of to company established norms, and to evolve a psychological "set," believing the "company" is wrong in most things it attempts to do. It also causes accidents due to inattention, disregard of safety rules, poor "attitude" toward the company and safety, etc. The normal management reaction to these symptoms is more control, more specialization, and more pressure. Management believes even more strongly that "they" must be controlled, i.e., treated like children. The problem becomes circular.

What can we do about it? Since we cannot feasibly change mature people into immature ones (nor would we want to), the only option is to look at the organization and see how we can change its characteristics. This leads us to organizations (and safety programs) with less control, less specialization, and fewer superior-subordinate relations. This theory leads us right back to McGregor's thoughts on participative leadership. It also leads us to a principle known as *job enrichment.* Job enrichment, however, is more generally thought of as stemming from the research of Frederick Herzberg.

THE MOTIVATION-HYGIENE THEORY

Herzberg[58] proposes that job satisfaction and job motivation are not on the same continuum (see Exhibit 3.1) as we often have viewed them. Whereas we could (we thought) made a dissatisfied worker a motivated worker by manipulating a variable or two (pay, environment), Herzberg says that there are certain variables that he calls *hygiene factors* and different ones that are *motivation factors.* By improving the hygiene factors (company policy, supervision, interpersonal relations, status), we can make a dissatisfied worker a satisfied worker, but this does not mean that worker will be motivated. Different factors motivate (achievement, recognition, the work itself, responsibility). The motivation factors have to do with the job itself, while the hygiene factors are peripheral to the job. As Exhibit 3.1 indicates, a worker can be high on the hygiene scale and low or the motivation scale, or vice versa, or high on both, or low on both. Herzberg views job enrichment as the answer to motivating employees. Enrichment does not mean enlargement (not just more work) but rather it means building some meaning into jobs: allowing workers to put themselves into their job, allowing workers to "manage" their own jobs.

In safety management, job enrighment is particularly applicable. In fact, safety can be used as one of the areas of enriching the worker's regular job. Some of the ways this can be done are by

1. Building safety into each job by assigning the workers problem solving, goal setting, developing measurement criteria, determining what controls are needed and job safety analysis tasks.
2. Allowing the workers to design the safe procedure, standard operating procedures, or safety rules.
3. Providing external and internal satisfaction through job recognition.

THE EMPLOYEE-CENTERED MANAGER

McGregor, Argyris, and Herzberg all strongly suggest that the close control of the worker does not get the results wanted and poses major problems in attitude or relationships. Rensis Likert's[88] research proves this. Exhibit 3.2 indicates the relationship he found between pressure and productivity. Exhibit 3.3 indicates the relationship between freedom and productivity. Likert describes the importance of what the supervisor thinks of subordinates and their productivity (see Exhibit 3.4). In short, Likert's findings seem to say that the *more* the supervisor supervises,

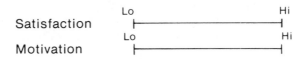

Exhibit 3.1 Herzberg's dual continuua.[58]

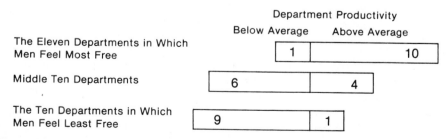

Department Productivity

	Below Average	Above Average
The Ten Departments Which Feel The Least Pressure	1	9
The Middle Eleven Departments	6	5
The Ten Departments Which Feel the Most Pressure	9	1

Exhibit 3.2 Relationship between pressure people feel for production and productivity of the departments.[88]

the *worse* the results, and the supervisor who lets subordinates alone to "do their own thing" gets the most production. While Likert does not speak directly to safety, it would seem the results would hold true there too.

MOTIVATING MANAGERS

The behavioral scientists have also looked at the subject of motivating supervisors and managers. Perhaps the best model of manager motivation was constructed (and fully tested) by Porter and Lawler.[135] This model explains in some detail the forces that go into supervisory performance (effort, ability, and role perception) and of the forces behind effort (amount of reward and probability that the reward will be received). This particular model is one of the best for direct use by safety management, explaining perhaps better than any other model why supervisors do or do not perform in safety.

OTHERS

We have touched (lightly to be sure) on only a few of the more famous behavioral scientists. All seem to have something to say to safety management. Others also speak clearly to us. For in-

Department Productivity

	Below Average	Above Average
The Eleven Departments in Which Men Feel Most Free	1	10
Middle Ten Departments	6	4
The Ten Departments in Which Men Feel Least Free	9	1

Exhibit 3.3 Relationship between freedom that people feel to set their own work pace and their productivity.[88]

Number of First-Line Supervisors Who Are

	Job-Centered	Employee-Centered
High-Producing Sections	1	6
Low-Producing Sections	7	3

Exhibit 3.4 Relationship between how the supervisor thinks of his people and their productivity.[88]

stance, you may wish to look at Odiorne's work[122] (Management by Objectives) which has been adapted to Safety by Objectives; Lewin's[87] (Group Dynamics); McClelland's[93] (High Achievers); Hovland's[61] (Communications); Mager's[103] (Training and Analyzing Performance Problems). These will all be discussed later, but this initial overview gives a flavor of behavioral science and indicates some of the more obvious ways it can help us with the most variable and unpredictable component of our system. In many cases, safety managers seem to be building programs and techniques that historically seem sound but in reality are in opposition to what we now know about people. We try to motivate workers and supervisors with things that demotivate. We try to communicate in ways that are irrelevant to the worker. We try to force and control workers into becoming safe workers. We try to "build a safe attitude" without knowing what an "attitude" is. Behavioral scientists obviously do not have all the answers and have not unlocked all the secrets of the human being. But they have found out a few things and safety managers can well use those few things.

In the 1980s managers became particularly enamored with more new approaches to people management. In the early 80s the "One Minute Manager"[18] philosophy emerged—a cute rewrite of the author's previously published textbook *Situational Leadership*. While cute, it had a solid philosophical base which will be discussed in some depth later. Then the "excellence" concepts became popular, first with *In Search of Excellence*,[124] and later with *A Passion for Excellence*,[123] followed by a host of different offshoots of the concepts.

HOW INDUSTRY USES BEHAVIORAL SCIENTISTS

Before looking specifically at the uses of the behavioral scientist in safety management (which is practically nil), we might look at how industry in general has the knowledge. Lennon and Hollenbeck[83] spell out the ten most common ways the behavioral scientist has contributed to management as:

1. *Survey Methodology.* Managers concerned about characteristics of their organizations have often used employee attitude surveys to identify problem areas, to assess the effects of organization change or policies on job satisfaction, and as one index of management performance. Early uses of employee surveys often took the form of one time "how are we doing?" looks, but current usage emphasizes comparisons over time or across organizational units.

2. *Selection and Assessment of Personnel.* The contribution of the behavioral sciences, and in particular of psychology, to improved selection, placement, and assessment of personnel is a most common use.

3. *Programmed Instruction.* This concept, derived largely from the work of **B. F. Skinner**, includes such sensible elements as behavioral definition of training goals, the analysis of learning tasks into manageable units, and systematic applications of principles of reinforcement to the learning situation.

4. *Rational Decision Making.* Systematic investigation of the decision-making and problem-solving process has been a frequent concern of behavioral scientists and has led to the conviction that the abilities involved can be appreciably improved through appropriate training.

5. *Human Relations.* This emphasis was not the product of any specific behavioral scientific research, though some think of it as associated with the work of Elton Mayo and Fritz Roethlisberger or William F. Whyte. Certainly it was not a tool or technique, yet it was a state of mind that influenced the direction of much of the behavioral research of the 1950s.

6. *Job Enrichment.* The general heading of "job enrichment" includes the modification of the content of jobs for the purpose of increasing employee satisfaction, interest, and pride in work, and thereby motivation and performance. The assumption is that as employees are given more responsibility and more control over their work, they will be more highly motivated to perform.

7. *Motivation Theory.* In the minds of some observers, the impact of the motivation theorists and the application of the practices they advocate constitute the major contribution of the behavioral scientists to industry in the past decade. Douglas McGregor, Frederick Herzberg, Abraham Maslow, Chris Argyris, and David McClelland are names as familiar to management as they are to other behavioral scientists. McGregor's Theory X and Theory Y, Maslow's Need Hierarchy, Herzberg's Two-Factor Theory, and McClelland's Need Achievement have made a notable impact on managerial style.

8. *Participative Management.* In line with the urgings of the motivation theorists has been the proposition that workers would be more fully committed if they had a voice in setting the goals and conditions of their work—the so-called participative management view, an outgrowth of the work of Kurt Lewin[87] on group dynamics. Participative management enjoyed widespread attention in the 1950s and 1960s. A primary proponent has been Rensis Likert, again, a name familiar to many in management.

9. *Organization Development.* Organization Development (OD) is a process of systematic, planned change in an organization. It incorporates a definite philosophy that organizational change should be in the direction of mutual trust, openness in communication, readiness to deal with feelings as legitimate data, and understanding intra- and inter-group relations. OD frequently uses as a basic tool experienced-based learning, most often sensitivity training.

10. *Industrial Counseling.* Clinical psychology continues to contribute to management.

11. *Use of One Minute Approaches.* A whole new concept emerged—a concept which says that management is something that happens between two people—a boss and a worker (not between a boss and a group of people). Management takes place one on one; it happens in the one to one interaction. Management occurs when a boss and a worker together define the key objectives of that worker's job, what is important in that job, and what is not. Then management becomes simple for the boss, and subordinates can be effectively managed with short little one minute contacts, both positive and negative (with the strong emphasis on positive).

12. *Excellence, et. al.* Another whole new concept that suggests companies become successful only when they put primary emphasis on what is good for their people. The people orientation is what makes them better.

Notable in this list is the fact that safety is not included. Also notable is the fact that safety managers have not been known as heavy users of any of the above techinques. It is the rare safety manager who has effectively used attitude surveys, programmed safety texts, job enrichment, participative safety programs, motivation theory, or even counseling. Our literature does not even discuss these things, even though management literature has been talking about them for many years.

It seems as if our newly formed and developed concepts in management are not in tune with our safety programs of today. Or to put it another way, our safety programs often run counter to the facts from research that lie behind our new management concepts. The examination of a few areas might illustrate.

Perhaps a good starting point might be with the book *In Search of Excellence,*[124] the non-fiction best seller by Thomas Peters and Robert Waterman, one of the most popular management books in years. The authors quite simply have looked at the companies in this country that consistently seem to be the best performers, the most effective in terms of bottom line performance. From this in-depth look at the best run companies, they have distilled some rather interesting keys to effective management. For instance:

— The best companies have "A Bias for Action." They are ready to innovate; they try new things; they even encourage failure. There are precious few innovators in our profession.
— The best foster autonomy and enterpreneurship down through the organization. Does the corporate safety manual, the rules and regulations, foster this?
— The best have hands-on, value driven executives; they are out on the shop floor. Management is not seen as the supervisor's exclusive task. The supervisor is clearly not the key person.
— The best utilize a simple organizational form: a lean staff indicating fewer safety staff people will get more line management action.
— The best have simultaneous loose-tight properties. They allow great autonomy while adhering to firm tight values. If safety is one of those tight values, coupled with great line flexibility, we'd have super safety programs. *In Search of Excellence* does challenge traditional safety thinking severely.

Dr. Rensis Likert did major research many years ago on the relationship between organizational climate and the bottom line indicators of corporate success. He found unbelievably high correlations between his defined, quantified climate and return on investment, profitability, and growth of the organization. Consider some of his climate areas:

— The amount of confidence and trust between the worker and management (both ways). His concept of team relationships clearly got the best results. "Them vs. Us" relationships clearly got the worst results. In safety this might mean that a "police vs. policed" relationship between management and the worker is counterproductive.
— Teaching workers how to solve their own problems was found to be much more successful than giving workers all the answers. Sometimes it seems that we in safety are much better answer-givers than teachers of how to solve problems.
— Information sharing was important in Likert's climate profile—not what people need to know, but rather what information they did not need but would simply like to know to better identify with the organization.
— Actively soliciting the ideas and opinions of the worker was extremely important in climate. It is of crucial importance in safety, for the worker is the greatest expert on the safety of that job.

— Recognizing accomplishments was also of great importance in climate—the positive recognition given when someone does something right or good. So many safety programs are built upon the negative approaches. We lie in the weeds to find the unsafe actors, so that we can zap them.

Likert's research showed clearly that certain climates are closely connected with organizational success. There is little doubt that a similar connection exists between safety success and climate.

Climate studies not only are valuable to safety professionals because of this connection, they also are valuable because our safety programs at times serve to build an unhealthy climate through policing, through answer giving, through ignoring the ideas of the worker, and through primarily negative approaches. If these are occurring, it is likely our safety programs are causing more accidents than they are preventing.

AUDITS OR CULTURALLY BASED PROGRAMS

Our profession has become enamored in recent years with the audit concept, the pre-identification of who is to do what and then checking on how well it is being done. I, for one, applaud the concept as a better way to measure results than our old numbers games of frequency rates, severity rates, OSHA incidence rates, claim costs, etc., being used as measuring sticks.

There are, however, many ways to build and perform audits. Some seem to make much more sense than others. We have a number of audits devised internally in an organization to define and measure location activity and success—an audit program devised for that organization. We have packaged audit programs—many of them published and for sale in the market place. National Institute for Occupational Safety and Health (NIOSH) has published some for certain industries and private concerns are selling others.

All of these should start from the knowledge of what should be the components of an effective safety program. What should be the elements of your company's safety plan? To answer this question much research has been performed by many people and organizations and some of it is conflicting. The National Safety Council in 1967 gave us research which resulted in a rank listing of 78 safety program elements, as judged by the best U.S. companies. Research at Michigan State University[147] and at the University of Nebraska[141] gave additional insights on what constitutes the essentials of a safety program. Stanford University has done similar research in construction. All of these give us somewhat different results in terms of what should be considered "essential."

Then we have other research studies which further muddy the water. Dr. Foster Rinefort's doctoral dissertation at Texas A&M University looked at nine different categories of Texas companies—three different sizes each of three different types of businesses. His results showed different rank order listings for each of the nine categories. It also gave a cost-effective break-even point for each of the nine. Items that were cost effective in one category were a waste of time in another. This suggests that in different organizations perhaps different program elements are needed and perhaps other typical elements might well be ignored.

Several years ago, the Association of American Railroads made an indepth study of all major U.S. railroad safety programs. They found little safety program uniformity and no elements really essential to program success. Their results suggest the safety program must be right for the specific organization. What is right and essential at one property is not right nor essential at another. There seems to be no simplistic solution to our complex organizational safety problems nor no one set of program elements that are essential for safety success. There is no one safety program that is right for all.

Our newest writings in management concepts suggest that the starting point of all organizational improvement is to first assess the corporate culture that exists, and then systematically improve on it by building on the cultural strengths, while phasing out the cultural weaknesses. What are cultural strengths? Probably those elements that Peters and Waterman identified in *In Search of Excellence,*[124] those elements of climate that Likert identified years ago. What are cultural weaknesses? Probably a "Them vs. Us" relationship, a lack of information sharing, a lack of positive recognition.

Current management theory and good safety research suggest that what is right for your company is unique. There are no effective packaged programs. It seems the research and theory say your audit program must look at the things that are right for your organization.

BEHAVIOR CHANGE OPTIONS VERSUS THE 3 Es

While it has been said before and should not be repeated, behavioral science theory based on over 60 years of excellent research has clearly told us to throw away the three Es of safety (engineering, education, enforcement). To say that worker behavior change can obtained only through education and enforcement is completely ludicrous. Worse than ludicrous, it is counterproductive. There are innumerable options to behavior change. The behavioral scientists suggest three broad categories of options:

- Behavior Modification, which can shape worker behavior through systematically reinforcing the right or desired behavior until the new habit (safe behavior) is formed. While we are only recently trying this in safety, it is a very feasible option for our use. Recent research has demonstrated that even when we simply give our positive reinforcements to workers working safely as often as we zap them for infractions, we markedly reduce the accident rate.
- Change attitudes so that later the behavior will follow the new attitude. It is complex and much more involved than preaching and teaching, but it works, and it is an option.
- Build a psychological environment so that a worker feels more comfortable working safely. This is sometimes called motivation. We must analyze the various pulls on the worker's behavior, and change those influences we can while recognizing those we cannot.

Where do the 3 Es fit in? This is not very clear. Education enables a person to work safely. Whether or not the worker chooses to use the knowledge is outside of education. Enforcement (sometimes known as discipline to a manager and as punishment to a worker) might increase the likelihood of future improvement in behavior; might decrease the likelihood of future improvement; might serve simply to make sure the person will not get caught next time; might aid in the process of alienating the worker from the organization; might cause the worker to want to "get back." We simply do not know what reaction will come.

Of all of the options available to us in management to affect the behavior of the worker, the behavioral scientists tell us that education and enforcement are the two least valuable.

SAFETY MEETINGS OR ONE-ON-ONES

I believe every safety program I have ever seen has as an integral component the safety meeting, either large or small, long or short, by safety pro or supervisor. Law often requires them and we all believe them to be helpful, if not effective. Are they? Perhaps, although a summary of the research several years ago came to the conclusion that there was no evidence to show that they were.

Here again the behavioral scientists seem to be suggesting other approaches might be even better. One offered for our consideration is the one-on-one contact between worker and supervisor. Consider the reasoning behing one-on-ones:

— In one-on-ones, a supervisor can break through the all important worker influence of the peer group. In the meeting setting, in fact, the influence is at its greatest.
— In one-on-ones, a supervisor can begin to understand the needs of the subordinate. This is absolutely essential to motivation, for motivation is defined as "a person doing whatever is necessary to satisfy current needs." In a group setting there is no way needs can be determined.
— In one-on-ones, a supervisor can achieve a "levelled" relationship with a worker, an equality, adult-to-adult relationship—another essential to motivation. In a group setting we stiffen the superior-subordinate (parent-child) relationship.

Perhaps the information from the behavioral people is worth looking at, maybe even worth trying, if we have that "Bias for Action" mentioned by Peters and Waterman.

We will attempt to relate some of these techinques to safety management in the following pages.

THE HUMAN APPROACH AND THE NEW BREED

The safety manager needs to consider the human approach to safety management more today than ever before. A recent survey discussed in *U.S. News and World Report*[159] indicated that compared with 20 years ago, today's workers are:

— A more diverse group—41 percent are women, 10 percent black, 23 percent under 35.
— More affluent—spendable income is up 9 percent; 63 percent of homes have two wage earners.
— Have more fringe benefits than ever before.
— Are better educated—17 percent of the work force are college graduates.
— Are largely in white collar jobs today—over 50 percent.
— Have more leisure hours.

Even with the above improvements, many, perhaps most, workers remain dissatisfied with their jobs. The University of Michigan's Survey Research Center showed worker dissatisfaction at the highest point in a decade:

— Sixty percent of the surveyed workers wanted different jobs.
— Thirty-nine percent thought they were underpaid.
— Thirty-six percent thought they were not using their skills.
— Fifty-five percent wanted more time off.

The survey further showed all percentages were up markedly over the same survey five years earlier.

People are disenchanted with work, or at least with the work they are currently doing. This disenchantment alters their behavior and in fact is another of the causes of unsafe behavior. The human approach to safety management attempts to better understand the worker and attempts to build a safety program based on that better understanding.

Part II

Motivating The Key People

Chapter Four

The Management Motivation Model

Theory is not an abstraction on the shelf but a daily tool, guiding us in all phases of our work, from conception of the problem to reporting the results.

B. Baxter, in a presidential address to the
American Psychological Association, 1965

In attempting to relate behavioral science knowledge to safety management perhaps the easiest place to start is with Heinrich's key person—the supervisor. It is easiest because there is a body of behavioral knowledge based on some excellent research that speaks directly to us about the motivation of and attitude of supervisors toward management's goals.

The attitude of the majority of supervisors today lies somewhere between total acceptance and flat rejection of comprehensive accident prevention programs. Most typical is the organization in which line managers do not shirk this responsibility but do not fully accept it either or treat it as they would any of their defined production responsibilities. In most cases their "safety hat" is worn far less often than their "production hat," "quality hat," "cost control hat," or "methods improvement hat." In most organizations, safety is not considered as important to the line manager as many, if not most, of the other duties performed. On what does a manager's attitude toward safety depend? The book *Managerial Attitudes and Performance* by Lyman Porter and Edward Lawler,[135] does a beautiful job of examining this and most of what we say here is based on what they have said.

Porter and Lawler build on motivation theory to construct a model of supervisory behavior. It is their basic model which we will use to describe supervisory safety behavior. Motivation theory, in general, attempts to explain how behavior gets started, how it is energized, sustained, directed, and stopped. While a great number of motivation theories have been proposed, there are only two theories that have been developed to a comprehensive state. The first of these is the *drive X habit theory* and the other is *expectancy X value theory.* The model constructed by Porter and Lawler that we are using is based on the expectancy X value theory of motivation.

Expectancy theory states that people have expectations or anticipations about future events. These take the form of beliefs concerning the likelihood that a particular act will be followed by a particular outcome. Such beliefs or expectancies could take values between 0 (no chance) and 1 (completely sure it will follow). Stating this basic theory in simpler terms, if line supervisors expect they will get something favorable for performing a safety function every time it is performed, they will be quite likely to perform it.

Porter and Lawler's examination of managerial attitudes is built around the theoretical model shown in Exhibit 4.1. The model suggests that managerial performance is dependent on three primary variables: abilities, role perception, and effort expended. All are important, and managers will not turn in the kind of performance we want unless all three are taken into account.

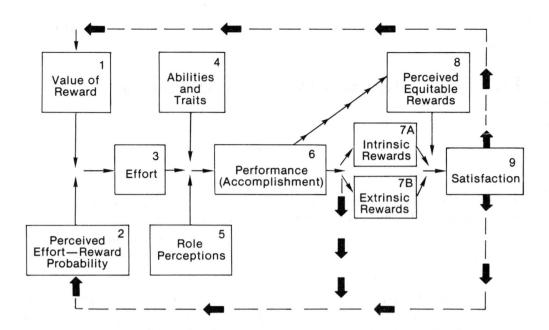

Exhibit 4.1 The Management model. [135]

There are two basic factors which determine how much effort employees put into their jobs. Their opinion of the value of the rewards and the connection they see between their effort and those rewards. This is true of a manager's total job, as well as of any one segment of it, such as safety.

VARIABLES IN THE MODEL

Value of Reward

This variable refers to the attractiveness of possible outcomes to individuals. If these outcomes are positive they can be thought of as rewards. The various rewards that a person might hope to obtain are the friendship of fellow workers, a promotion, a merit salary increase, or an intrinsic feeling of accomplishment. A given potential reward means different things to different individuals. For example, the friendship of peers (i.e., workers at the same job level) might be highly desired by one worker and be unimportant to another. A promotion might have very little positive value for one because of lack of desire to take on increased responsibilities, but for a middle manager in a large corporation, a promotion might be a reward of extremely high value. Thus, value of reward refers to how attractive or desirable is a potential outcome which results from action.

In safety, the manager looks at the work situation and asks, "What will be my reward if I expend effort and achieve a particular goal?" If the value of the reward which management will give for achieving the goal is considered to be great enough, the manager will decide to expend the effort. In effect, the line manager looks at the rewards that management offers and makes a judgment about whether those rewards are great enough. Chances are that the manager will decide they are if the rewards are in terms of advancement and additional responsibility, rather than some of the lesser enticements that management too often selects. In safety, just as in other areas, management's chosen rewards are often too small and too unimportant to entice the line manager.

Effort-Reward Probability

Effort-Reward Probability refers to an individual's expectations concerning the likelihood that the rewards depend on effort. Such an expectation can be divided into:

— The probability that reward depends on performance.
— The probability that performance depends on effort.

In safety, a manager asks the following questions when assessing how probable it is that the rewards really depend on effort:

1. Will my efforts here actually obtain the results wanted or are there factors involved beyond my control? (The latter seems a distinct possibility.)
2. Will I actually get that reward if I achieve the goal?
3. Will management reward me better for achieving other goals?
4. Will it reward the other manager (in promotion) because of seniority, regardless of my performance?
5. Is safety really that important to management or are some other areas more crucial to it right now?
6. Can management really effectively measure my performance in safety or can I let it slide a little without management's knowing?
7. Can I show results better in safety or in some other area?

Effort

Effort is a key variable in this model and it should be clearly understood as being different from performance. Effort refers to whether or not the supervisor tries; performance refers to whether or not he or she succeeds.

Abilities and Traits

This refers to the supervisor's characteristics, personality traits, intelligence, and manual skills. Is the individual able to perform? In safety, this means our ensuring, through selection and supervisory training, that the line manager has sufficient safety knowledge and abilities to control the people and the conditions under which they work. In most industries lack of knowledge is not a problem. Usually the line manager knows far more about safety than is ever applied. Many people agree that a manager can achieve remarkable results on accident records merely by applying management knowledge, even with little safety knowledge. If a manager does not have adequate safety knowledge, the problem is easily handled through training.

Role Perceptions

This refers to the direction of effort—the kinds of activities the supervisor believes necessary to perform the job successfully. If the supervisor's role perception corresponds to those of management then effort will be applied where it will count the most for successful performance as defined by the organization. If the supervisor's perceptions are wrong (do not correspond to those of management), then a great deal of effort may be expended without successful performance. In safety, role perception means: "Does the supervisor know management's desires on accident control?" and "Does the supervisor know what the duties are?"

In the area of role perception, we should be searching for answers to some questions about our organization and about each line manager in our organization. These questions should concern the content and effectiveness of management's policy on safety, the adequacy of supervisory training, the company safety procedures, the systems used to fix accountability, etc.

Performance

Performance is the end result of effort. It is that which we measure.

Rewards

This refers to desirable outcomes provided by the supervisor (intrinsic) or by others (extrinsic), either monetary or psychological. In safety, there is always an intrinsic reward, but perhaps not as often an extrinsic reward. Rewards are desirable states of affairs that a person receives from either personal thinking or the action of others (management). For predicting future performance, the most important things to know about rewards are their perceived size and their perceived degree of connection to past performance.

Perceived Equitable Rewards

This refers to the level or amount of reward that the supervisor expects as the result of a given level of performance. In safety, this often boils down to a simple assessment of the amount and kind of reward the supervisor thinks would be correct for safety effort. In many cases today the supervisor does not expect to receive much reward for safety effort, so therefore does not give much. This, however, becomes a real opportunity for management. If we can improve the reward structure for safety performance to the point where the supervisor will be pleasantly surprised rather than disappointed, it will reinforce continued safety effort.

Satisfaction

This is defined as the extent to which the rewards actually received meet or exceed the perceived equitable rewards. The comparison the supervisor makes between what is perceived as an equitable reward for safety performance and what is actually received (both extrinsic and intrinsic) for safety performance determines satisfaction, dissatisfaction, or pleasure with results of the effort. This level of satisfaction then serves to influence effort in the future.

Feedback Loops

Exhibit 4.1 shows three different feedback loops. The first of these goes from performance (accomplishment) (after box 6) back to the perceived effort reward probability (box 2). What this loop says is that achieving what is set out to be achieved will help to convince the supervisor that it can be achieved. Successful accomplishment will tend to increase perception of the probability that an attempt to do it will bring success.

The second feedback loop goes from satisfaction (box 9) back to perceived effort reward probability (box 2). This loop says that if the supervisor is satisfied after comparing the reward received with the expected reward (perceived equitable rewards), then a higher probability will be perceived the next time.

The third feedback loop goes from satisfaction (box 9) back to the value of the reward (box 1). This says that satisfaction from the last experience reinforces the belief that the value of the reward is there. The supervisor will be "more sure" of the reward than before.

This then is the model. While it may look complex, it really is not. The model states that the determinants of performance are abilities and traits, role perception, and effort. It further states that effort is dependent on the reward and the probability that reward is connected to effort. Upon accomplishment, the supervisor receives intrinsic and extrinsic rewards. These rewards are compared with the perception of what is equitable and determines satisfaction or dissatisfaction. The level of satisfaction influences the perception of rewards and the effort-reward probability for the future.

APPLICATION TO SAFETY

The model is very applicable to safety management and perhaps merely asking these kinds of questions on each box in the model will best summarize this applicability:

Box 1   What is the reward for safety performance? Compared to the rewards for production, quality, etc.?

Box 2   Does the supervisor believe this probability is high? Will attempts bring success? Will success bring rewards?

Box 3   The reward and the effort reward probability determine the amount of effort. How much effort will your supervisor give in safety?

Box 4   Abilities and traits are a function of your selection and training. What are their levels in your company in safety?

Box 5   Role perception is a function of safety policy and the supervisors' observations. Do they perceive safety as their function?

Box 6   Performance is a function of ability, role perception, and effort. What performance are you getting?

Box 7   The reward received is both internal and external. Is the external reward there? In what forms?

Box 8   The supervisor expects a certain reward for accomplishment. What is the expectation?

Box 9   The supervisor compares the reward to the expectation. Satisfaction (or lack of it) influences 1 and 2 which influence efforts from this point on. What comparison have your supervisors made?

Chapter Five

The Determinants of Supervisor Safety Performance

Exhortation is used more and accomplishes less than almost any behavior-changing tool known to man.

Robert Mager[102]

As previously discussed, supervisory safety performance, as viewed in the Porter and Lawler model, is determined by three broad areas: the supervisor's abilities and traits, the perception of safety as part of the total role in the company, and effort. All three of these can be influenced to a large degree by management and by the safety manager through management. While we will look in some detail at each of the three, we can introduce this examination with an overview of some fo the variables involved:

1. The supervisor's abilities and traits as they relate to safety seem to be determined to a large degree by the management tools of supervisory selection and supervisory training. Management cannot alter a supervisor's traits to a large degree but it can select supervisors who have the traits that the tasks seem to need. Of course, the supervisor's ability level can be changed markedly through training.
2. The supervisor's role perception in safety is almost entirely controlled by management. Role perception is a function of how clearly the boss's wishes are known as indicated by the safety policy and by how well that policy is reinforced.
3. Supervisory effort, as the model has shown, is dependent on the value of the reward for performance and on the probability or likelihood that the reward will actually follow effort. These variables are controlled to a large measure by management.

It would appear that supervisory performance is dependent on management as evidenced by the safety manager and the company's policy and systems (the safety program).

TRAITS AND ABILITIES

Traits

What traits are needed for safety supervision? We have no idea. There has been no research directed toward this question; however, there has been much written on leadership generally. Many of the early books dealing with leadership included lists of the psychological characteristics which are essential to leadership—one had nineteen traits, one had thirty-one traits, and one had ten traits. The traits most frequently noted were intelligence, noted by ten studies; initiative, by six; extraversion and sense of humor, by five; enthusiasm, fairness, sympathy, and self-confidence, each by four. One study of Boy Scouts showed a high positive correlation between a rating of leadership and specific traits. Rating of leadership correlated with: intelligence, 0.87; athletic prowess, 0.62; dependability, 0.87; acceptable and pleasant voice, 0.54; appearance, 0.81. General ratings of leadership seem to be consciously or unconsciously based on physique, appearance, and the like, so that such high correlations with specific traits may be somewhat misleading.[73]

Many studies have been undertaken to ascertain the personality traits of the leader. When personnel managers are discussing with other executives the kind of person who can fill a management vacancy, certain terms, irrespective of function or level, keep cropping up with monotonous regularity, e.g., analytical, decisive, determined, or acceptable, thus indicating that most managers still accept the personality-trait approach. Nevertheless, many researchers into leadership today have pretty well rejected the personality-trait approach.

In general, results in this field would appear to support the notion that while certain minimal abilities are required of all leaders, these traits are also distributed among other members of the group. Many social psychologists hold the view that leadership traits may exist, but if they do, they have not been recognized. In contrast to this, most managers (including many personnel managers) still believe that they have the ability to select good leaders on the basis of a brief interview. This optimistic if somewhat naive attitude springs from the belief that a manager is a person who has some of the following characteristics (which ones and how many depends on the prejudices of the individual making the selection): analytical, intelligent, "not too bright," keen, enthusiastic, aggressive, capable of maintaining smooth interpersonal relationships, persuasive, dominant, personally acceptable, tactful, extraverted, well-balanced, having the need to succeed, ambitious. Different selectors combine such lists of personality traits in their own ways according to their own particular bias. But is is just as well to remember that social psychologists are in general agreement that the numerous studies of the personalities of leaders have not produced any consistent or valid pattern of traits which distinguishes leaders from other group members.

The Psychologically Distant Manager.   Professor F. E. Fiedler,[73] (University of Illinois), has shown that the leaders of more effective groups, when evaluating subordinates, indicate that they maintain greater psychological distance between themselves and their subordinates than do leaders of less-effective groups. The explanation for this appears to be that a manager cannot properly control and discipline subordinates if there is too close of an emotional attachment. Likewise, a manager emotionally dependent on a subordinate cannot afford to alienate that subordinate for fear of losing support.

Two distinctly different portraits emerge from Fiedler's studies. The psychologically distant manager is more efficient and has the following characteristics:

1. Tends to formalize role relationships both with superiors and with subordinates.
2. Tends to be somewhat reserved and withdrawn in relationships within the firm.
3. Has a preference for formal staff consultation rather than seeking opinions informally.
4. Accepts and rejects subordinates on the basis of performance.
5. Though reserved, is still sufficiently skilled to ensure smooth interpersonal relationships.
6. Does not develop deep friendships with colleagues.
7. Demands and gets considerable freedom of action from superiors.
8. Expects subordinates to make mistakes and plans accordingly.
9. Prefers ambitious subordinates.

The Psychologically Close Manager.   The psychologically close manager is relatively inefficient compared to the distant manager and has the following characteristics:

1. Does not seek to formalize role relationships.
2. Behavior is more concerned with ensuring good human relationships within the organization often at the expense of efficiency.
3. Prefers informal discussion with subordinates rather than regular staff meeting.
4. Is inclined to select friends from colleagues and subordinates in the firm.
5. Is inclined to dominate and possess subordinates.
6. Will only delegate on relatively minor matters and insists on frequent individual consultation.

In other words, the psychologically distant manager is a task specialist, and the psychologically close manager is a human relations specialist. Fiedler lays out three important dimensions of the total situation which structure the leader's role:

1. Leader-member relations or the extent to which the leader enjoys the confidence and loyalty of subordinates.
2. Task structure or the extent to which the task represents an "order from above."
3. Position power or the power inherent in the position of the leader.

Another factor to consider in looking at the traits of supervisors whom we intend to have some control over through selection is the actuality of promoting a person into supervision. Professor Dalton,[32] a sociologist (University of California, Los Angeles), presents a fascinating analysis of managerial promotion in the "Milo Fractionating Center," a firm with a total work force of about 8,000. He analyzed the promotion scheme not in terms of formal company policies but in terms of actual patterns of interactions. The company's official policy indicated that ability, honesty, cooperation, and industry were important criteria for advancement. In addition, criteria such as age, employment background and service, formal education, and relevant training are often believed to be of importance. Dalton, however, found that none of these factors were really pertinent in actual promotion at Milo. Analysis of the occupational data, for example, revealed the results shown in Exhibit 5.1.

Managerial group	Data categories	Mean	Median	Range
First-line foremen	Age at appointment	36.0	37.0	16-58
	Years service at appointment	16.2	17.0	1-31
	Years education	11.0	11.0	6-20
	Current age	48.5	48.0	31.65
General foremen	Age at appointment	44.4	44.5	26.62
	Years service at appointment	16.2	17.0	1.31
	Years education	11.8	12.0	8.16
	Current age	50.0	50.0	35.65
Superintendents	Age at appointment	41.4	41.5	25.58
	Years service at appointment	19.8	19.0	3-35
	Years education	13.8	14.0	9-19
	Current age	48.7	49.0	35-65
Staff group	Age at appointment	36.6	33.0	24-54
	Years service at appointment	13.0	10.0	3-35
	Years education	15.2	16.0	9.19
	Current age	42.9	41.0	29.61

Exhibit 5.1 Mill manager data.[32]

Dalton found that several unofficial requirements were really the key factors in managerial promotions. These were: being a Mason, being of Anglo-Saxon or German ancestry, being a member of the local yacht club, and being a Republican. The ethnic requirement is particularly revealing if one considers that people of Anglo-Saxon and German origin made up only 38 percent of the surrounding community. Yet they controlled 85 percent of Milo's advisory and directive forces. Dalton also studied management promotion in three other companies and found that similar unofficial requirements were at play.

Thus, there do not seem to be known ideal traits for a supervisor. Even if there were, according to Dalton's work, it is doubtful if managers would actually be selected on the basis of these traits. Where does that leave us in looking at the traits aspect of the traits and abilities determinant of performance? Perhaps only looking at the supervisor's past record and performance in safety (as either a worker or supervisor). That record usually will indicate to us whether or not there is any interest in success at safety. This suggests that a "trait" of "interest in safety," or "past success in safety" might be a factor to look for or to use as one criterion of promotion to supervision in the future.

Abilities

For our purposes in safety supervision, we can use the term *abilities* in our traits and abilities determinant to mean knowledge, knowing "what to do" to get safety results. This becomes primarily a function of our supervisory training in safety.

Training is a powerful influence and motivator in safety, just as it is in many other areas. In training supervisors you give them only two things: knowledge and skill. You get three things back: knowledge, skill, and motivation. This has been shown over and over again in business; the value of training has been proven. Industry today is engaged in more training than ever before in history. Training is, however, often terribly misused. It is too often considered a panacea for problems in industry. Too often management perceives a performance problem and, without diagnosing the problem, decides some training is needed when the problem may actually be in poor procedures, poor selection, or poor management. Training is effective when aimed at defined needs, when analysis shows the problem to be lack of knowledge or lack of skill. It is a waste of time when the problem is elsewhere in the organizational system.

The job training method used to be a single four-step process: tell them, show them, observe them, correct them. Industrial training today is too complex to be boiled down to a simple four-step process, but it does start with one essential step: *define the objectives of your training.* Objectives should always be spelled out before content and method are decided on. Training objectives should be stated in terms of desirable terminal behavior. They should state how the participant should perform once training has been completed. For instance,

As a result of this course, the participant should be able to:

1. Investigate any injury accident to determine five causes, and record the information properly on the accident investigation form.
2. Transmit management's policies on safety to employees.
3. Orient each new employee in safety.
4. Make a job safety analysis of each job supervised.

Actually the above objectives have not been taken far enough. To properly define training objectives you should:

1. Define the terminal behavior by name. In other words, what will be accepted as evidence that the trainee has achieved what you desire? For instance,

   "As a result of this training, the supervisor will be able to investigate an accident."

2. Define the desired behavior further by describing the important conditions under which the behavior will be expected to occur, such as when, how, and where it must be achieved? For instance,

   "As a result of this training, the supervisor will be able to investigate any accident *which occurs in the department and complete the company accident report within 24 hours.*"

3. Specify the criteria of acceptable performance by describing how well the learner must perform to be considered acceptable, or what is acceptable performance? For instance,

   "As a result of this training, the supervisor will be able to investigate any accident which occurs in the department and complete the company accident report within 24 hours, *identifying at least five contributing causes and suggesting at least five corrective actions to be taken by either the supervisor or by management.*"

When your objective statements are this detailed, you will find it much easier to put the training presentations together, for you will know very precisely what to teach and, more importantly, you will find it much easier to determine whether or not you have succeeded in your teaching, for you will know precisely how to test the trainee to see if the terminal behavior is what you have spelled out so clearly.

Training in safety, particularly supervisory training, must be directed not merely at disseminating knowledge but rather at telling participants what management wants done and how to do it. After objectives are defined, content generally comes easily. The last step should be method, which is too often used as a first step. Today there is tremendous choice in method of instruction. In past years a trainer could lecture or discuss; today there are audiovisual aids of every description: films, slides, filmstrips, television programs. Furthermore, there are many programmed instruction texts, management games, etc.

Training should be given to supervisors when there is a defined need, that is, when they do not know how to do what you are asking them to do. The content of a training program is what management wants employees to know. More specifically, "Teach them how to carry out the responsibilities management has assigned." Actually, content, or what to train, is decided as objectives are identified. If objectives are stated in terms of what a participant should be able to do on completion of the training, selection of content is simple. For instance, if one objective is that the participant should be able to make a job safety analysis of each job, one segment of the training will consist of an explanation of a job safety analysis and another segment will consist of the actual training to use that tool.

There are many sources of help for supervisor training, including packaged self-study, programmed and discussion courses. They are available from universities, vocational schools, television, government agencies, insurance carriers, and trade associations. Some courses are excellent, some are good, and some are poor. Some may be a waste of time. The primarily question in selecting outside help should be whether the intended program will fit your defined objectives. Again, the first step is to determine whether or not training is the answer to your defined problem. If it is, the second step is to define the objectives of that training. Only then are the decisions as to content and method made.

ROLE PERCEPTION

One of the three determinants of supervisory safety performance is role perception. Role perception depends first on knowledge of management's stated (and reinforced) desire that safety be a part of the supervisor's job. Management influences the supervisor's perception of the role through training, through systems of fixing accountability, and through policy. It all starts with policy as explained in training and as reinforced through accountability.

Policy

Our thinking starts with written safety policy: a definition of management's desires regarding safety. In discussions regarding policy, we often bog down trying to sort out policy from procedure, from standard operating procedures, from rules. The only important thing is whether or not management's interest is accurately communicated. One dictionary says that policy is "a settled course adopted and followed by a body." When top executives determine and announce a "settled course," they are affirming a shared purpose or one in which they want to enlist the voluntary cooperation of every member of the organization. They are giving out a guide for thinking, and this guide is to be used by the other members who have been delegated authority to make decisions in keeping the company on a given course while realizing company purposes.

A safety policy is management's expression of the direction to be followed. It is management's first step in organizing to accomplish its desires. It is important that management's safety policy be in writing, in almost all cases, to ensure that there is no confusion concerning direction and assignment of responsibility (role perception). Why do we urge written safety policy even when other company policy is not in writing? Why do we propose that safety policy be publicized for all employees when other policies are merely put in the policy manual? Because safety policy, more than most other management policies, requires some action from each individual in the organization, from the president to the lowest rated worker. Safe performance of an organization requires that a decision be made by each person in the organization. The single greatest factor that each person will consider in making a decision for or against safety is, "What does the big boss want from me?"

What is included in the safety policy may vary from company to company. No one policy is right or wrong. Each should be written specifically for the organization it serves. It would seem, however, that we could outline as a minimum some areas which ought to be touched on in any safety policy:

1. *Management's Intent.*   What does management want?
2. *The Scope of Activities Covered.*   Does the policy pertain only to on-the-job safety? Does it include off-the-job safety also? fleet safety? public safety? property damage? fire? products safety?
3. *Responsibilities.*   Who is to be responsible for what?
4. *Accountability.*   Where and how is it fixed?

5. *Staff Assistance.* If there is a safety function, how does it fit into the organization? What should it do?
6. *Safety Committees.* Will there be committees? What will they do?
7. *Authority.* Who has it and how much?
8. *Standards.* What rules will the company live by?

Writing a policy does not solve any problems of fuzzy role perception in supervisors. It only begins the solution. A policy is like a law or a safety rule. If it is not reinforced regularly in some way, it becomes meaningless. In many plants with a "policy" of wearing ear protection when the noise level is over 90 decibels, no one wears that protection in that circumstance. Is that really the policy of the company? Obviously not. Writing words on paper does not keep plugs in ears. Writing words on paper does not make supervisors responsible for safety. And writing a safety policy does not get supervisory safety performance. But, if those words are reinforced by actions, then they become policy and they help a supervisor to perceive that safety is, in fact, a part of the job.

Role Perception Defined

Role perception then is nothing more than whether or not each supervisor perceives, or rather believes, that doing something in safety is really a part of the job. Role perception includes the interpretation of management's priorities. Is there belief, from management's past actions, that safety is as important as, or not as important as, production?

Accountability

The average company, even when it writes a safety policy, does not define any procedures to fix accountability, that is, to measure the safety effectiveness in performance of line managers. Until such procedures are designed, safety policy cannot exist. Fewer than 10 percent of the companies in the United States have written safety policies. A study of 74 safety policies of companies in all business types throughout the United States revealed that only one even mentioned the idea of fixing accountability to the line organization. Based on these figures, only 13 companies in every 10,000 actually do fix accountability for safety. Instead of simply preaching for fifty years that the line has responsibility, we should have been devising procedures to fix such accountability. When someone is held accountable (is measured) by the boss for something, responsibility will be accepted. Without accountability there is no accepted responsibility. Effort will be put in the direction where the boss is measuring, for this is the role perception (by the boss's measurement system).

Effort

The primary way that management influences the supervisor's effort is by determining the value of reward for safety performance and effort-reward probability. This is done through policy consistently and constantly followed up by measurements (systems to fix accountability).

REWARDS AND EFFORT-REWARD PROBABILITY IN SAFETY

Rewards for safety performance are no different from rewards for performance in any other area where management asks for performance from the supervisor. While our model focuses primarily on positive rewards (peer acceptance, subordinate approval, enhancing the likelihood of promotion, merit salary increases, higher bonus, intrinsic feelings of accomplishment, pat-on-the-back, compliment from the boss), it also could mean negative reward (chew-out, lower and harder pat-on-the-back, reprimand).

The main difficulty in developing a reward system is not in determining what the rewards will be for performance but rather in determining when the reward should be given, that is, when the supervisor has "earned" it. For in safety we have precious few good measurement tools that tell us when a supervisor is performing.

MEASURING TOOLS

In safety work, there are three ways of measuring supervisory performance. We can measure the activities of the line, we can measure the results of those activities or we can use a combination of both. The most used measurements seem to be based on results.

Results Measurements

Exhibit 5.2 shows a partial listing of the things that we might consider measuring for results. One of the simplest means of doing this is to charge accidents to the department in which they occurred. If supervisor A has a man suffering a disabling injury, it shows up on supervisor A's record. An adaptation of this is to charge the claim costs back to the line. Here we are measuring the line supervisor in terms of dollars, which is a far better measuring stick than any other that we have today. The dollar sign means something to that supervisor. Every accident can go directly into the departments profit and loss statement; the cost will come out of the working budget.

SYSTEMS
1. Charge accidents to departments
   a. Charging claim costs to the line
   b. Including accident costs in the profit and loss statements
2. Prorate insurance premiums
3. Put safety into the supervisor's appraisal
4. Have safety affect the supervisor's income

RESULTS MEASUREMENTS
1. Number of incidents
   a. Accidents
   b. Injuries
   c. Other
2. Costs
3. Frequency and severity indicators
4. Estimate costs
5. Loss ratios
6. Costs of damage
7. Number of unsafe acts
   a. Sampling

Exhibit 5.2 Measurement of results.

Prorating insurance premiums merely means to charge back to departments their "share" of the insurance premium dollar in the same rate as their "share" of the total corporate dollar losses. Putting safety into appraisals is effective for when line supervisors are appraised on safety records as well as on production records, they generally becomes far more interested in accident prevention and begin to do something about it. This is often overlooked in the appraisal of line managers. Also, when a line supervisor's paycheck is in some way influenced by the accident records as well as by the production records, safety becomes a primary thing to worry about and consider. This has been very effectively used in some construction companies. When a job superintendent must subtract accident costs from the profit figure for a job and when a personal bonus for the job is reduced or wiped out because of it, accident prevention becomes an important factor on the next job.

Also shown in Exhibit 5.2 is a list of results that might be measured. Most items listed are self-explanatory.

Two of the techniques utilize the dollar as the measuring stick instead of measuring supervisors in terms of number of accidents, number of days lost, or the commonly used frequency rate or severity rate. Many people today believe that the dollar is a far better measuring stick than any other in safety, and many companies are beginning to utilize it effectively. However, it is difficult to utilize the dollar for the following reasons:

1. In serious accidents, actual claim costs are not available for a long period of time, in some cases years after the accident has happened.
2. There is no way to convert a frequency rate or a severity rate into dollars.
3. If a company operates in several states, the actual claim costs will be an unfair measuring stick because the benefits vary so much from state to state (although this is now changing).

To overcome these difficulties, it is possible to use a system in which the costs are estimated by a predetermined formula based on previous costs for similar accidents in the state where the company operates. Thus, when an accident does occur, an estimate of the final cost can be computed readily using this formula. (Such a system is described in *Techniques of Safety Management.*[127])

Loss ratios are seldom used, but are a feasible measurement. They can be computed by dividing the dollar loss (real or estimated) by a computed departmental premium (man-hours times a rate obtained from the carrier). An advantage of loss ratios is that there is a rough standard built in by virtue of the fact that because there is a "break-even" loss ratio in that rate. Find out what it is (often around 60 percent). It becomes an arbitrary point each supervisor must stay under. Costs of property damage should also be included if they result from incidents that you are attempting to control.

Activities Measurements

Exhibit 5.3 lists some of the items that management might measure the line organization against to determine what they are doing to prevent accidents from occurring. This is perhaps more important than the measurement of results because it measures line effort in controlling losses before the accidents have happened.

Management can measure line supervisors to see whether or not they are utilizing such techniques of accident control as toolbox meetings, job hazard analyses, inspections, accident investigations, incident reports, safety committees, and safety meetings. Management may require line supervisors to submit activity reports. When management measures these activities, it is set-

```
ACTIVITIES TO BE MEASURED
1. Safety meetings that supervisor holds
2. Tool box meetings
3. Activity reports on safety
4. Inspection results
5. Accident investigations
6. Incident reports
7. Job hazard analysis

         SYSTEMS
1. Regular reports
2. Sampling
3. SCRAPE
4. Performance rating
```

Exhibit 5.3 Measurement of activities.

ting up a system of accountability for activities. It is emphasizing to the supervisor that management wants performance in safety. It is, in terms of our model, raising the supervisor's perception of the probability that effort will be rewarded.

Exhibit 5.3 also mentions some systems. Regular reports required from supervisors is a simple system. An example of such a report is shown in Exhibit 5.4. Sampling is a possible system and it could be statistical sampling or a simple sampling of acts and conditions made periodically by management using a form like the one shown in Exhibit 5.5.

Safety sampling is a tool which has the potential of telling management which line supervisors are doing their job in safety and which are not. Safety sampling is a method of systematically observing workers in order to determine what unsafe acts are being committed and how often they are occurring. The results of these obsevations are then used to measure the effectiveness of line safety activities (see Appendix A).

In utilizing the tool of safety sampling, first a list of unsafe codes is prepared. The most common unsafe acts are listed on a form. Next, a sample is taken by walking rapidly through the operation and observing each employee quickly. An immediate decision is made whether or not each employee is working safely. If it is noted that the employee is working safely, it is checked off on a theater counter as one safe observation. If the employee is working unsafely, the observer will record this as one unsafe act. The third step is to validate the sample statistically to determine whether there are enough observations to constitute a representative sample. The fourth and final step is to prepare a report for management. This shows each supervisor's rating expressed as a ratio of safe to unsafe acts. This can be compared with past records and with the records of other departments. Management then can judge line performance and apply whatever action is necessary. Safety sampling measures supervisory performance.

SCRAPE (System of Counting and Rating Accident Prevention Effort) is a systematic method of measuring accident prevention effort. The SCRAPE rate indicates the amount of work done by a supervisor and by the company to prevent accidents in a given period. Its purpose is to provide a tool for management showing before accidents whether or not positive means are being used regularly to control losses (see Appendix B for details). Performance rating is similar in that it attempts to quantify the supervisor's effort (as well as role perception and ability). This also is covered in Appendix B.

Built into the above systems are some of the regular elements of our normal safety programming. These tools should be reexamined in the light of how they are to be used. They should be surveyed from new angles—not necessarily from the same perspective in which they are usually

REPORT OF SUPERVISOR'S SAFETY ACTIVITIES

Supervisor _____ DEPARTMENT _____

Date _____ This Report Covers _____ To _____

Inspections Made

Date of inspection _____No. Hazards Corrected ____No. Recs. to Mgmt. _____

Date of inspection _____No. Hazards Corrected ____No. Recs. to Mgmt. _____

Date of inspection _____No. Hazards Corrected ____No. Recs. to Mgmt. _____

Comments:

Meetings Held

Tool Box Meetings   Date _____ No. Employees _____ Subject _____

Date _____ No. Employees _____ Subject _____

Date _____ No. Employees _____ Subject _____

Other Meetings (Explain)

Accidents Investigated

Number of Accidents Investigated This Period _____

Number of Hazards Corrected _____

Number of Recommendations to Management _____

Comments:

Employee Contacts

New Employee Safety Orientation

Name _____ Date _____ Name _____ Date _____

Name _____ Date _____ Name _____ Date _____

Name _____ Date _____ Name _____ Date _____

Other Employees

Name	Date	Subject	Name	Date	Subject

Use of Safety Materials

List Materials Used this Period

Accident Record	This Period	Year to Date
Number First Aid Cases		
Number Doctor Cases		
Number Lost Time Cases		
Man Hours Worked		
Frequency Rate		
Severity Rate		
Comments:		

Exhibit 5.4 Report of supervisor's safety activities.

MANAGER'S SAFETY SURVEY

PART 1. UNSAFE ACT SAMPLING

A. Number of Safe Observations

B. Number of Unsafe Observations

Unsafe Act	Dept.				
No Safety Glasses/Unauthorized Glasses					
Not Using Machine Properly					
Machine Unguarded/Guard Not Adjusted					
Not Using Tools, Jigs, Pushsticks, etc.					
Working Near Tripping Hazard					
Improper Use of Air Nozzles					
Improper Use of Hand Tools					
Loose Clothing Near Machine					
Improper Lifting/Positioning					
Climbing on Racks					
Unsafe Loading/Piling/Storing					
Using Defective Equipment					
Other (Specify)					

C. Percentage

		Last Per.		Last Year	
Dept.	%	Dept.	%	Dept.	%

PART II. UNSAFE CONDITION INSPECTION

D. Violations	Dept. ⟶				
Machine Guards — Transmission					
— Point of operation					
— Missing					
— Not properly adjusted					
Electrical — Cords					
— Grounds					
Flammables — Amount					
— Use					
— Extinguishers					
— Exits					
Falls Floor surfaces, objects					
Air Nozzles					
Hand Tools — Guards					
— Grounds					
Other (Specify)					

E. Number

		Last Per.		Last Year	
Dept.	No.	Dept.	No.	Dept.	No.

Exhibit 5.5 Manager's safety survey.

seen. In the past, we have tended to use inspection for the purpose of seeking out hazards. We have used accident investigation for the purpose of identifying an unsafe act or an unsafe condition and we have used record keeping to compute frequency and severity rates. Inspections have been used to spot conditions, but seldom to spot acts. Investigations have been used to unearth symptoms more often than causes. Records have been used to tabulate accident types, accident agencies, and injury types more often than accident causes. Let us look at each briefly.

1. *Inspections.* The single most important reason for management making inspections is seldom mentioned. It is to measure the supervisor's performance in safety. If this inspection is used as a measurement tool, the line manager will inspect more often to ensure that conditions remain safe and that fewer unsafe acts occur and not wait until the safety specialist comes around to do the inspection job.

   It is generally agreed that responsibility for conditions and for people belongs to the line supervisor. Thus so should responsibility for the primary safety inspection. By *primary safety inspection* we mean the inspection intended to locate hazards. Any inspections performed by staff specialists then should be only for the purpose of auditing the supervisor's effectiveness and be a direct measurement of safety performance and effectiveness.

2. *Investigation.* The primary accident investigation function has always been the supervisor's. The tools that we give the supervisor ought to lead to determination of some of the many underlying causes. It is proper that the line supervisor should investigate and be allowed to determine what really happened. The form in Exhibit 5.6 would force identification of at least five causes and encourage dealing with all of them. If we, as management, are going to measure performance in investigation, then we must routinely rate the supervisor (see Exhibit 5.6).

3. *Injury Records.* Injury records should be designed so that they measure the line manager and to measure the results of the line manager's safety performance, they should be set up so that:
   a. The accident records are kept by supervisor (by department).
   b. They give some insight as to how the accidents seem to be happening (agency, body part, event, etc.).
   c. They are expressed eventually in terms of dollars by department (by supervisor).
   d. They conform to any legal and insurance requirements.

      The past performance of any group is the best standard to use as a guide for present performance. The people in the group understand it and will accept it, whereas they are often reluctant to accept an arbitrary standard set by someone outside the group. Often they believe that competing against a different group is unfair, since no two groups face the same challenges, hazards, or situations. Management is interested principally in improving its own company.

When past performance is used as a standard, it is often difficult to know whether or not the changes that have occurred are truly significant, reflect a different supervisory performance, or are merely chance happenings or random fluctuations. There is a technique now in use which helps us to determine this, called the *Safe-T-Score* (described in *Techniques of Safety Management*).[127]

These then are some of the measurement tools and techniques which allow us to find out who is performing. We then can direct those rewards that we do control to those supervisors who are performing. In doing this we will ensure future effort in safety since the supervisor will know better the value of the reward and will have a better perception of the probability that a known reward does follow effort. Similarly, we will know through those measurements where any negative rewards should be directed.

SUPERVISOR'S REPORT OF ACCIDENT INVESTIGATION

Name of Injured _____ Date of Accident _____ Time _____
Seriousness: Lost Time   Doctor   First Aid Only   Near Miss
Nature of Injury _____
What Happened? _____
_____
_____

What acts and conditions* were involved? What causes them? How were they corrected? *At least five (use back also).

Unsafe Act/Condition/Symptom	Possible/Probable Cause	Correction/Suggested Correction
1.		
2.		
3.		
4.		
5.		

Supervisor _____ Department _____

**INVESTIGATION RATING
SHEET**

		Circle One	
1. Was it on time?		Yes-5 pts.	No-0 pts.
2. Was seriousness indicated?		Yes-5 pts.	No-0 pts.
3. Does it say where it happened?		Yes-5 pts.	No-0 pts.
4. Can you tell exactly what the injury is?		Yes-5 pts.	No-0 pts.

	Circle One					
5. How many acts and conditions are listed?	5	4	3	2	1	0
6. How many causes are identified?	5	4	3	2	1	0
7. How many corrections were made or suggested?	5	4	3	2	1	0
8. How many of the listed corrections would have prevented this accident?	5	4	3	2	1	0
9. How many corrections are permanent in nature?	5	4	3	2	1	0
10. In how many of the corrections listed is the supervisor **now** doing something differently?	5	4	3	2	1	0

Total of Circled Points _____
Multiply X 2 _____

Reviewed by _____
General Foreman   SCORE _____

Exhibit 5.6 Supervisor's report of accident investigation.

Chapter Six

Evaluating Performance Problems

Solutions to problems are like keys in locks; they don't work if they don't fit.

Robert Mager[103]

The final area we will look at in this section on motivating the key person assumes at the outset that this key supervisor is *not* doing all we want done in safety. We start out assuming there is a safety performance problem and then we try to determine why there is a problem. Should we expend our effort on clarifying the role perception? Or should we concentrate on the reward and the perception of how that reward relates to expended effort? Or should we concentrate on abilities and bring them up to a higher level?

Let us look first at the latter and, and we so often do in real life, let us start out assuming lack of performance, and therefore, probable need for training. The safety manager spends a good deal of time in training activities. Training is suggested an utilized as a solution to safety problems more than any other type of solution. It is true that training is a powerful influence over employee behavior. It is true that when you train you not only provide knowledge and skill but also motivation. However, there are some cautions to be observed. One of the better texts on industrial training (McGehee and Thayer's[96]) warns that:

> Proper utilization of training in modern industry and business requires that it be put in its proper context. It is not an end in itself, but a means to an end.

They further state that:

> To be effective, this management tool must be used when and where it is needed and not as window dressing to impress visiting firemen with the alleged personnel-mindedness of an organization. We suspect that the effectiveness of a training program may be an inverse function of the elaborateness of the lithography, and the multiplicity of forms and manuals which are shown visitors.

It seems particularly true in the safety profession that training is often used as the solution selected regardless of the problem defined (or without any definition of the problem). The purpose of this section is to look at the identification of training needs, or rather the definition of the problem, for the purpose of determining whether or not it is a training or some other type of problem.

Peter Pipe and Robert Mager[103] recently published one of the simplest and finest books available on problem definition entitled *Analyzing Performance Problems* (or *You Really Oughta Wanna*). The entire book is written around the flowchart shown in Exhibit 6.1. The

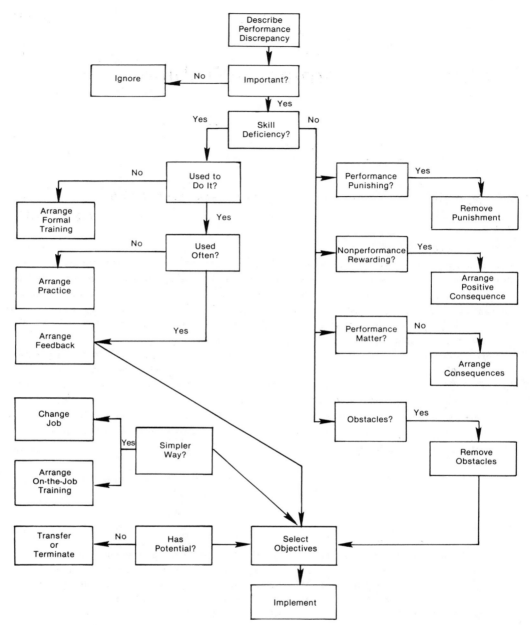

Exhibit 6.1 Mager's basic model. [103]

original flowchart in included here for its value in organizational nonsafety problem analysis, but our main purpose is to adapt this chart to the field of safety management and to use it as a tool to help us identify and analyze safety problems as a prestep to objective setting. We could look at a flow chart or model such as this in two ways: (1) from the standpoint of a safety manager (or middle manager) analyzing the performance discrepancies of first-line supervisors or (2) from the standpoint of a first-line supervisor analyzing the performance problems of the workers.

ANALYZING A SUPERVISOR'S PERFORMANCE DISCREPANCIES

Placing ourselves in the position of the corporate safety manager with a job of assisting the first-line supervisor, we find the original model slightly adapted as shown in Exhibit 6.2. We start by a simple description of a performance discrepancy such as, "Joe is not adequately investigating accidents," or "George is not teaching his new employees the safety rules."

Our second step is to ask the question, "Is the discrepancy important?" The central idea here, of course, is whether or not the alleged problem is really a problem and whether or not the discrepancy is really worth spending our time on. For instance, we may have cited George's "attitude" as a problem. Every time we survey his department he gets highly abrasive and irritated with us for interfering with "his" safety program. Is this a problem? Maybe. Maybe not. It could be a serious problem if George is also not doing a job (not performing) in safety. It may well be no problem if he is performing and, in fact, does not need any help from us. Assuming, however, that the discrepancy is important, we go to the next step which is the determination of whether or not the performance discrepancy is due to a skill deficiency. This

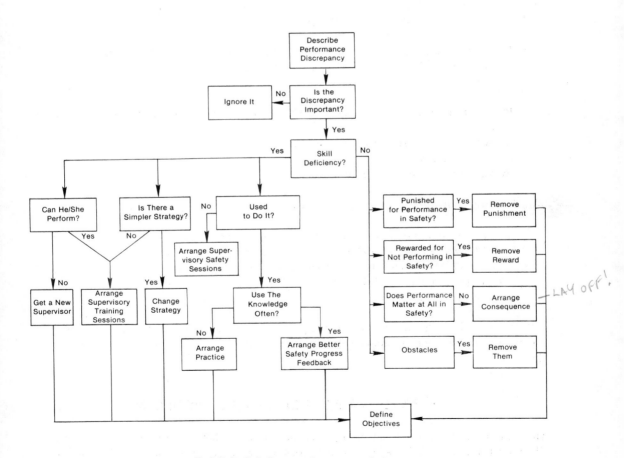

Exhibit 6.2 Supervisory model.

is the key branching point in the model for it leads us to either a training or a nontraining solu-tion. How do we make this determination? Mager and Pipe suggest a simple question: "Could he do it if his life depended on it?" If the answer is no, there is a skill deficiency—it is a training problem. If the answer is yes, it is not a training problem.

Let us assume we answer this key question "yes" and branch left on Exhibit 6.2. We might then ask whether or not he ever knew how to perform satisfactorily. If the answer is no, we must provide some real initial training. If the answer is yes, he only needs practice and we should arrange situations where he can "brush-up" on his unused skill.

A second question might be, "Is there a simpler or different strategy or approach he might use to get results? Would 5-minute safety talks to the group achieve his purpose instead of in-dividual employee contacts? Would delegating the inspection task to a lead person serve the purpose? If the answer is yes, a simple change may suffice. If no, we are back to supervisory training to improve his performance in the needed strategy. We must, for instance, train him to properly investigate accidents.

A third question is, "Once trained, can he perform satisfactorily?" If we answer this no, we have the wrong man. If we answer it yes, we again come back to supervisory training. After answering these three questions, we are now ready for the definition of our training (or non-training) objectives. In other words, we now are ready to sit down and specify what we are going to do. Actually to properly define our training objectives, we have a somewhat more com-plex task than just specifying what we are going to do. We will briefly discuss this task later.

There are four questions we ask on the right branch of the model:

1. Is he punished for performing in safety? This may sound unreal, but actually it is quite real in safety. There are several ways a supervisor could be punished for performing in safety.

   By a manager who rewards only for production, thus punishing a supervisor for spending time only on safety.
   By peers who ridicule safety performance.
   By subordinates who scoff at him for his efforts.
   By management which undermines his safety efforts.

If such punishment exists it should be identified and attempts made at removal.

2. Is he rewarded for not performing in safety? Is he rewarded more for performance in other areas? He will spend time and effort where there is potential for reward.
3. Does safety performance matter at all? Or does nothing happen to a supervisor for either good or poor safety performance?
4. What other obstacles lie in the way for safety performance?

Either branch of the model ends up at a point where we are ready to define objectives. This supervisory model might be summarized by the checklist (Exhibit 6.3) of key issues and ques-tions to ask.

ANALYZING A WORKER'S SAFETY PROBLEM

The same general approach can be used for diagnosing worker safety performance problems. The model in Exhibit 6.4 has been adapted for this. This model can be used by first-line super-visors as a prestep to setting objectives for worker improvement. In the employee model we ask almost the same questions in the same order as for the supervisor model. We initially describe what the employer is doing wrong, determine if it is really hazardous, and then determine whether or not it is a skill deficiency by our question, "Could the employee do it if a life depended on it?"

If we determine that it is a skill deficiency, we ask three questions similar to those asked of the supervisor: Did the employee use to know? Is there a simpler way? and Does the employee have potential? The answers to these questions lead up to giving workers more feedback, more practice, more training, job changes, or transfers.

If we determine that we do not have a skill deficiency, we then ask the same four questions as before:

I. The supervisor isn't doing what he should be doing in safety. I think I've got a training problem.
    1. What is the performance discrepancy?
        ____ Why do I think there is a training problem?
        ____ What is the difference between what is being done and what is supposed to be done?
        ____ What is the event that causes me to say that things aren't right?
        ____ Why am I dissatisfied with the situation?
    2. Is it important?
        ____ Why is the discrepancy important?
        ____ What would happen if I left the discrepancy alone?
        ____ Could doing something to resolve the discrepancy have any worthwhile result?
    3. Is it a skill deficiency?
        ____ Could he do it if he really had to?
        ____ Could he do it if his life depended on it?
        ____ Are his present skills adequate for the desired performance?
II. Yes. It is a skill deficiency. He couldn't do it if his life depended on it.
    4. Could he do it in the past?
        ____ Did he once know how to perform as desired?
        ____ Has he forgotten how to do what I want him to do?
    5. Is the skill used often?
        ____ How often is the skill or performance used?
        ____ Does he get regular feedback about how well he performs?
        ____ Exactly how does he find out how well he is doing?
    6. Is there a simpler solution?
        ____ Can I change the job by providing some kind of job aid?
        ____ Can I store the needed information some way (written instructions, checklists) other than in someone's head?
        ____ Can I show rather than train?
        ____ Would informal (i.e., on-the-job) training be sufficient?
    7. Does he have what it takes?
        ____ Could he learn the job?
        ____ Does he have the physical and mental potential to perform as desired?
        ____ Is he over-qualified for the job?
III. It is not a skill deficiency. He could do it if he wanted to.
    8. Is desired performance punishing?
        ____ What is the consequence of performing as desired?
        ____ Is it punishing to perform as expected?
        ____ Does he perceive desired performance as being geared to penalties?
        ____ Would his world become a little dimmer (to him) if he performed as desired?

Exhibit 6.3 Checklist of key questions. [103]

9. Is nonperformance rewarding?

_____ What is the result of doing it his way instead of my way?

_____ What does he get out of his present performance in the way of reward, prestige, status, jollies?

_____ Does he get more attention for misbehaving than for behaving?

_____ What event in the world supports (rewards) his present way of doing things?

(Are you inadvertently rewarding irrelevant behavior while overlooking the crucial behavior?)

_____ Is he "mentally inadequate," so that the less he does the less he has to worry about?

_____ Is he physically inadequate, so that he gets less tired if he does less?

10. Does performing really matter?

_____ Does performing as desired matter to the performer?

_____ Is there an undesirable outcome for not performing?

_____ Is there a source of satisfaction for performing?

_____ Is he able to take pride in his performance as an individual or as a member of a group?

_____ Does he get satisfaction of his needs from the job?

11. Are there obstacles to performing?

_____ What prevents him from performing?

_____ Does he know what is expected of him?

_____ Does he know when to do what is expected of him?

_____ Are there conflicting demands on his time?

_____ Does he lack the authority?

the time?

the tools?

_____ Is he restricted by policies or by a "right way of doing it" or "way we've always done it" that ought to be changed?

_____ Can I reduce "competition from the job" — phone calls, "brush fires," demands of less important but more immediate problems?

IV. What should I do now?

12. Which solution is best?

_____ Are any solutions inappropriate or impossible to implement?

_____ Are any solutions plainly beyond our resources?

_____ What would it "cost" to go ahead with the solution?

_____ What would be the added "value" if I did?

_____ Is it worth doing?

_____ Which remedy is likely to give us the most result for the least effort?

_____ Which are we best equipped to try? Which remedy interests us most? (Or, on the other side of the coin, which remedy is most visible to those who must be pleased?)

Exhibit 6.3 (Continued)

Does the employee get punished for working safely (by the supervisor due to less production and by peers through ridicule)?

Does the employee get rewarded for working safely (by anybody)?

Does it make a difference to anybody if the employee works safely?

Are there any other obstacles?

It is perhaps easier to see how these four questions can be very real factors in safety performance.

With this model as with the others, we end up at the point where we are ready to define objectives. This definition of objective is a relatively easy task if we have followed the right hunch of our models and have a nontraining situation, or if in following the left hunch, we end up with a nontraining solution. It is, as mentioned earlier, somewhat more complex when we are discussing training objectives.

PREPARING TRAINING OBJECTIVES

According to Mager, in his book *Preparing Instructional Objectives,*[102] an objective is an intent communicated by a statement describing a proposed change in a learner—a statement of what the learner is to be like after successfully completing a learning experience. It is a description of a pattern of behavior (performance) we want the learner to be able to demonstrate. Mager further spells out what objectives really are and emphasizes their importance:

> The statement of objectives of a training program must denote measurable attributes observable in the graduate of the program, or otherwise it is impossible to determine whether or not the program is meeting the objectives.
>
> When clearly defined goals are lacking, it is impossible to evaluate a course or program efficiently, and there is no sound basis for selecting appropriate materials,

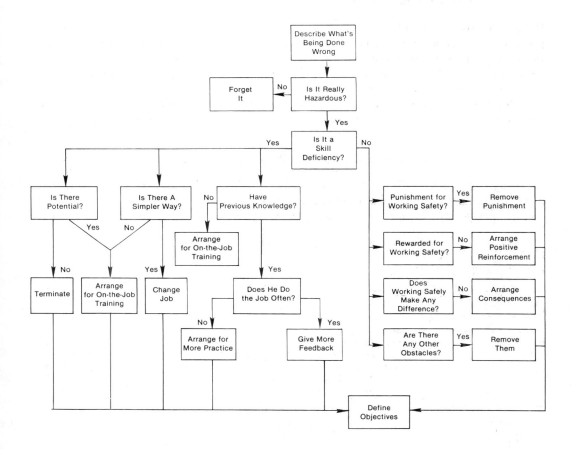

Exhibit 6.4 Employee model.

content, or instructional methods. After all, the machinist does not select a tool until he knows what operation he intends to perform. Neither does a composer orchestrate a score until he knows what effects he wishes to achieve. Similarly, a builder does not select his materials or specify a schedule for construction until he has his blueprint (objectives) before him. Too often, however, one hears teachers arguing the relative merits of textbooks or other aids of the classroom versus the laboratory, without ever specifying just what goal the aid or method is to assist in achieving. I cannot emphasize too strongly the point that an instructor will function in a fog of his own making until he knows just what he wants his students to be able to do at the end of the instruction.

Finally, Mager describes his method for writing objectives similar to those discussed earlier. They describe the desired behavior of the learner:

First, identify the terminal behavior by name; you can specify the kind of behavior that will be accepted as evidence that the learner has achieved the objective.

Second, try to define behavior further by describing the important conditions under which the behavior will be expected to occur.

Third, specify the criteria of acceptable performance by describing how well the learner must perform to be considered acceptable.

When you have completed this process you have adequately written your training objectives (set your terminal behavior). Having completed this, you are now ready to decide how to get the learner from the present state of performance to your written statement of desired terminal performance. The difference between these two levels of performance is what must be taught.

While this process may sound like excessive work and very time-consuming, many trainers find that when they actually properly develop training objectives in this manner the best instructional methods become obvious, and the training content is a long way toward completion.

A CONCEPTUAL TRAINING FRAMEWORK

According to Dr. Fred Luthans[91] (University of Nebraska), supervisory training may be thought of as a planned effort to stimulate *change,* either in the supervisor or in performance goals of the organization. A conceptual framework for supervisory training can be depicted as in Exhibit 6.5.

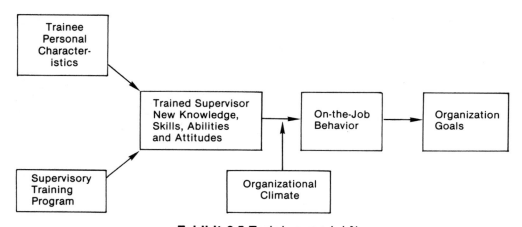

Exhibit 6.5 Training model.[91]

In this model, the inputs are the trainees' present knowledge, skills, abilities, and attitudes and the training program (the techniques and methods) used. The output of the interaction between these two is new knowledge, new skills, new abilities, and new attitudes. But a disparity often exists between what the supervisor learns in the training sessions and what is applied back on the job. The existing organizational climate largely determines if job performance is actually changed. The role expectations of superiors, peers, and subordinates are the most influential aspects of this climate. For actual job behavior to change, the role expectations of significant others must agree with the knowledge, skills, abilities, and attitudes learned by the supervisor in the training session. Congruency between role expectations of others and what is taught in the program is necessary for a permanent change in job behavior.

The last link in the sequence depicted in the model is between job behavior and organizational goals. A change in job behavior can affect the goals of the organization.

TRAINING METHODS

There is relatively less emphasis on method today in behavioral science literature. This is understandable, for as trainers in industry spend more time and effort on defining behavioral objectives (determining where they are going) and on evaluating training effectiveness (determining if they got there), they find method is almost unimportant or, rather, it just seems to fall into place. We should, however, look very briefly at what we do know about the relative effectiveness of some of the more widely used training techniques:

Lecture Method

This is probably the oldest, cheapest, and least effective technique available to a trainer. The lecture generally consists of one-way communication, with the instructor presenting information to a group of passive listeners. There is little or no opportunity to clarify meanings, obtain feedback, or account for individual differences. It is generally recognized that this lecture method is of minimal value in promoting attitudinal or behavioral change (McGehee and Thayer).[96]

The Conference Method

In conferences, unlike lectures, trainees play an active role. The emphasis is on small group discussions, and the leader provides guidance and feedback rather than instruction. Criticisms center around the inability to cover much material in a reasonable length of time using the conference technique.

Case Method

This technique assumes that improvement can best be attained through the study of cases. The trainees are presented with cases and then requested to state their solutions or recommendations to the group. Active participation is encouraged. This allows an individual to obtain feedback regarding his or her own solutions and to learn how others would approach the same problem. Some feel that the case method is unable to teach general principles and is unreal since it lacks real-life pressures.

Role Playing

Role playing involves the acting out of parts assigned to individual trainees. Role playing attempts to simulate actual organizational situations. Trainees spontaneously "act out" solutions to problems as would the persons whose roles they are playing. Role playing has proven to be an effective technique for promoting attitude change. Janus and King[66] found that a person playing a role shifts private opinion in the direction of that role. However, trainees often feel the technique is childish, and it is still an artificial situation where the results do not count. Also, it is time-consuming and expensive (Bass and Vaughan[11]).

All of these methods are commonly used in supervisory safety training. Since the real value of traditional supervisory training techniques is difficult to measure, many organizations assume that their safety training program is doing a pretty good job and is achieving its stated objectives. This is probably not true. In general, empirical studies have raised some serious questions about the value of such training. If nothing else, they show that management must not automatically assume that just any training will have desirable goal attainment effects on supervisors.

Most traditional training programs have been concerned with attitude change. Several of the studies have lent support to the notion that such attitude changes can be taught to supervisors. However, the crucial question becomes the degree of transfer of these attitudes of actual job behavior. This question is the most critical consideration in training supervisors. If the ultimate goal of supervisory training is to accomplish the goals of the organization, then training that does not transfer to job behavior is of no value and may even have negative value. In other words, a simple relationship between attitude change and a change in performance does not exist. Having a supervisor's attitudes changed by a training program is of little help unless it can be shown that this attitude change affects performance.

Another major purpose of supervisory training and development is to develop certain job behavior. In several studies, information was collected from supervisors, peers, and subordinates concerning the behavior of supervisors after training. It is rather convincing that such training does induce behavioral changes, but there is still the problem of determining whether these behavioral changes have contributed to job performance. Research on training indicates that it has not been proven that improvements in skills and attitudes have any effect on job performance. Research shows that no simple relationship exists between measures of attitudes and measures of job behavior, performance, and effectiveness.

According to Dr. Luthans,[91] the traditional approach to supervisory training and development is based on the following set of assumptions:

1. The supervisor can be taught new knowledge, skills, abilities, and attitudes.
2. These new inputs will produce a significant alteration in his motivational pattern.
3. These new supervisory inputs will be retained even when the supervisor leaves the training program.
4. The supervisor will be able to apply the new inputs to job behavior, performance, and effectiveness for goal attainment.

He states that these traditional assumptions no longer seem valid. A new set of assumptions is needed for effective supervisory training.

An interesting study was made in 1976 by Dan Norem,[121] while at California State University, Fresno, that sheds some light on the effectiveness of safety training. This study looked at safety training practices in companies with high versus low accident experience. The survey was on selected industrial firms in a four county area of the San Joaquin Valley, CA, to determine if significant differences could be found between companies with low accident experience versus those with high accidents in their utilization of safety training. The findings:

1. Both high and low accident frequency groups used written guidelines for safety training, with the low group developing a higher percentage of guides for specific job hazards.
2. Low accident companies reported more time spent in safety orientation of new employees than the companies in the high group.
3. The high group indicated a significantly higher use of a checklist to cover all points, but did not display inclusion of orientation tours or personal protective equipment demonstrations to the extent of the low group.
4. The high accident group indicated greater overall training for supervisors.
5. The high accident group had more extensive employee safety training practices than the low group, particularly with respect to handling hazardous materials, changing job responsibilities, new equipment, processes, layouts, and vehicle operations.

The conclusions drawn from this study were that:

1. There is no indication that training practices are significantly different between high versus low accident experience companies.
2. The findings infer that low accident companies provided more extensive orientation safety training than did high groups.
3. Differences in workforce makeup (turnover, seasonal employment, and educational levels) might be more of a factor than training.

While there are some confusions in the results, it is fairly clear that safety training is not the simple answer that we have believed to all of our safety problems.

It appears from most research and from what the behavioral scientists are telling us that we cannot use training as a panacea any longer. It just does not answer all (perhaps hardly any) of our problems. Mager tells us to define the problem first and infers most of the time that "they could do it if their life depended on it." Luthans suggests that our traditional approaches and the assumptions which lie behind training are no longer valid. And all of this leads us right back to the Porter and Lawler model suggesting we concentrate our efforts on those determinants of performance that influence role perception and effort and worry less about traits and abilities.

Part III

The Key-Chain Holders

Chapter Seven

Top Management's Role

Company management has the power to directly alter the attitude of its employees to make them receptive to our posters, literature, films, etc. Management does this every day to get what it wants from its employees—except in accident prevention.

D. C. Petersen[127]

We started in our look at how safety management might better use behavioral science knowledge by looking at Heinrich's key man. We now will look briefly at those people in the organization who hold the key chain; those people who can most influence the key person. There are a number of key-chain holders, and we will look at them in these categories:

1. *Top Management.* Those people who decide what they want done (set direction or policy).
2. *Middle Management.* That group located in the organization between the policy makers and the first-line supervisors.
3. *Staff People.* The various specialty people who have a function of assisting the top in setting policy, of working with the middle in some rather ill-defined ways, and of influencing the first-line supervisors in a number of ways to get them to want to do what they (and management) wants done.

THE SURVEYS

In the eyes of most safety professionals the first-line supervisor carries out the most critical functions. The second most critical functions are top management's. A 1967 National Safety Council Survey[132] was made that brought out this and other rather interesting points. The purpose of the survey was to determine which factors are considered most important to a comprehensive industrial safety program.

One hundred and forty-eight safety experts took part in the survey by completing a questionnaire which rated the importance of various safety activities. A total of seventy-eight activities in eight safety program areas were included. The eight "major program areas (MP)" covered by the questionnaire are shown in Exhibit 7.1 as well as some of the activities within each major area. Exhibit 7.2 gives an example of the complete list of activities in the MP area as they appeared in the questionnaire. The experts rated the importance of the eight major areas as well as the groups of activities within each area. Exhibit 7.3 shows the rank order of the major safety program areas. The first five areas represent what might be considered the basics of a safety program, while the rest are somewhat peripheral. Exhibit 7.4 shows the top ten subitems; Exhibit 7.5 shows the bottom ten.

Supervisory Participation (SP)
Enforcing safe job procedures
Setting an example by safe behavior
Training new or transferred employees
in safe job procedures

Middle Management Participation (MP)
Setting an example by behavior in
accord with safety regulations
Restating management's position on
safety
Using safety as a measure of
management capability

Top Management Participation (TP)
Setting an example by behavior in
accordance with safety regulations
Assigning someone to coordinate safety
on a full or part-time basis
Publishing a policy expressing
management's attitude on safety

*Engineering, Inspection,
Maintenance (EIM)*
Specifying guards on machinery before
it is purchased
Setting up a formal lockout procedure
Establishing a system of preventive
maintenance for tools, machinery,
plant, etc.
Inspecting tools and equipment
periodically

Screening and Training of Employees (ST)
Making safety a part of every new
employee's orientation
Including safety in supervisory training
courses
Including safety requirements in job
procedures based on job safety
analysis

Coordination by Safety Personnel (CSP)
Advising management in the formulation
of safety policy
Analyzing the safety program to
determine its effectiveness
Assisting and advising other
departments on various safety-related
matters

Forming a Record Keeping System (R)
Requiring the department supervisor to
conduct investigation of disabling
injuries
Using a standardized injury investigation
form
Including recommendations in injury
statistics reports

*Motivational and Educational
Techniques (ME)*
Providing for the employees a list of
general safety rules
Establishing a procedure for disciplining
violators of safety rules
Holding workplace safety meetings

Exhibit 7.1 Top-rated activities in each major area. [132]

The emphasis is on supervisory and top management participation, which reflects that most voters saw the' supervisor as the crucial link directly affecting employee behavior. This, of course, is from a survey of *safety directors,* people brought up under the concepts of the key person being the supervisor. Even so, note the number of TP (top management participation) items in the top ten. This suggests that top management must provide the initial push; the supervisors must maintain program momentum daily; and middle management participation is necessary to create the chain of communication and command. (You do not produce safety with middle management, but you cannot produce safety without it.)

This study from the National Safety Council is included here even though it is quite old because it clearly underlines the importance of each level of management in safety success. This also has been brought out by many other studies.

NIOSH funded a study done in 1978 by Smith, Cohen, Cohen and Cleveland:[28]

This research was a companion study to an earlier questionnaire study by Cohen, Smith, and Cohen (1975) in which safety program practices of matched pairs of low and high accident rate plants were compared to determine factors that might

The following items refer to middle management participation. Please rate them according to their importance in a total program.

Middle management restating management's position on safety	1	2	3	4	5
Middle management attending safety meetings	1	2	3	4	5
Middle management making periodic reviews of safety performance of employees	1	2	3	4	5
Middle management establishing checks to ensure adherence to safety program goals	1	2	3	4	5
Middle management setting an example by behavior in accord with safety regulations	1	2	3	4	5
Middle management using safety as a measure of management capability	1	2	3	4	5
Middle management serving on committees investigating serious accidents	1	2	3	4	5

Exhibit 7.2 Middle management participation section of questionnaire.[132]

Element	Rank
Supervisory participation	1
Top management participation	2
Engineering, inspection, maintenance	3
Middle management participation	4
Screening and training of employees	5
Records	6
Coordination by safety personnel	7
Motivational and educational techniques	8

Exhibit 7.3 Rank order of major safety program area.[132]

Item	Major program area	Rank
Enforcing safe job procedures	SP	1
Setting an example by safe behavior	SP	2
Middle management setting an example by behavior in accord with safety requirements	MP	3
Training new or transferred employees in safe job procedures	SP	4
Making safety a part of every new employee's orientation	ST	5
Top management setting an example by behavior in accordance with safety regulations	TP	6
Top management assigning someone to coordinate safety on a full or part-time basis	TP	7
Including safety in supervisory training courses	SP	8
Top management publishing a policy expressing management's attitude on safety	TP	9
Advising management in the formulation of safety policy	CSP	10

Exhibit 7.4 Top ten activities.[132]

Item	Major program area	Rank
Sending safety material to employee's homes	ME	69
Instructing supervisory personnel in first aid basics	ST	70
Investigating all accidents	CSP	71
Top management reporting safety performance to stockholders annually	TP	72
Keeping employee safety records	SP	73
Using posters to promote safety	ME	74
Setting up employee safety committees	ME	75
Offering incentives for good safety records by departments	ME	76
Offering prizes to individual employees through safety contests	ME	77
Enforcing safety regulations	CSP	78

Exhibit 7.5 Bottom ten activities.[132]

account for the difference in safety performance. The questionnaire study surveyed 42 pairs of plants in 6 industries which were matched by work force size, industrial category, and location with one member of the pair having a high accident rate and the other having a low accident rate. In the present study, on-site surveys were made of a sample of 7 pairs of the questionnaire respondents in order to expand on the results of Cohen et al. The on-site surveys resulted in findings similar to the questionnaire study but reported additional differences in safety program practices that could account for plant safety performance. The data indicated that the low accident companies differed from their matched high rate partners in the following ways: (1) Greater management commitment and involvement in the safety program and safety matters; (2) a more humanistic approach in dealing with employees stressing frequent positive contact and interaction; (3) better employee selection procedures; (4) more frequent use of lead workers to train employees versus supervisors; (5) a much greater degree of housekeeping and general plant cleanliness; (6) better plant environmental qualities; and (7) lower turnover and absenteeism among a more stable workforce.

A 1983 study at the University of Nebraska[141] looked at 143 firms in Nebraska representing 37,541 employees and found that the firms that had top management actively involved in promoting safety awareness had 469.8% fewer accidents than those firms that did not experience this active role (based upon a 99.9% confidence level).

A 1973 study conducted at Michigan State University by Shafai-Sahrai[147] tested a number of hypotheses:

The first hypothesis concerning occupational safety is based on managerial attitude.

In the firms with lower work injury frequency and severity rates, top management is highly interested and involved in the company's overall safety programs and actively participates in and supports safety activity.

For analysis of the data collected for testing this hypothesis and those that follow, the Wilcoxon Matched-Pairs Signed-Ranks Test primarily was employed. In some cases where data type were not amenable to evaluation by that test, the sign test was used. Instances where the latter test was employed will be indicated in reporting the results of testing. Where no such specification is made, the Wilcoxon test was used.

In the case of the above hypothesis the results of testing provided positive and strong support for confirmation since a significance level of 0.05 was obtained.

Exhibit 7.6 shows, in summary, how top management's support and involvement in safety differed in LO-IN and HI-IN firms. Although differences exist in most of the activities listed in the table, three areas represent the largest difference. It appears that top management of LO-IN firms puts more emphasis on personal audit and inspection than does top management of HI-IN firms. Also, in LO-IN firms it shows more interest in plans for achieving certain safety objectives and holds review and analysis sessions to ascertain that those plans are being carried out properly and according to projected objectives. Obviously, this practice enables top management to take timely and constructive corrective measures. In the HI-IN firms this practice is not being given much emphasis and, as is shown in Exhibit 7.6, considerable difference exists between the two groups of firms in this matter. The third large difference is in the area of including safety reports, figures, and achievements on the agenda of board meetings. This is being practiced in the LO-IN firms to a greater extent than in the other group. It appears that improvement in the above three areas may contribute to filling the existing gap between firms with markedly different work injury rates.

Exhibit 7.7 shows the scores of the matched pairs of firms with regard to top managements' support and involvement in safety. These are the actual scores which were used to test the hypothesis and determine its significance level. The most interesting and important observation in this table is that where great differences exist between matched firms' scores, a great difference is also observable in their respective injury frequency rates. This point can be clearly observed in the firms in industries 201, 265, 346, 354, and 371.

Description of Top Managements' Activity with Regard to Safety	LO-IN Firms (N = 11)		HI-IN Firms (N = 11)	
	Yes	No	Yes	No
1. Does he attend any safety meetings in the company?	8	3	6	5
2. Does he chair any of these meetings?	3	8	2	9
3. Does he regularly receive safety reports?	11	0	11	0
4. Does he personally conduct any safety audit or inspection?	9	2	4	7
5. Is he a member of any safety organization?	0	11	0	11
6. Does he regularly attend any safety meetings or conferences outside the company?	2	9	0	11
7. Does he emphasize plans for achieving certain safety objectives?	10	1	8	3
8. Does he actively participate in execution of safety plans?	10	1	7	4
9. Does he hold review and analysis sessions in order to compare the results of carrying out safety plans with projected objectives?	9	2	3	8
10. Are safety figures, reports, and achievements included on the agenda of company board meetings?	8	3	3	8
Total Score	70	40	44	66

Exhibit 7.6 Differences in top management's involvement in safety within sample firms.[147]

SIC Code	LO-IN Firms		HI-IN Firms	
	Score	Injury Frequency Rate	Score	Injury Frequency Rate
201	7	12.9	2	51.7
202	9	25.2	7	37.0
203	2	19.8	2	61.9
251	4	25.2	4	89.4
265	7	14.5	2	45.1
332	6	15.5	5	69.67
335	4	34.1	6	45.0
339	8	41.7	7	71.1
346	7	16.0	4	65.2
354	9	44.2	1	173.0
371	7	51.6	4	128.0
Total	70		44	
	(Mean = 6.3)		(Mean = 4)	

Exhibit 7.7 Total scores of sample firms on top managements' involvement in safety.

In brief, a statistical testing of this hypothesis shows a very strong significance level, and a comparison of the firms' scores while considering their injury frequency rates provides evidence that leads strongly to the belief that top management's earnest support and involvement in company safety is undoubtedly a determinate in the achievement of good safety performance. Where this support and involvement is missing, firms tend to suffer greatly from high work injury frequency rates.

This study and many others illustrate that top management participation is highly important in functioning safety programs. This is not exactly a new or exciting thought. Safety people have spent a large amount of time thinking and talking about how to get management's backing, how to enlist its support, or how to get its interest. They have little to show for their efforts for the most part. There is precious little behavioral research to help them in learning how to enlist their executives in their safety programs or, for that matter, in even understanding the executive. Those readers presently in management may be interested to note that they are objects being studied and may be interested in the results of some of these studies. For safety staff, the results of these studies offer some insights in how to get management's attention and how to communicate with it.

A major difficulty in the study of executive life is the problem of framing questions that are simple enough. Questions relevant to the study of executive behavior include: What does the executive do? How is time allocated to different functions? How much time is spent alone and how much communicating with others? What methods of communication are used? It is one of the oddities of managerial literature that we know so little about the behavior of the executive in any significant detail. Direct studies of executives are few in number.

The most famous research in this field is the study by Sune Carlson of nine Swedish managing directors and one French managing director, which is reported in his book *Executive Behavior*[27] Carlson first summarized the literature on executive management. He found that literature was more concerned with general speculation regarding the work of the executive than with actual descriptions of the behavior in which they engaged, that is, the literature was prescriptive rather than descriptive. He then started his research by collecting data under four headings:

1. *Place of Work.* By recording the place of work Carlson was able to form some idea of what the managing director was doing or not doing. As Carlson observes, if a managing director spent all his time at the office, he could scarcely have been carrying out a plant inspection.
2. *Contact with Persons and Institutions.* If it is accepted that the essence of management is working through others, then recording whom the managing director meets becomes a major focus of research. Carlson found that his managing directors spent between 65 and 90 percent of their total working time in contact with people.
3. *Technique of Communication.* This category refers to the methods used by the executive to get in touch with other people and describes methods of obtaining information and conveying ideas and intentions. Under this heading Carlson also included the technique of getting information by plant inspection.
4. *Nature of the Question Handled.* Questions were classified according to the field of activity: whether the question was one under development (planning) or one of current operation, and whether the question was one of policy or operation.

HOW AND WHERE THEY SPEND THEIR TIME

Exhibit 7.8 shows what was found out about how executive time is spent. The median figure was around 9¾ hours per day at work. According to this study the executive spent a considerable proportion of time outside the plant. Much of this time was spent at meetings with government officials, unions, or employers' associations representatives, etc. Inspection tours took 10 percent of the executive's time; inspection of the home plant outside the office, 7 percent; and time with subordinates in the office, 3 percent. From conversations with them it became clear to Carlson that the executives regarded inspection tours as extremely important. Further, there was a general feeling of regret that inspection tasks could not be attended to sufficiently because of the lack of time. Carlson makes the point, nevertheless, that most of them seem to be

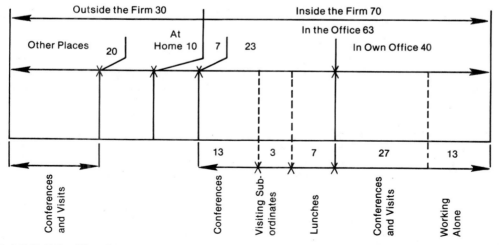

Exhibit 7.8 Distribution in percent of the working time during the weekdays, except Saturdays, when the executive was at the office.[27]

mistaken as to the amount of inspection work they really did. Wishful thinking led them to believe that things were not as bad as they really seemed to be. When the executives were asked how frequently they made a plant inspection the typical answers were, "Once every three weeks." But the evidence from Carlson's research did not support such assertions.

Carlson also refers to the "diary complex" which apparently is very common.

> There is a tendency for business executives to become slaves to their appointment diaries—they get a kind of "diary complex." One can seldom see two business executives talking together without their diaries in their hands, and they feel rather lost unless they know that they have these diaries within easy reach. When they start their working day they will look up what they have to do, and whatever is in the diary they will fulfill punctually and efficiently. If one wants to be sure of getting something done by this group of people, one has to see that it gets into their diaries. One should never ask a busy executive to promise to do something, e.g., "next week" or even "next Friday." Such vague requests do not get entered into the appointment diary. No, one has to state a specific time, say, Friday 4:15 p.m., then it will be put down and in due course done. The more exactly the time is specified, the more certain it will be that the task will be attended to.

Carlson's figures seem to suggest some things:

1. The executives that must lead the safety programs are extremely busy persons who often are busy in the less important things. They may well be highly interested in safety and in safety performance from their people, but just have not had the time to formulate thoughts on safety. For instance, the executive's alone periods (that is the time free from interruption, from visitors or telephone calls) had an average length of 10 or 15 minutes. "Alone intervals" of 10 minutes are obviously of little value for any depth of thought.
2. The executives would like to spend time in the plant seeing how things are going. What better way than a safety survey?
3. The safety manager ought to set a definite time in the diary as the other managers do and the executive ought to ensure that this happens.

COMMUNICATION AND NATURE OF QUESTIONS

Subordinates rarely contacted the chief executive directly and were reluctant even to ask for appointments. The chief executives received many communications but sent only a few letters each week. The telephone was especially useful for getting information quickly as well as for following up decisions. Carlson also makes the point that many executives are reluctant to make policy.

> The absence of clear policies may be the result of a defensive position; and it may be the outcome of the feeling that too rigid policy decisions may hamper the necessary flexibility in the firm's activities. Even when the chief executive sees the necessity of policy decisions, he may postpone action because of lack of time. A policy decision is generally much more difficult and time consuming to make than a decision regarding some matter of detail. But by doing so he becomes more and more over-loaded by details to be able to take on policy decisions. The very reason why he is over-loaded by details is the absence in the organization of established policies.

A fixed safety policy relieves the executive of daily policy decisions.

Carlson is one of the few who have attempted to classify or understand the executive. Such research is very important to our success.

Some other research has been done. For instance: Professors Wald and Doty[165] carried out an extensive "depth" study of thirty-three highly successful corporation officers who represented twenty-nine different organizations in order to discover whether there exists a definite pattern of background and personality that leads to executive competence. They found this profile of the executive:

1. The successful or likely-to-be-successful executive has experienced a happy home life in his earlier years which has been conducive to the development of security and self-confidence.
2. He is extremely interested in and feels very much attached to his present family unit.
3. The educational level completed by the typical executive is far above the average of the general population.
4. He takes full advantage of varied educational opportunities.
5. He is an active participant in and leader of social organizations during childhood and throughout his career as a worker.
6. He is interested in religion as a force toward developing high moral and ethical standards.
7. He has experienced and continues to experience good health.
8. He is interested in people—particularly in selling them on the idea of fundamental co-operation. He is interested in the written and spoken word as a means of communicating ideas. He is not preoccupied with the technical phases of his work, but rather with promoting harmonious human relationships.
9. He possesses very superior mental and analytical ability.
10. He is serious and conscientious in his approach to work. He is willing to take risks only after full consideration of the available facts.
11. He is forceful and intense, actively seeking new work to be done and new methods of doing it.
12. He is objective in facing his personal problems, frank and straightforward in his dealings with people, and spontaneous in his interpersonal relationships.
13. He is ambitious and able to identify his ambitions with those of his company to an outstanding degree.

Items 6, 8, 11 and 13 would indicate that an executive probably would have a sincere interest in safety.

Zygmunt A. Piotrowski,[131] a psychologist at Jefferson Medical College of Philadelphia, and Milton R. Rock, managing partner of Edward N. Hay and Associates in Philadelphia, studied top executives of business and manufacturing corporations in order to determine which, if any, traits differentiated between success and failure. The 110 subjects, presidents and vice presidents, took verbal tests and the Rorschach test. At the end of a follow-up period of four years, the executives were classified into three groups: successes, failures, and intermediates. The chief result was that self-trust, initiative, stamina, and other personality traits were the critical factors contributing to success, and not intelligence, although a minimal IQ of 120 seems to be a prerequisite.

What does all this mean to safety? It seems to mean this. The executives needed so badly for success are extremely sharp, intelligent persons who appear very normal, well-adjusted, and highly interested in the people who work for them. Time limitations seems to be the main reason for inactivity in the safety program. On the other hand, they seem to be spending time on some less important things. They seem shielded more by our reticence to contact them than from any other reason, and they have indicated (at least in research) that they would like to know more about what is going on. All of this does not fit with some safety professionals' past approach of pleading (to deaf ears) for management's support. Management is interested in safety but in lots of other things as well. Those who occupy top management positions are extremely busy people. They do first what they feel is most important and this is based on the information they get. According to Dr. Barenklau:[8]

> Perhaps one of the reasons top management appears to need motivating is that safety seldom is part of the executive purview. Safety specialists may have themselves to blame. Top management tends to look at things through an economic screen. Although this is common among safety professionals also, there might be a tendency to underemphasize economic considerations in favor of humanitarianism or other social goals. The latter are important, yes, but they tend to fall outside the ordinary range of executive concern.
>
> Safety professionals agree that compared to most things safety is hard to measure. If an organization is building internal-combustion engines, for example, and 80 are supposed to come off the line each day, there's an obvious problem if only 70 come off the line on a given day. One of management's prime goals is production—and production rather readily gives itself to accurate and instantaneous measurement.
>
> But if top management asked the question, "Have we been safe today?" how would we answer? We could say we haven't had any accidents, or we didn't have any near misses today, but it's pretty hard to say that we had a safe day as such. We may have had several close calls that could have resulted in injury or property damage, but since the problem didn't materialize, the day appeared to be relatively safe. Safety people should develop better measures, and indeed many of us have.
>
> While it is necessary to get a commitment from top management in order to achieve job safety, several roadblocks stand in the way: (1) Corporate safety goals might be fuzzy. (2) Incidents (accidents) are rare. The etiology of the safety movement within the organization might be somewhat obscure. ("What are we trying to do with our safety program?")
>
> Another roadblock to obtaining a management commitment might well be the fact that safety professionals over the years have tended to treat symptoms rather than causes. For anyone who has spent many years in safety, this is a hard thing to admit. But accidents are only symptoms, the result of operational errors. When we have accidents we should immediately try to determine the error which allowed the accident to take place. It is not an easy job; often there are multiple causes for accidents and injury. This fact alone tends to obscure the safety effort—and so in the midst of his busy day, the boss tends to overlook the obscurity and thus the safety effort itself.

Management is first of all interested in how the safety professional's ideas relate to the profits of the organization. That is, what will management get in return for the money it is being asked to spend? Thus, safety people ought to be dollar-oriented when talking to management. Even if management understands the language of frequency and severity rates, dollar indicators ought to be used instead. Here are some possibilities:

1. Dollar losses (claim costs) from the insurance company.
2. Total dollar losses (insurance direct costs) plus first-aid costs not paid by insurance.
3. So-called "hidden costs".
4. Estimated costs.
5. Insurance loss ratio.
6. Insurance premium.
7. Insurance experience modification.

8. Insurance retrospective premium.
9. Sales needed to make up for losses incurred, etc.

Invariably management is financially oriented. Dr. Barenklau states:[8]

> Top management will go along with things that are economically feasible, that
> return a bit more than a dollar for each dollar spent. Top management wants to
> participate in profit improvement programs—even if we call them "safety programs."
> When the boss asks, "What will it get me?" we can say, "Boss, it will get you five
> things:
>
> 1. Money—money which accumulates from lack of accidents. Money which
>   transfers directly from overhead to profit on the company balance sheet.
> 2. Money—money that would ordinarily be spent for fire and equipment losses
>   were our safety program less adequate.
> 3. Money—money saved through reduction of workmen's compensation pre-
>   miums because the boss wants and gets a safe workplace and holds others
>   accountable to achieve it.
> 4. Money—money that would be spent paying cargo claims, filling out papers,
>   and doing things over.
> 5. Money—money spent replacing injured employees with unskilled substitutes
>   who in themselves might be a cost until they become trained."

OSHA AS AN ATTENTION GETTER

If it accomplishes nothing else, OSHA has focused executive attention on safety more than ever
before. While this attention starts out negatively (fear and disgust) and while it misdirects the
executive (to physical conditions only), nevertheless it is executive attention for which the
safety manager did not have to plead. It would be a shame to waste it.

Now seems an opportune time to obtain executive interest in *safety* as well as in compliance,
to strive for executive understanding of safety performance and the importance of personal input to the program. As misdirected as OSHA seems to be when we consider the true causes of
accidents, perhaps its real value is as an attention getter to safety in general. Safety managers
would be remiss not to capitalize on the executive attention.

WHAT DO YOU WANT THE EXECUTIVES TO DO?

Perhaps the real key to our past problems of getting management's "backing" is that the few
times we have gotten their attention we have not been sufficiently clear about what exactly we
wanted done. We have shown our figures which either spell out that we are doing a good job, or
that "they" (the line) are not doing their job. But we have not stated exactly what we want
management to do differently from what has been done in the past.

This is partly due to the measurement difficulties referred to earlier. We are hesitant to ask
the big boss to "zap" a line manager for lack of performance in safety when we really are not
sure (have no crisp measure) of what has or has not been done in safety. We are hesitant to have
the big boss "zap" a line manager for a poor safety record when there could be a large degree of
luck involved. Similarly, we are hesitant (and should be) to reinforce with rewards from the top
when we do not know if the wanted behavior has been there. Perhaps one reason for our management measurement difficulties is that we are not too clear when telling the boss what we

want done. About the clearest thing we do is ask for issuance of a safety policy. But even in doing this we fail to include a procedure for follow-up so there is no way of knowing whether or not anyone is following the directive. And even with the best of intentions it is often not possible to make personal plant visits often enough to know what is happening.

We should spell out in more detail exactly what we suggest management's role should be in our safety program. For instance, we should suggest that:

1. A safety policy be issued.
2. A reporting system be instituted on who is and who is not performing in safety by some predetermined criteria of performance.
3. A way should be initiated to indicate positive or negative rewards to immediate subordinates (middle managers).

Exhibit 7.9 shows one simple reporting form by which top management would receive monthly information on one single sheet that tells which first-line supervisors are performing in those areas believed to be important (in this case, in timing and quality of investigations, in meetings held, in inspection made and in any other area). Any circled item would indicate unacceptable performance and the middle manager would be negatively rewarded. Similarly, a notation on the report of notably good performance would be positively rewarded by the executive receiving the monthly report. This particular report is put together from information on the Supervisor's Activity Report shown in Exhibit 5.4. On a day-to-day basis this may be the only executive input needed in the program. At times, of course, policy decisions might be necessary, but these would be handled by examples as any other management policy decision might be handled.

Another way of informing the executive of what is going on and of explaining exactly what action you want taken is through your annual report. Every safety manager ought to utilize this tool. If management does not ask for such a report, it should be submitted anyway. It would certainly be peculiar for management to appoint a person to handle the safety staff function (whether full or part time) and then never ask what the company is getting for the money spent.

Whatever the safety person wants management to know should go into the annual report. For example, the report might answer these questions:

1. How did we fare last year? (Give results expressed in management's terms.)
2. What did we accomplish last year? How are we stronger than we were before?
3. What are our objectives for next year? How will we be stronger at this time next year?
4. What do we need from the executive?

This report is crucial to a safety professional's relationship with management. Future goals included in the report tell management in what direction you intend to move. If management approves, it is committed to that direction.

Next to the first-line supervisor the top executive is crucial to the success of your safety program, perhaps even more crucial (regardless of the study mentioned earlier) for it is the executive that holds the key chain. Our key person will not perform without the executive wanting this performance.

VISIBLE TASKS

Management thinking tends to support the concept of delegation of responsibility to the lowest possible level. While this concept is excellent, there are some real hazards in the concept when

MONTHLY SAFETY REPORT

Supervisor	Accident Investigation		Optional Strategies (choose 2)					
			Meetings		Inspections			Other
	No Not On Time	Quality Rating	On Sched-ule	No. Reach-ed	On Sched-ule	% Safe	No. Hazards	Comments OK — Not OK

Comments:

_____

Any Item Circled Above Is Below Acceptable Performance.

Exhibit 7.9 Monthly safety report.

applied to safety. When all responsibility in safety is delegated, it leaves upper levels of management with little to do about safety, other than the traditional role of authorizing policy. Too often this lack of involvement on a regular basis can be construed by the individuals in the organization as saying that safety is unimportant to the executives, that it has a low priority, that it is not an important enough issue on which executives should spend their time.

It is, therefore, extremely important that safety responsibilities not be delegated totally to the lowest possible level. All levels of management, including the top level, must retain some specific predetermined tasks that are visible—that can be seen by all workers in the organization as an indication of the importance of safety to that executive.

What are these tasks? There is a large variety depending on what is comfortable for the personality of the executive. Here are some examples:

— Some begin their regular weekly or monthly supervisory meetings with safety as the first item on the agenda. It is made quite clear that all subordinate supervisors present follow suit in this demonstration of management interest in safety.
— The *"dial for a safety message"* program of one company utilizes it's top officials to record terse key safety messages that have proven to be most effective.
— Executive management personnel frequently conduct final hearings on major accidents or disabling injuries. Although it appears that practices vary greatly, it seems to be a common practice to have subordinate key management personnel present, i.e., the division manager, the department superintendent, the injured worker's supervisor, etc.

Many have latched on to the "Management By Wandering Around" concept from *In Search of Excellence,* and use it to look for safety problems and to discuss safety with workers.

THE EXECUTIVE ROLE IN SAFETY

As indicated earlier, the role of upper management is threefold. They must

1. Ensure the performance of subordinate managers (middle managers) in safety—an accountability function.
2. Ensure the quality of that subordinate's safety performance—a quality control function.
3. Personally engage in some specific safety-related function that can be seen as a demonstration of priorities—a visibility function.

Chapter Eight

The Middle Managers

There are members of the management team at each level: in the headquarters office, at regional offices, and at the specific plants or installations. It is through these team members that safety must gain functional recognition; become identified as an integral part of the system.

W. Pope and T. Cresswell[134]

UNDERSTANDING THE MIDDLE MANAGER

The Industrial Safety Study[132] mentioned earlier also showed the importance of middle management participation. Exhibit 7.3 indicated that the role of the middle manager was more important to safety success than the screening or training of employees, our record keeping function, the coordination done by the safety manager, and all of our motivational and educational techniques. In that survey the third most important single item on a list of seventy-eight items was "middle management setting an example by behavior in accord with safety requirements." That survey also spelled out exactly what the role of the middle manager might be in safety. Exhibit 7.2 indicated it is the role of the middle manager to restate policy, to participate in safety meetings, to review employee safety performance, to establish checks to ensure adherence to safety program goals, to set an example, to utilize safety performance as a measure of management capability, and to serve on investigating committees, in short, to be an active participant in the program transforming the executive's abstract policy into supervisory action.

We have already stated that there is very little behavioral research to help us to understand and to work successfully with our executives. There is even less discussion of the middle managers. And what discussion there is leads us to question rather pointedly some of our historical managerial principles and beliefs. The general conclusions of these studies challenge the established view of the middle managers. The traditional picture of middle manager behavior is one which is abstract and static, where managers are seen as having nicely bounded, carefully defined, and compartmentalized roles. They are given authority which is supposedly commensurate with responsibility and they are seen as exhorting, directing, or cajoling subordinates. The middle managers receive instructions from management and transmit them to subordinates; by this process they make sure subordinates know what they are supposed to do and that they will be held responsible for doing it. The distinguishing marks of this system of organization are that every individual has a job which is clearly defined (it has a clear beginning and end), and that every member of the organization is required to make discrete decisions within the framework of "delegated" authority. This approach is characterized by organization charts and by terms such as *delegation, line,* and *staff.*

Empirical studies show clearly that successful managers in modern organizations operate in a manner very different from that suggested by the traditional view of management. The facts of organizational behavior make it perfectly plain that the modern manager operates in a dynamic context where things are not as simple as the *principles of management* might suggest. The manager's job today is *not* a job with neatly defined authority and responsibility. The traditional idea that organizations can be characterized as networks of roles connected by single lines of authority is very far from being the case; it is naive. There is a multiplicity of relationships between managers in an organization.

Leonard Sayles, in *Management Behavior,*[143] discusses some of the "old wives' tales" of management theory and includes the following theories:

1. A manager should take orders from only one man, his boss. Most managers, in fact, work for, or respond to many people (customers or those in a position to make demands upon them).
2. The manager does no work himself; he gets things done only through the activities of his subordinates. Actually the manager himself must carry on many relationships with others and participates in activities of all kinds to get things done.
3. The manager devotes most of his time and energy to supervising his subordinates. Actually the manager is away from his subordinates a significant portion of the time.
4. The good manager manages by looking at results. Actually, methods of continuous feedback are required.
5. To be effective, the manager must have authority equal to his responsibility. Actually, a manager almost never has authority equal to his responsibility; he must depend on the actions of many people over whom he has not the slightest control.
6. Staff people have no real authority since they are subsidiary to the line organization. Actually, staff groups have very real power.

Since the middle managers' jobs are not neatly defined, it would be naive to spell out the safety responsibilities in a neatly defined manner. Perhaps the items mentioned in that Industrial Safety Study[132] are sufficient definition and any further refinements should be left to the individual company. While it seems unwise to tightly define the middle managers' safety responsibilities, this by no means suggests they should be left out of the program. Obviously they must be an integral part.

THE FUTURE OF MIDDLE MANAGEMENT

Most of the above comments are based on thinking about our traditional kind of industrial organization as shown on the organization charts with boxes and lines. However, these boxes and lines are actually quite meaningless. Just because there is only one line into a middle manager's box does not mean there is only one boss to please.

Because of the reasons already stated and because of the impact of technological change, our traditional organization structures are changing also. For instance, organizations which have automated their production and information systems have experienced organizational structure changes and an expansion of top management. Those organizations which have planned to make changes in structure as a result of automation have generally found that the traditional pyramid structure of organizations has been changed. Studies have generally reached these conclusions:

1. The number of levels of authority in an organization increases with increasing technical complexity.

2. The ratio of managers and supervisors to total personnel increases with technical complexity.

It is agreed by many that new technology, in the form of automation or cybernetics, will alter the organizational structure and expand the bureaucracy. (Cybernetics studies communications and organizations and their automation.) The most obvious impact on organizational structure will be the decline of the concept of middle management as we now know it. The organization of the future will eliminate a large number of administrative functions and decision making will be based on the information flow through the organization.

Exhibit 8.1 shows the traditional pyramid organizational structure. The precybernetic organizational structure requires a middle management to perform specialized tasks of data analysis for top management. The future organizational structure in a cybernetic society is shown in Exhibit 8.2. In the cybernetic organizational structure, computers will reduce the use of middle management in performing routine tasks as indicated in the exhibit. The replacement of middle management by computers does not remove those people from the organization. Their functions change and many of the people will be absorbed by top management to assist in decision making.

Present concepts of organization will undergo drastic changes with an increase in automation: departmentalization and specialization will be reduced; future organizations will become more adaptable to handle changes in structure and will be better able to adjust to given situations; team participation in decision making will be more prevalent; and shared administration by many executive-level administrators supported by a total information system for technical decision making will result. The present pyramid organizational structure of top, middle, and lower management will give way to one in which there will be only a top and lower level of

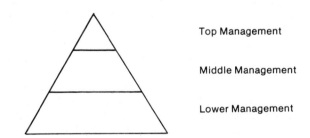

Exhibit 8.1 Precybernetic organizational structure.

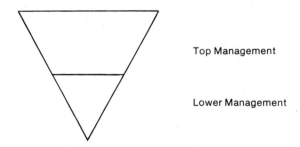

Exhibit 8.2 Cybernetic organizational structure.

administration. The present middle management functions will be replaced by computers. In turn, middle management will be absorbed by the top or lower management levels.

While middle managers may well be a casualty of technology, they are not yet. We still must deal with them. We *do* need them. They are still of considerable importance to our safety program, but we need to motivate them or to "turn them on" in safety.

WHAT MOTIVATES THE MIDDLE MANAGER?

First of all, everything that has been said about motivating supervisors using the effort-reward probability model could be restated here. Performance failures of middle managers are fully explainable by that model. Role perception is perhaps the box in the model that best explains the lack of middle manager performance. Whereas we strive in our policy to pinpoint responsibility of the first-line safety supervisor, we seldom pinpoint middle manager responsibilities (or duties) as crisply. The total effort portion of the model also needs close examination to determine whether or not we reward middle managers for safety performance and how likely it is that this reward does follow performance.

There are other ways we might look at motivating middle managers in safety. One of these is offered by Sidney R. Wilson,[177] who states:

> It is an interesting phenomenon that the effectiveness of efforts to motivate company employees is inversely proportional to their salary and stature in the organization. Motivation at the shop floor level has long been a highly developed science, largely a matter of sophisticated analysis of such factors as incentive scales, piece rates, and other variables. Motivation of field salesmen also involves a dispassionate application of incentives such as commissions, bonuses, conventions, and prizes. Performance at the management level has the most profound effect on a company's fortunes. Yet motivational efforts are least effective at this level. For the most part, we have been unable to apply the principles that are known to work so effectively at lower levels. We seldom specify what performance standards are required and, even when we do, we seldom make incentive awards contingent on achievement of those standards.
>
> The difficulty with motivation at the management level arises principally from the failure to recognize that executives respond to incentives exactly as machinists or salesmen do. We seem to feel that the performance of decision makers can be improved only by influencing directly the executive's motivation.
>
> The problem is illustrated by the dictionary's definition of the word "motive" as "some inner drive, impulse, intention, etc., that causes a person to do something or act in a certain way." We deal directly with performance against known standards and objectives at the lower levels of an organization, considering a person's inner drives irrelevant as long as the job gets done. On the other hand, we consider that executive performance depends on these inner drives, and we ignore the direct relationship between incentives and performance. Thus, our efforts to motivate managers are directed mainly toward changes in their attitudes with the inevitable decrease in effectiveness that must result from trying to influence something that cannot even be observed.
>
> The typical process of executive motivation can be summarized in a sequence of events: This sequence presents some obvious difficulties in trying to motivate executives:
>
> First, there is a heavy emphasis on evaluations of verbal performance. Job performance is more difficult to observe and evaluate objectively at the decision-making level: it is even difficult to agree on what an executive does. However, verbal per-

formance is easy to observe, so we tend to base our conclusions about an executive's attitudes more on the basis of what he says than what he does.

Second, all judgments of executive performance are filtered through these vague and imprecise conclusions about attitudes.

Third, advice on job performance tends to take the form of communications about organizational objectives. We tell an executive the company needs more profit, or more products. But we usually tell him little about the type of executive performance needed and fail to help him rearrange his priorities.

Fourth, we also spend a good deal of time counseling executives concerning their attitudes. Such counseling may take the form of talks with colleagues and superiors, inspiring messages from the president, or suggested attendance at useless seminars on the role of management. Admittedly, these counseling activities may have a considerable impact on verbal performance: the executive learns quickly to say the right words at the right time. This is extremely gratifying to the counselor, and the evaluation of verbal performance tends to receive even more emphasis.

Job performance, however, is not likely to be affected noticeably.

The fact is that all observations and measures of attitude are highly unreliable, no matter who makes them. Objectivity is impossible. The key to the achievement of performance objectives is to eliminate this mysticism and to concentrate on the performance itself. This is the advantage that can be gained from a well-planned program of incentive management.

When we make deliberate use of available incentives in a management development program, the sequence of events changes considerably to the process described in Exhibit 8.3.

There are certain important differences between this process and the one described earlier. First, no attention is paid to an executive's verbal performance unless it is a part of, or heavily influences, his job performance.

Second, all observations concern factors that are, in fact, observable.

Third, since the evaluation of what the executive says has no place in the incentive management process, one is forced to examine only the performance itself. Finally, the necessity to describe the incentives that will be awarded for particular performance calls attention to the way in which they are used and tends to decrease the frequency with which they are applied to encourage irrelevant or even undesirable performance.

Actually, Wilson is utilizing one of the more modern psychological techniques, more frequently referred to as *behavior modification.* This will be treated in some detail later.

Let us apply some of these thoughts to safety for our middle managers. One of Wilson's key points is that in incentive management we provide rewards only for observed behavior, not for verbal behavior. In safety this means that we must know whether or not our middle managers are giving performance in safety. We must know, or measure, performance, and not depend on anything less. We must, first of all, decide what behavior we want from them and then measure to see if we get it. Wilson actually is saying what our effort-reward probability model said earlier. We must decide what we want at each level, measure to see if we are getting it, and then make sure rewards (incentives) are provided consistent with and contingent upon performance.

It becomes our job as safety managers to set up a system in our company that achieves these things. In establishing our system, the first step is defining desired behavior. We said earlier that we cannot specify behavior here; such behavior depends on each company's own system. We can, however, reiterate the common performances of the Industrial Safety Study:

1. To restate policy.
2. To participate in meetings.

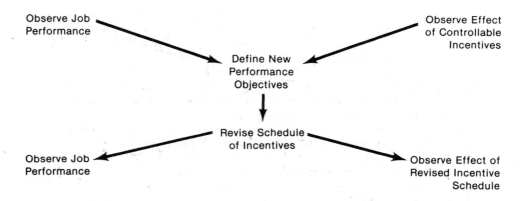

Exhibit 8.3 New process of management motivation.[171]

3. To review employee performance.
4. To establish checks to ensure adherence to the program.
5. To set an example.
6. To use safety performance as a measure of management capability.
7. To serve on investigating committees.

These and others can be quantified, observed, and measured. Incentives (rewards) can be structured for each type of behavior wanted and a system devised to link the reward consistently to the behavior. Later, several such systems will be described.

THE MIDDLE MANAGEMENT ROLE

Middle managers (anyone between the executive on the top and the supervisor) must become an integral part of the safety program for program success. Traditionally we have very effectively built them out of our programs. Since their subordinates (the supervisors) will respond and react primarily to their wishes, their wish list must include safety performance. If results in safety is not perceived by middle managers as their problem, it is unlikely they will put the pressure on those beneath them. If middle managers are not required to carry out some performances related to safety, it is unlikely they will see subordinate safety performance as important. And if middle managers do not engage in some regular visible safety related performance, it is unlikely their subordinate workers will see safety as a high priority item.

Therefore, the role of middle management in the safety program is:

1. To ensure subordinate performance (Accountability).
2. To ensure quality performance (Quality Control).
3. To engage in some visible activity (Visibility).

THE PLANT MANAGER

In the last two chapters we've talked about top management's role and middle management's role. Sometimes it is difficult to ascertain exactly which a person is. For instance, in many organizations with many locations, there is a location manager, i.e. a plant manager. Is a plant manager top or middle management? In effect, both. Kurt Frank,[39] writing in *Professional Safety* magazine, offers some help to that individual:

When a manager takes over the location's top position he realizes rather soon that he owns it all. It is his plant, his equipment, his people; it is his safety program. His initial reaction and responsibility should be to give support to and:

- *Continue the existing program.* Whatever is going, keep it going for the time being, give it your support. When the newness of the assignment has worn off and other high priority items addressed and direction agreed to, then,
- *Evaluate and critique the inherited safety program.* Where are we? Where are we going? After considerable analysis and soul searching, cut out the ineffective and reinforce the residual. A trimming process. At this point,
- *Introduce and implement changes.* Freshen up, repackage the program. Improve safety awareness and interest of subordinates to achieve even better safety performance.

Now that the plant manager has modified and improved the safety program for the location to his satisfaction, what role does he play in its implementation? The plant manager is interested in every portion of safety, but there are many opportunities for his personal involvement. Let us examine areas where the plant manager can be involved, in what way and how much of his time, on the average, might this take.

Promotional programs—½ hour per week

- Appoint a steering committee representing all location personnel, including safety, to conceive and develop the promotional and monthly programs.
- Be personally involved in the kick-off meeting to demonstrate that it is his program.
- Plant manager schedules one general monthly safety meeting per year with all employees.

Plant appearances—2 hours per week

- Make self visible in the plant for any reasons—whatever the variety.
- Production/safety contacts with hourly personnel. Do not overlook the first line supervisor, he needs a pat on the back too.
- Periodic safety messages (bulletins, fliers, concerns) from the plant manager should be posted or distributed.

Relationships—1½ hours

- To ensure interaction with hourly personnel, set up small on-going committees and special meetings to allow intimate discussions with key employees with safety responsibilities (accident investigation teams, plant safety committees, luncheons, etc.).

Executive safety committee—¼ hour

- Schedule meetings once a month in a definite time slot and maintain.
- Chair the committee, suggest topics and agree to finalize agenda with safety manager.

Policy setting—½ hour

- make certain that policy is set on *purpose*, rather than *by accident*.
- Review all existing safety related procedures (in order of priority) with the objective being to get *his* name (i.e., "I have read and accept") on them.

Example setting—½ hour

- By wearing proper protective equipment.
- Challenging unsafe acts in plant.
- Questioning unsafe conditions.
- Observing closely some job operations.

OSHA inspections—¼ hour

- The plant manager is expected to participate in the initial meeting.
- Participation in the closing conference.
- Report monthly to OSHA office on citations.

Involvements in new safety training programs—1½ hours

- Be part of the kick-off (as a minimum), but as a participant is even better.

Safety Indoctrination—¼ hour

- The plant manager greets each new employee the first day or week on the job and says a few words and sets safety priority.

Audits—½ hour

- Ensure that in-depth operation audits are challenging and a proper number are scheduled each year.
- Review the completed audio with the team to determine the audio quality, assign priorities to concerns to demonstrate the genuine interest of plant manager.

Emergency action—½ hour

- Participate in appropriate public relations and headquarters communication during emergencies.
- Visit the area soon after incident, to demonstrate concern and to offer comfort.

Accident investigation—1½ hour

- Set climate (and attitude) to make certain all significant accidents are investigated.
- Make a personal visit to accident area.

Personal trait or theme

Each plant manager has an operating mode, a habit, a technique which serves as his trademark. Examples are:

- A special telephone number which in effect, is a "hot line" to the plant manager.
- Each day at precisely 10:00 a.m., he is at a plant work station.

Management by Objectives/Results program—¼ hour

- Have a major personal safety goal.
- Establish a format for supervision which emphasizes activities for a basic rating, plus a result (like statistics), to achieve an outstanding rating.

Evaluation of safety program

- Follow-up (completion) on commitments serve as indicators.
- Conduct a one-on-one survey with each subordinate.
- Self evaluation (available from corporate safety).

Chapter Nine

Staff

For each of us, the duties that we perform are the duties we have set for ourselves. The job of staff safety specialist is very much a self-defined job.

D. C. Petersen[127]

THE ROLE OF STAFF

We would be remiss not to talk about the role of safety staff, even though the Industrial Safety Study of the National Safety Council[132] rated coordination by safety personnel seventh in a list of eight. Regardless of this ranking, it is safe to say that where safety staff is involved in the organization, whether full or part time, much of the progress obtained depends on that staff. Let us first of all look at what the role of safety staff is in an industrial organization.

The goal of management is efficient production—production which maximizes profit. To obtain this goal management has four basic resources: employees, facilities, equipment, and materials. Management brings many influences to bear on both of these resources. Upon the company's personnel, management applies such influences as training, selection and placement process, employee health programs, and employee relations. Facilities, equipment, and materials are influenced by maintenance, research, and engineering. These influences and basic resources are brought together through various standard operating procedures. The function of safety staff is to:

1. Build safety into these procedures.
2. Continually audit the carrying out of these procedures to ensure that the controls are adequate.

Safety staff accomplishes these tasks by asking *why* certain acts and conditions are allowed and asking *whether* certain known controls exist.

Safety is not a resource; it is not an influence; it is not a procedure; and it certainly is not a program. Rather, *safety is state of mind,* an atmosphere that must become an integral part of each and every procedure that the company has. This, then, is what we mean by *built-in* or *integrated safety*. It is the only brand of safety that is permanently effective.

Since any accident, any unsafe act, or any unsafe condition indicates a system failure, the safety professional must become a systems evaluator. Just stating this, however, does not tell the safety manager what to do, what the actual duties of the staff safety specialist are, or whether he has a full- or part-time position. Actually we cannot. For each of us, the duties that we perform are the duties we have set for ourselves. The job of staff safety specialist is very much a self-defined job. The duties will vary, depending on the size of the organization, the number of locations, the operations themselves, the people above and in line management, the problems presently facing the company, the other staff people and specialists available, and where the safety specialist fits into the organization.

The American Society of Safety Engineers (ASSE) has at least spelled out the areas in which the specialist should work:

1. Identification and appraisal of accident- and loss-producing conditions and practices and evaluation of the severity of the accident problem.
2. Development of accident prevention and loss control methods, procedures, and programs.
3. Communication of accident and loss control information to those directly involved.
4. Measurement and evaluation of the effectiveness of the accident and loss control system and the modifications needed to achieve optimum results.

Beyond these broad guidelines the safety staff job is self-defined. To help in this definition the safety manager might refer to Exhibits 9.1 and 9.2.

Exhibit 9.1 briefly outlines the major kinds of activities in which the safety manager will engage. First of all, a system of measurement must be structured so that accountability can be fixed, and so that rewards can be applied properly to the right people and at the right time in order to reward or reinforce the desired behavior. Some of these systems might incorporate estimated costing, sampling, rating effort, SCRAPE, and performance measurements as described in Appendixes A and B. Secondly, in safety programming the safety manager must ensure that safety is included in orientation, that safety training is provided where needed, that safety is a part of supervisory development, that the organization's attention is focused on safety, that safety is included in employee selection and in the medical program. Thirdly, the safety manager must function as a technical resource: know how to investigate in depth, know where to get technical data, know the standards, and know how to analyze new products, equipment,

```
1. Measuring safety performance, using
     Results Measures
     Activity Measures
     Audits
     Records
     Statistics
     Inspections
     Sampling
2. Safety program development, including
     Orientation
     Training
     Supervisory department
     Motivation
     Gimmicks
     The selection process
     Medical Controls
3. Being a technical resource
     In investigation
     On standards and regulations
     On consumer products
     On new equipment purchases
4. Being a Systems Analyst
```

Exhibit 9.1 Safety staff's role.

1. Management's policy
   - What does the boss want from supervisors? From middle management? From plant management? From you?
2. Accountability
   - What means will there be of measuring safety performance? How will management know?
   - For results
   - For activities
   - Methods
     - The safety audit
     - Statistics and records
     - Inspections
     - Sampling
3. What is safety staff's role?
   - Safety programming. What elements should there be?
   - Technical, advice, assistance or control
   - Systems: developing, installing, controlling
     - Records, follow-up, identifying problems, new equipment, etc.
4. Where does safety fit:
   - Employee selection and placement?
   - Training of employees, new and ongoing, job and special
   - Supervisory training
   - Motivation, creating and maintaining interest
   - Medical program
5. Are you organized for OSHA?
   - Who does what?
   - Recordkeeping requirements
   - Knowledge of the rules and standards
   - Documentation

Exhibit 9.2 Areas to consider in organizing the staff function.

and problems. And fourthly, the safety manager must function as a systems analyst, searching for why and whethers.

Exhibit 9.2 says much the same thing. It suggests that the safety manager should concentrate initially on five key areas and find answers to questions in these five areas. These questions, or similar ones, might suggest those weaknesses in the system that would be the most important on which to begin working.

WHAT IS STAFF?

We should probably define more clearly what we mean by safety staff. Staff work consists of assisting an *executive* in the discharge of executive functions, under the supervision of, on behalf of, and frequently in the name of the executive. The staff does the work that the executive would do with adequate *time* and the *specialized knowledge*. This work includes planning, supervising, directing, coordinating, organizing—all, however, under the supervision of the executive. Obviously, since the line has primary safety responsibility, the safety director must be staff. As staff there is no responsibility for the safety record or the results; responsibilities involve activities which help the line achieve its goals in safety. At first, the safety specialist has no authority over the line, but may have a great deal of influence, which is quite different from authority. How much influence or power will depend on the organization and on the personality of the individual in the line position. The safety specialist in any organization obtains results by using one of two methods: (1) making a recommendation to an executive in the line chain of command and that executive issues an order or (2) obtaining acceptance for suggestions voluntarily from line supervisors without taking the chain-of-command route. More often than not, the safety specialist achieves success by the second route and uses the first route only for rare emergencies.

Most line managers realize that the safety specialist does have stronger influence than that shown on the organization chart. This staff specialist is an expert, has certain status, has management's interest and backing, and, if worst comes to worst, has some influence on management's appraisal of the line manager, and hence on that manager's future. So, although the safety specialist has no authority, there is an element of power. There also are situations where the safety specialist is given degrees of authority over the line. It may be stated in the policy that under certain situations the safety specialist must be consulted and must give approval, or can even temporarily step in and assume command (stop the operation). This granting of temporary authority to the staff safety specialist in industry is common and correct. Even so, we must keep in mind that basically the work is accomplished by influencing the line and not by directing it.

WHERE TO INSTALL SAFETY

To whom should the staff safety specialist report—a line manager of a staff executive? These questions have been debated for years. A survey of their members by the American Society of Safety Engineers published in 1960 showed that 48 percent of the respondents from industry reported to either the personnel manager or the industrial relations director. This study showed that while a reporting relationship was perhaps a factor in accident prevention results, it apparently was not among the more significant factors and that because each company in each industry is different, there can be no hard and fast rule about the exact position of the safety function in an organization. Today we are sure of only one thing: there is no one right answer. It depends on the organization and on the personalities of the people involved. We can, however, offer some criteria in assessing the right place for safety in an organization:

1. *Report to a Boss with Influence.* In part this is a personal evaluation, in part it is an evaluation of the structure. If the boss is line, authority must encompass the hazards to be controlled. If staff, there must be access to an executive whose command will buttress necessary action.
2. *Report to a Boss Who Wants Safety.* This follows inevitably. Problems arise when a chief executive wants results, but the voice of the safety director is muffled by an immediate boss whose attention is devoted to other problems.
3. *Have a Channel to the Top.* Management properly sets the priorities between production results and safety results, between sales expansion and elimination of unsafe driver-sales-people, between security of confidential research and the prying eyes of the safety specialist. This is not to say that safety must be placed in the upper echelons, but it does assert that channels to the upper echelons must exist so that all parts of the organization can reach their ears. Too often the only channel is a bypass with all its frictions.
4. *Install Safety Under the Executive in Charge of the Major Activity.* The safety function in this case serves as staff to that executive.

So we see a staff job that is pretty well self-defined and whose influence depends on personalities as well as location. Who might best fill this kind of job? This also has no easy answer.

In 1969 Bracey[29] made a study of the effectiveness of safety directors. He attempted to test these hypotheses:

1. Safety directors who are more effective have attitudes about certain organizational, technical, and behavioral concepts that are different from the attitudes of the less effective safety directors. Also, the more effective safety directors tend to have stronger convictions about these concepts than do the less effective safety directors.
2. Safety directors with nonengineering backgrounds have different attitudes about these concepts than those with engineering backgrounds. Engineers value and have stronger convictions about the technical concepts than nonengineers. Also, there is a relationship between the college degree held and managerial effectiveness.
3. Experience affects attitudes towards these concepts and the effectiveness of the safety directors.
4. The organizational position of the safety director affects the amount of power possessed. Also, the amount of power of and the position of the safety director have an influence on effectiveness.

Bracey sent 583 questionnaires (with 42.5 percent returned) to safety directors in four industries: petroleum refining, chemical, electrical manufacturing, and transportation equipment. The questionnaire included a semantic differential test of seven organizational, technical, and behavioral concepts, and asked questions on organizational characteristics, educational data, experience of the safety director, and accident statistics. The returns were divided into effective and ineffective classifications based on the accident frequency rates by industry. He found:

1. That the effective safety directors did have different and stronger convictions about the concepts relating to accident prevention.
2. That no correlation existed between education and the accident record. A test of attitude differences indicated that education played only a minor role in causing differences in attitudes. It was concluded that engineers and nonengineers make equally good (or bad) safety directors.
3. That the portion of the third hypothesis dealing with experience and attitudes was accepted while the portion concerning experience and effectiveness was rejected. Experience may be helpful in attaining strong attitudes, but experience does not seem to be related to effectiveness.
4. No apparent relationship was found between the level of the safety director and the amount of power possessed. No correlation was found between position, power, and accident frequency rates. More powerful safety directors did have stronger convictions about the concepts than the less powerful.

Thus it is apparent that there is no one best kind of safety director, just as there is no one best kind of manager or executive. The best safety manager is the one who manages to assess situations, the company, and what needs doing, and then gets it done through the people in the organization that must do it.

We suggested earlier that the safety manager start out by assessing the situation and defining the role. Exhibit 9.1 suggests the role will include four basic areas. We discussed areas 1, measuring, and 4, systems analysis, earlier. We will not deal with area 3 to any real degree here as this is the highly technical aspect of safety and there is a great deal of information available from all sources. We will, however, deal with area 2, safety programming, in the next sections. We will look at company styles and leadership patterns and how these might be assessed and how your choice of safety programs ought to depend on your analysis. We will look at workers and what kinds of safety programs they will buy today. And we will look at some programs that work.

Change Agent

The bottom line is that the safety professional's task is to effect change in the organization. How can he or she effectively do this? Exhibit 9.3 suggests the change is a slow process, which starts from the top—the executive who feels some pressure to do better and reacts. Executive pressure in safety management is at an all time high due to the legal environment, legislative changes, high costs, and other factors.

How that executive reacts to the changes you propose is illustrated in Exhibit 9.4. The clearer we can make the changes and the more trust we have earned, the more positive the perception of the changes. The executive then evaluates the personal impact of the changes, which dictates reaction.

SAFETY STAFF EFFECTIVENESS

William English, the Director of Loss Control for Marriott Restaurant Operations, wrote in the *National Safety News* on the effectiveness of the safety manager. He quoted Peter Drucker as saying efficiency is doing things right, but effectiveness is doing the right things. Therefore, the effective safety manager is the one who is able to analyze needs, coordinate the formulation of

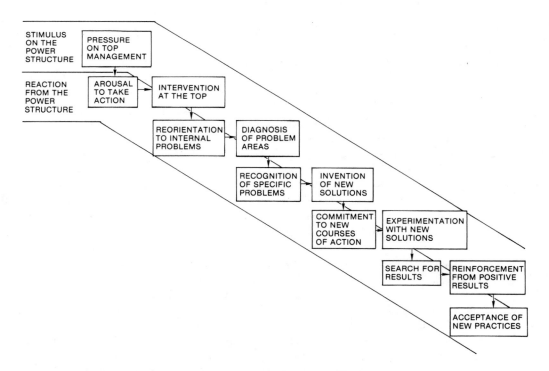

Exhibit 9.3 Process of change in attitudes toward safety.[31]

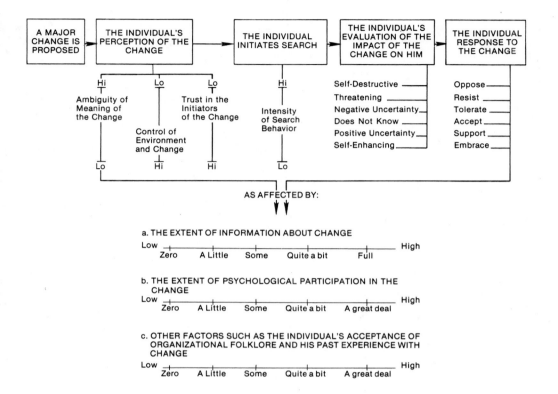

Exhibit 9.4 Model for understanding an individual's response to change.[13]

remedies, sell the remedies to the organization, and help the organization implement the prescribed programs. These functions require judgment, planning, and enough industrial experience to understand the organization and its function.

It is no doubt true that the most effective safety managers are the ones who in fact do do the right things (that is advise, develop solutions, communicate those plans to the line, and monitor results), and the ineffective safety managers are the ones who concentrate on doing the things that the line managers refuse to do.

Part IV

The Company And Its Style

Chapter Ten

Climate

If you want to motivate your employees toward a strong commitment to the goals of your company, you must have a positive philosophy that is effectively communicated throughout the organization.

Burt K. Scanlon[144]

Before examining the elements that might go into a safety program, it seems wise to look first at the company itself—at the environment the program will live in, the climate of the organization, and the climate of the safety program itself.

SAFETY PROGRAM CLIMATE

There seems little question that the way the worker sees the company safety program strongly influences not only behavior on the job but ability to learn from and to respond to safety materials. Workers in different companies characterize their safety programs quite consistently in any given company. This was brought out in a study[86] a number of years ago made for Employers Insurance of Wausau by Social Research, Inc. They found these general company types:

The Overzealous Company

This is the kind of company where a great deal of safety equipment must be worn, machines are so guarded that they are difficult to get near and work with, and the tone of the safety program is heavy. Such a company is likely to impose harsh punishment (a three-day layoff, for example) for some minor infraction of safety procedures. There seem to be endless meetings, films, manuals, and preachments about safety, often not involving the worker directly in any personal way. In such companies response to safety education materials is not lively and workers feel overexposed to it.

The Rewarding Company

This is the kind of company that might offer prizes for safety records or for entering into safety slogan competitions. While the prizes are relatively small, there is a sense of competition for them. Employees in such companies feel that safety programs are important and that the company generally feels responsible about safety matters.

The Lively Company

This is the kind of company that has a safety program which stimulates competition among the various plants, offers plaques, has boards to record the number of hours passed without accident, or posts a continuing safety record at the plant entrance. These are companies who teach the workers to identify with safety goals and the employees are proud of their record. Safety in these companies becomes one of the lively aspects of the job and means more than avoiding risk or accident; it is a concrete symbol and goal.

The Negligent Company

This is the kind of company which seems to have programs only after the fact, that gets busy about safety only after a major accident happens. The people feel that the company does not really care, that it passes out safety equipment because it is current custom and in order to protect itself, and that it passes out safety information material in the same way.

This study further reported that the ideas about the friendliness of safety people, the manner in which safety is talked about and promoted, and the efficiency with which matters are attended to vary with the concept of the company as described above. In the overzealous company, the safety specialists usually seem loud and harsh, overly watchful, too quick to criticize. In plants where programs for safety are livelier and more rewarding, safety people are usually seen as friends, people much like the workers. In the more lax organizations, the complaint is that there are not enough safety officials around and that the workers doubt reporting to them does much good, since little happens.

Despite these various attitudes, it seems evident from the comments of these people that they see safety as a peopled and policed program, one to which particular individuals and duties are assigned. These conclusions are the result of detailed in-depth interviews with workers to assess the need for and worth of safety media. The types of climates described are common in industrial safety programs.

CORPORATE CLIMATE

The safety program climate is only one aspect, another is total corporate climate. This is a major influence on the behavior of both managers and employees. According to Burt Scanlon,[144] of Oklahoma University, one point of concern is that the employees' perception of the organization's climate and philosophy may be different from that which is intended. Two possible reasons may contribute to this difference in desired versus actual perception. First, perhaps not enough effort has been expended in comunicating the guiding philosophy down the line. The second factor may be a discrepancy between what is professed and what actually occurs. The individual's closest point of contact with the organization is the immediate superior. If the superior's actions do not reflect the organizational philosophy, a perception discrepancy occurs.

Requirements for Individual Performance

According to Scanlon, some of the basic climate requirements for maximum individual performance in an organization are:

1. There must be central overall goals or objectives toward which the organization is striving.

2. The objectives must be communicated down the line with the idea of getting commitment to value, reasonableness, feasibility, etc.
3. Functional areas, departmental units, and individuals must also have specific goals to obtain. These must be derived from the central goals and interrelationships must be perceived.
4. The interdependency of all subunits within the organization in the accomplishment of results must be clearly established and a framework for interunit cooperation must exist.
5. Meaningful participation on the part of the individual should be the keynote. This means participation in the sense that the individual has a "real" part in determining job objectives.
6. There must be freedom to work in the sense that a person has an opportunity to control and adjust performance without first being exposed to authority and pressure from superiors. In addition, support and coaching from superiors is critical.

CLIMATE FACTORS

M. Scott Myers[114] notes a number of specific climate factors which are necessary before motivation can occur. Exhibit 10.1 spells out the climate factors as well as the formal programs for motivation he has used at Texas Instruments. He explains the climate factors:

> The first of the climate factors is *expansion.* If you are fortunate enough to live in a growth organization, it is hard to be a bad manager simply because, as the organization expands it offers increasing opportunity to achieve, to grow personally and professionally, to acquire a sense of responsibility. In addition, change itself makes Theory X more bearable. Joe, down on the assembly line, thinks "I can't stand the boss, but things are changing so fast I'll soon be rid of him." Of course, when the organization is stable, this kind of boss becomes more oppressive.
>
> Another climate factor to be considered is *delegation.* If every manager were Theory Y in his assumptions, delegation would happen naturally. Unfortunately all managers are not. But it is possible to encourage delegation by the manner in which you organize your business. Some of the desirable characteristics of the small organization can be retained within a large company by organizing the business into blocks, products, or families of products of income-producing services, or of staff functions.
>
> Our third climate factor is *innovation.* Exhibit 10.2 shows the relationship between level of motivation and utilization of suggestions of our managers. Among some 400 highly motivated managers, 73% said that their ideas were usually adopted. Of the approximately 400 poorly motivated managers, 69% said their ideas were not used. You can argue that the low group has half-baked ideas, and that's probably part of the reason they are poorly motivated. But part of the reason

Climate	*Formalized programs*
Expansion	Planning
Delegation	Performance review
Innovation	Educational assistance
Fluid communication	Compensation
Goal orientation	Attitude surveys
Stability	Work simplification

Exhibit 10.1 Media for motivation.[114]

Utilization of Suggestions

Level of Motivation	Ideas Not Used	Ideas Used
High	27%	73%
Partial	42%	58%
Low	69%	31%

Exhibit 10.2 Utilization of suggestions vs. level of motivation.[114]

the upper group is highly motivated is that they exist in a climate that is adjusted to change.

Fluid Communication is the fourth requirement for our climate, necessary to the job of getting the superior product or service out of the door. But as our organization grows, a formalizing of communication usually follows which can actually result in a break-down of communication. In a small shop, Joe yells something across the room to Harry; information is exchanged, and that's all that's needed. But in our growing company we dictate memos, which are typed in duplicate, filling up cabinets and taking up floor space. Boys are hired to deliver the memos to distribution points, from which they go into baskets, tying managers to their desks reading and answering these communications. We are caught up in a growing mountain of paper work which worsens until some agile competitor comes along who doesn't have this paper problem and threatens us by giving superior service at a lower cost. We clean up our systems, discovering that we can throw out 90% of what we have in our files. And gradually they grow fat again.

We aren't saying you should quit writing memos. Business, the law and contracts require documentation. We do say that you should think of the best way of communicating, using an informal system. Speak face to face. Use the telephone wherever possible. You pay a price for using the informal system; you need to keep in mind who else needs to have the information—your boss or his boss or someone else down the line. If you pay this price, you're in business; if you fail to pay it, your penalty is the "paper umbrella."

The fifth element of climate is *goal orientation*. This differs from authority orientation. When you ask a person why he is doing something and he says, "Because the boss told me to," that's authority orientation. But if, in answer to your question, he explains the job and what it will achieve for the organization—that's goal orientation.

One aspect, one condition, of goal orientation merits discussion here: the so-called status symbol. First of all, status symbols are inescapable. Whenever two or more people get together, such symbols emerge. And what are the typical status symbols of industry? Office size, varying according to rank, office furnishing, parking privileges, named parking lots, the executive dining room, manner of dress, the first name basis among equals, identification badges reflecting rank, and so on.

Why are we careful about this? We believe that status symbols of the kind listed, based on rank, increase social distance, and when you increase social distance you inhibit communication. Now, you never forget who your supervisor is, but you never are as close to him as you would be were he not your supervisor. You don't need any symbols to remind you of this. The lieutenant doesn't go up to the colonel and ask him to go out for a cup of coffee. That big eagle on the colonel's shoulder says the lieutenant had better not. But if the lieutenant and the colonel aren't wearing special identification, if their ranks aren't flaunted, they can talk as individuals.

We aren't discussing status itself, but the symbols of status. There's a good deal of difference. Some people have greater capability, greater responsibility, earn greater salaries. But flaunt the differences before those with unequal opportunities—special privileges, a carpet on the floor, a rubber plant in the corner—indulge in these—and you pay a high price. The symbols separate you from the people who do the work, placing the emphasis on authority rather than goal orientation. It's well to remember that as managers we *achieve our goals through motivated people.*

The final organizational climate factor is *stability.* It means many things, one of which is job security. This isn't to say that every individual hired has assurance of a life-time job. Individuals who don't perform well are let go. But in talking about job enrichment we do feel that where people manage their own work, possibly eliminating their own jobs, they must feel assured that there will always be another job for them. This kind of stability is essential.

SYSTEM 4

Rensis Likert,[89] one of the most famous of the behavioral scientists, also discusses climate to a large degree when he describes his *system 4* type company. He has isolated three variables which are representative of his total concept of *participative management.* These include (1) the use of supportive relationships by the manager, (2) the use of group decision making and group methods of supervision, and (3) the manager's performance goals. A supportive relationship is shown by the degree to which the manager has:

1. Confidence and trust.
2. Interest in the subordinate's future.
3. Understanding of problems and the desire to help overcome problems.
4. Trained and helped the subordinate to perform better.
5. Taught subordinates how to solve problems rather than giving the answer.
6. Given support by making available the required physical resources.
7. Communicated information that the subordinate must know to do the job and needs to know to identify more with the operation.
8. Sought out and attempted to use ideas and opinions.
9. Approachability.
10. Recognized and credited accomplishments.

The use of group decision making and supervision does not mean that the group makes all decisions. The emphasis here is on the involvement of people in the decision-making process to the extent that their perceptions of problems are sought, their ideas on alternative solutions are cultivated, and their thoughts on implementing decisions which have already been made are solicited. The participative process can be applied on either an individual or a group basis.

Likert measures the relationship of the above to productivity. He states there is strong evidence to suggest that the organization which exhibits a high degree of supportive relationships, and which utilizes the principles of group decision making and supervision where there are high performance aspirations, has significantly higher levels of achievement. He makes a distinction between what he calls casual, intervening, and end-result variables.

The *casual variables* refer to different management systems characteristics as follows:

System 1, Eplorative Authoritative.
System 2, Benevolent Authoritative.
System 3, Consultative.
System 4, Participative Group.

Likert explains each in detail in his book, *The Human Organization.*[89]

The *intervening variables,* such as loyalty; performance goals of subordinates; degree of conflict versus cooperation; willingness to assist and help peers; feelings of pressure; attitude toward the company, job, and superior; and level of motivation are of key importance.

The *end-result variables* refer to tangible items such as volume of sales and production, lower costs, and higher quality (results). An authorative approach may initially achieve improvements in the end-result variables, but at the same time the intervening variables begin to disintegrate. Turnover, absenteeism, and a progressive deterioration in the end-result variables themselves will eventually occur. A participative group approach leads to a gradual upgrading of the intervening variables and long-run permanent gains in end-result variables. The critical factor is time. Can you wait for your results?

Participation is one of the main ingredients in gaining employee commitment on an overall basis. It can lead to less need for the use of formal authority, power, discipline, threat and pressure as a means of getting job performance. Thus, participation and its resultant commitment become a positive substitute for pure authority. Commitment may be much harder to achieve initially, but in the long run it may prove much more effective.

Later we will examine in some detail how participative management works in safety programming when we talk about participative safety programs. Likert's description of participation, however, is really a discussion of .climate. An assessment of climate is fundamental to safety program success. Climate builds the employee attitudes. Acceptance of our program is a function of corporate climate, and our communication with employees depends to a large degree on our credibility. Climate, while hard to define, is important.

What builds climate? Myers[115] (Exhibit 10.1) suggests six components and Likert[89] suggests ten. Scanlon[144] spelled out six different ones. Each of us might be able to identify others from our experience.Anyone who has worked in an organization knows climate exists and knows its importance. I have worked in and for companies with vastly differing climates, from permissiveness to tight control, from climates of freedom to climates of guardedness, from climates with creativity to climates that stifle creativity, from climates that allow you to work the hours that you choose to climates that are run by real, but nonexistent time clocks, etc., etc., etc. And from climates that say safety is an important, integral part of the job to climates that say it is not. Similarly, we have all seen or experienced differing climates generated from safety programs as described earlier by Levy and Greene[86] in the Social Research report. The climates described are real. They are a product of both the corporate climate and the functioning safety program.

As we structure a safety program, climate must be considered. Better yet, before we structure our safety program, the corporate climate should be assessed and our program should be structured so as to create a safety climate.

What shapes a Company's Culture?   The researchers say five basic elements shape an organization's culture:

1. *Business Environment.* The "marketplace reality" of what a company does, its competitors, technology, regulations, etc.
2. *Values.* The basic concepts and beliefs of a corporation that define for employees "If you do this, you too will be a success."

3. *Heroes.* People who personify the culture's values and provide tangible role models.
4. *Rites and Rituals.* Systematic and programmed routines of corporate life. 'In mundane manifestations they show employees the kind of behavior expected of them. In ceremonies, they provide visible and potent examples of what the company stands for."
5. *Cultural Network.* An organization's primary means of communication, where 90 percent of the real business goes on.

In Search of Excellence[124] suggested eight attributes that make up a culture that spells success:

1. *A bias for action.* "These companies may be analytical in their approach to decision-making," but they are not paralyzed by it. Standard operating procedure is 'Do it, fix it, try it.' "
2. *Close to the customer.* "These companies learn from the people they serve. . . .Many. . .got their best product ideas from customers. That comes from listening, intently and regularly."
3. *Autonomy and entrepreneurship.* These companies foster leaders and innovators—"champions"—by encouraging practical risk-taking and supporting good attempts, even when the attempts fail.
4. *Productivity through people.* "Excellent companies treat the rank and file as the root source of quality and productivity gain."
5. *Hands-on, value driven.* Excellent companies pay "explicit attention" to values and "their leaders have created exciting environments through personal attention, persistence and direct intervention—far down the line."
6. *Stick to the knitting.* "The odds for excellent performance seem strongly to favor those companies that stay reasonably close to the businesses they know."
7. *Simple form, lean staff.* "The underlying structural forms and systems in the excellent companies are elegantly simple. . . .Top-level staffs are lean. . . ."
8. *Simultaneous loose-tight properties.* "The excellent companies are both centralized and decentralized. . .they have pushed autonomy down to the shop floor or product-development team. On the other hand, they are fanatic centralists around the few core values they hold dear."

DETERMINING SAFETY CLIMATE DIMENSIONS

Dov Zohar,[176] (Israel Institute of Technology), recently identified some of the dimensions of safety climate in an article in the *Journal of Applied Psychology*. His purpose was to define organizational characteristics that differentiate between high versus low accident rate companies. He identified these characteristics:

1. One consistent finding was that factories having successful safety programs invariably had a strong management commitment to safety. There was little consistency in how this commitment was exhibited, but the hallmark of successful companies was active visible personal executive involvement.
2. In low accident rate companies, safety was given a high priority in company meetings and accidents were treated as symptoms of design failures.
3. In better companies, safety personnel were seen as having a higher status.
4. In better companies, safety training was an integral part of new worker training.
5. In better record companies, there tended to be open communication links between workers and management.

6. A good record was invariably accompanied with good plant housekeeping.
7. Better companies tended to have a stable work force with lower turnover and older workers.
8. Finally, successful companies had distinctive ways of promoting safety. These included guidance and counseling rather than enforcement and admonition.

A similar recent study was performed in the coal mining industry by the Naval Weapons Support Center to determine the relationship between climate and safety results. Their data suggested that:

1. When decisions are decentralized, when management is flexible and innovative in trying new procedures and programs, and when morale is high, disabling injuries decrease.
2. As disabling injuries increase, feedback, continued employee development, and consistency of orders improve, which then appears to decrease the injuries.
3. Production pressure appears to lead to an increase in disabling injuries, which in turn results in a decrease in production pressure.

Chapter Eleven

Management Styles

The classification of the behavior of managers, bureaucrats, or administrators has occupied the attention of political and administrative scientists, sociologists, and psychologists for some years. It is widely recognized that such typologies may enable organizational, group, and even personal life to be better understood and changed.

W. J. Reddin[138]

Management reading in recent years has been sprinkled with articles and books which describe various styles of management. Most of these writings tend to classify management into styles according to two orientations: how much the manager is oriented either to the tasks that must be performed or to relationships and the feelings of subordinates. Different authors use different classifications, and different labels for the types of managers.

THEORY X VERSUS THEORY Y

Douglas McGregor[98] used as his basis of classification some basic assumptions the manager makes about human nature.

McGregor summarizes[98] the conventional conception of the management task in terms of three propositions. He labeled this set of propositions *Theory X*:

1. Management is responsible for organizing the elements of productive enterprise—money, materials, equipment, people—in the interest of economic ends.
2. With respect to people, this is a process of directing their efforts, motivating them, controlling their actions, modifying their behavior to fit the needs of the organization.
3. Without this active intervention by management, people would be passive, even resistant, to organizational needs. They must, therefore, be persuaded, rewarded, punished, controlled. Their activities must be directed. This is management's task. We often sum it up by saying that management consists of getting things done through other people.

McGregor stated that behind this conventional theory there are several additional beliefs:

   a. The average person is by nature indolent; he works as little as possible.
   b. He lacks ambition, dislikes responsibility, prefers to be led.
   c. He is inherently self-centered, indifferent to organizational needs.
   d. He is by nature resistant to change.
   e. He is gullible and not very bright.

At the time of his writing (1960), management was fashioned from propositions and beliefs such as these. Even today, organizational structures and managerial policies, practices, and programs reflect these assumptions.

The findings from the social sciences challenge this whole set of beliefs about human nature and about the task of management. The social scientist does not deny that human behavior in industrial organization today is approximately what management perceives it to be, but believes that this behavior is not inherent nature. It is a consequence rather of the nature of industrial organizations and of management practices. The conventional Theory X management approach is based on mistaken notions of what is cause and what is effect.

MCGREGOR'S NEW THEORY OF MANAGEMENT

For these reasons, McGregor proposed a different theory of managing people based on more adequate assumptions about human nature and human motivation, which he called *Theory Y.* Theory Y is based on these assumptions:

1. Management is responsible for organizing the elements of productive enterprise—money, materials, equipment, people—in the interest of economic ends.
2. People are not by nature passive or resistant to organizational needs. They have become so as a result of their experience in organizations.
3. Motivation, a potential for development, a desire to assume responsibility, and a readiness to direct behavior toward organizational goals are all present in people. It is one responsibility of management to make it possible for people to recognize and develop these human characteristics for themselves.
4. The essential task of management is to arrange organizational conditions and methods of operation so that people can achieve their own goals best by directing their own efforts toward organizational objectives.

McGregor's Theory Y is a process of creating opportunities, releasing potential, removing obstacles, encouraging growth and providing guidance.

Since one of our fundamental tenets is that safety should be managed like any other function, all of the above applies to safety management. The line manager can be either Theory X or Theory Y oriented in safety management. Theory X safety management is directive, highlighted with rules, assumes "they" want to get hurt and must be controlled, and believes that "they" do not know enough to spot a hazard. Theory Y safety management attempts to harness and utilize what people know and allows them to determine their own rules.

THE MANAGERIAL GRID

Another approach to managerial style typology is the Managerial Grid. Blake and Mouton[77] classify management styles similarly by drawing a *Managerial Grid* as shown in Exhibit 11.1. The two axes of the grid are Concern for People or Relationships Orientation, and Concern for Production or Task Orientation. With the grid we can refine further the management style identification and recognize that a single manager can be high in both concern for people and concern for production or job achievement. Conversely, the manager can be low in both, or high in one and low in the other. Shown on the grid are five management types:

1. *The 9,1 manager,* the production person who is not interested in the employee's feelings.
2. *The 1,9 manager,* the human relations specialist who is not interested in getting any work done.

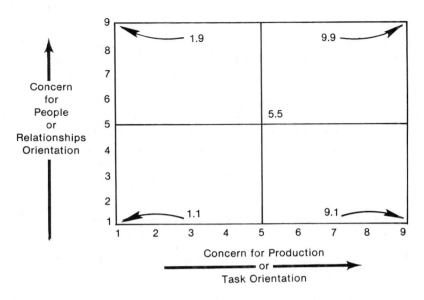

Exhibit 11.1 The managerial grid.[17]

3. *The 1,1 manager,* who believes workers are so naturally lazy that one can expect practically no production from them and actually expects none.
4. *The 9,9 manager,* who expects the best from people and gets it, because they will be happier as they do achieve goals on their job.
5. *The 5,5 manager,* who goes halfway each way, compromising both concerns and gets half-happy people and mediocre production.

MANAGEMENT STYLES AND SAFETY

The safety manager needs an understanding of Theory X versus Theory Y and of the types of managers shown on the Managerial Grid because success depends on a number of line managers that must be influenced and motivated into action. Each of these managers is located in a different place on the grid. One supervisor may be a 9,1 manager, while the next department may have a 1,9 manager. How do we as safety staff specialists motivate a 9,1 manager? Certainly differently than a 1,9 manager. And how does each of these types attempt to control the behavior of their people? Certainly quite differently. Their respective results will be quite different, and their results are our results.

When the 9,1 manager faces a safety problem, it is seen as a threat to production. Probably this very threat is the best form of motivation. If safety training is not an integral part of job training, it will not get done, because job training is seen as important to getting the job done and safety training could be perceived as a waste of important production time.

When the 1,9 manager faces a safety problem, it is seen as a threat to the welfare of the people. Reaction probably will be to provide any needed safety training and to maintain physical conditions. A weakness may be in the enforcement of safety rules. Our best sales point for the 1,9 manager would seem to be to emphasize that employees could be injured, rather than discuss improved efficiency or lowered costs.

The 9,9 manager is sharp enough to see the relationship between safety and production, so it it usually sufficient just to identify the problem.

The 1,1 manager is a problem to us as well as to top management, but it is highly unlikely we will have many 1,1 managers with whom to deal as they do not last long in industry.

The 5,5 manager is also a problem. As a compromiser in both human relations activities and production, safety enforcement also will be halfway, which may not be far enough. As a laissez faire manager who will give lip service, but no real commitment to either subordinates or management, this manager will be difficult to change.

The Managerial Grid and the Theory X-Theory Y approaches to management style typology could help the safety manager to understand line managers and to know how to motivate them. The theories infer that to successfully motivate line supervisors, there is probably no single "right" approach. Each individual supervisor, depending on personal task and relationship orientation, will react differently to each motivational approach used.

EFFECTIVENESS OF OUR STANDARD MOTIVATORS

Consider, for instance, some of our standard supervisory motivators in safety. We use these kinds of things to motivate our supervisors:

1. *Management's Policy.* We use an expression of management's will in writing as a motivator for supervisors in the belief that once the supervisor knows that the boss wants safety results, the supervisor will try to give the boss those results. Obviously, the higher the concern for production, or task orientation, the more effective policy will be. If the supervisor perceives safety as part of the defined task and has a high task orientation, this policy will be effective as a motivator.

2. *Accountability Systems.* These systems make policy live. The stronger the task orientation, the more effective these systems will be. If management is measuring the supervisor in terms of results and places safety results on a par with production results, then the 9,1 and 9,9 managers will be powerfully motivated by the measurements.

   Accountability systems could be a good stimulus for the relationships-oriented managers also. If the systems measure the activities the supervisors are performing (training activities, inspection, employee contacts, etc.), they are measuring areas that are perceived as important to the 1,9 and 9,9 supervisors. They will see these measurements as making them look good. They will react to these measurements by doing more safety training, inspecting, etc., which should bring further results. The managers with a high concern for people, however, may not react to accountability systems that measure only results, particularly if they cannot translate these results (dollars, frequency rates) into things that are good for the people.

   In short, accountability systems that measure only results are effective for task-oriented (high concern for production) managers. Accountability systems that measure only activities are effective for relationships-oriented (high concern for people) managers.

3. *Training.* Providing needed knowledge and skills to the supervisors usually will result in motivation. It may, however, depend on the training content. The 9,1 managers may consider safety training a waste of time unless it also shows them how to improve production performance. For the task-oriented managers, safety training should be an integral part of job training.

   The 1,9 managers may consider safety training more important than some job training. Relationships-oriented managers would certainly react more favorably to the term *safety education.* These managers would react better to supervisory training that builds supervisory skills as it improves safety performance than to pure safety training which teaches only safety skill.

4. *Special Emphasis Campaigns.* The high task-oriented (9,1) managers will no doubt see these activities, as well as all use of safety educational media, as a waste of time. They might, however, react well to a context or competition with other department managers. The task-oriented managers like competition. This might explain the successes of some safety contests.

The high relationships-oriented (1,9) managers on the other hand, would react favorably to safety educational media, to special emphasis campaigns, and to most training or visual aid materials. They might, however, not react well to competition with peers.

It would seem that it might be useful for the safety managers to try to identify the styles of line management people as it may well help in plotting strategy for future safety activities. If, for instance, the managerial styles in company tend to be the predominantly high relationships-oriented, they might rely heavily on accountability systems that measure activities rather than results. They might also choose to concentrate further on supervisory safety training and make use of more special emphasis campaigns. On the other hand, if the styles are predominantly high task-oriented, they might well emphasize policy more and rely more heavily on accountability systems that measure results. They would do well to continue striving to get top management to place heavy emphasis on the fact that safety performance is a key part of the manager's job.

The safety managers might also evaluate carefully the management style of top management since this will probably be an indication of what the predominant management style (climate) will be.

LEADERSHIP STYLES

As we indicated earlier, climate and managerial style seem to be closely interdependent—to a large degree climate is dictated by style. It seems to take all kinds of managers to keep a company going. Each kind of leader seems to create a human climate which has specific effects on followers. Although the number of different methods used by leaders must be enormous, it is easy to classify most leaders according to one or another style of leadership. The classification which has been used most is autocratic style in contrast with democratic style. There are more recently named styles, which partly overlap the democratic and autocratic. Although these create about the same climates as the democratic or autocratic, each does emphasize some variation which has practical significance:

Authoritarian	in contrast	Equalitarian
Dictatorial	with	Facilitative
Leader-centered	in contrast	Group-centered
Production-centered	with	Worker-centered
Restrictive		Permissive
Job-centered		Employee-centered

Renis Likert[88] uses the last two terms above and describes the relative effectiveness of the two styles:

1. Supervisors with the best records of performance focus their primary attention on the human aspects of their subordinates' problems and on endeavoring to build effective

work groups with high performance goals. These are referred to as *employee-centered leaders.* Exhibit 11.2 presents the findings from one study. It illustrates the pattern of results from several different studies in widely different kinds of work.

2. The performance goals of supervisors are also important in affecting productivity. If a high level of performance is to be achieved, it appears to be necessary for a supervisor to be employee-centered and at the same time to have high performance goals. Exhibit 11.3, however, shows that there is a marked inverse relationship between the average amount of "unreasonable" pressure the workers in a dependent position feel and the productivity of the department. Feeling a high degree of unreasonable pressure is associated with low performance.

3. General rather than close supervision is more often associated with a high rather than a low level of productivity. This relationship, found in a study of clerical workers, is shown in Exhibit 11.4 for supervisors. Similar results were found for nonsupervisory employees. Supervisors in charge of low-producing units tend to spend more time with their subordinates than do the high-producing supervisors, but the time is broken into many short periods in which they give specific instructions, "Do this, do that, do it this way."

4. Genuine interest on the part of a superior in the success and well-being of subordinates has a marked effect on their performance. For example, Exhibit 11.5, shows that high-producing supervisors tend either to ignore the mistakes their subordinates make, knowing that they have learned from the experience or to use these situations as educational experiences by showing how to do the job correctly. The supervisors of the low-producing sections, on the other hand, tend to be critical and punitive when their subordinates make mistakes.

	Job-Centered	Employee-Centered
High-Producing Sections	1	6
Low-Producing Sections	7	3

Exhibit 11.2 Employee-centered supervisors' production vs. job-centered supervisors' production.[88]

	Department Productivity	
	Below Average	Above Average
The Ten Departments Which Feel the Least Pressure	1	9
The Middle Eleven Departments	6	5
The Ten Departments Which Feel the Most Pressure	9	1

Exhibit 11.3 Relationship between pressure felt by people and their production.[88]

	Number of First-Line Supervisors Who Are:	
	Under Close Supervision	Under General Supervision
High-Producing Sections	1	9
Low-Producing Sections	8	4

Exhibit 11.4 Relationship betweeen closeness of supervision and production.[88]

	Foremen's Reaction to a Poor Job (As Reported by Their Men)	
	Punitive-Critical	Nonpunitive: Helpful
High-Producing Foremen	40%	60%
Low-Producing Foremen	57%	43%

Exhibit 11.5 Relationship between the manager's reaction to a poor job and production.[88]

While Likert's work is aimed primarily at relating managerial style to production and not safety, it seems to have a message for safety managers. On the surface, at least, it would seem that employee-centered supervisors would generate a better safety record than job-centered supervisors. The Levy and Green[86] study referred to earlier infers this. If the relationship shown in Exhibit 11.2 holds true for safety as well as for production, a logical conclusion would be that the relationships shown in Exhibits 11.3 and 11.4 might also be true in safety.

DISCIPLINE EXPERIMENT

There is a study which shows that the relationship punishment and production, shown in Exhibit 11.5, also holds between discipline and safety results. Maier describes this study in his text *Psychology In Industry*.[105]

Disciplinary action is often associated with accidents or safety violations so that regardless of how a superior feels about punishment he may be involved in administering penalties. What does a supervisor do when he finds a man breaking a company regulation for which disciplinary action is specified?

Many foremen have reported to the author the dilemmas they face when a good worker commits a violation. They know that laying a man off often creates hardships, destroys friendly relations, and lowers morale. Sometimes grievances are filed, and when this occurs, their decisions are frequently reversed. They also know that they can get into trouble if they ignore violations because all guilty persons are supposed to receive the same penalty. It is not uncommon for higher supervisors to demand strict enforcement of company regulations. Some companies go so far as to have the safety department police the job because foremen are too lax. When they take this action they remove safety from the foreman's duties. Campaigns and training programs are then instituted to make foremen more safety conscious. Foremen disillusioned by such experiences resolve the dilemma by not 'seeing' the violations.

Let us take a specific situation in which a foreman thinks he has found a lineman working on top of a utility pole without his safety belt. Should he attempt to determine whether a violation has occurred? If he does, he motivates the lineman to lie or to defend himself, which leads to unpleasant relations and poor cooperation. If he ignores the incident, he shirks his duty and may bring on trouble for himself with his superiors. Yet he is expected to build morale and carry out company regulations.

An experiment using 154 pairs of real foremen in a simulated situation was performed to determine what foremen do when confronted with this problem. One member of each pair acted as the foreman, the other acted as a lineman who had neglected to follow a safety regulation, but engaged his belt when he saw the foreman approaching. The foreman suspected a violation.

The results are shown in Exhibit 11.6. Interestingly, 25 per cent of the foremen did not discuss the violation, but instead talked generally about the company safety drive. The rest brought up the question of the violation. In these pairs, 40 per cent of the linemen denied the violation and were not laid off, while 60 per cent admitted it. Admission required the foreman to make a further decision. Of these, 85 per cent decided not to lay off the worker as required by the company regulation, while 15 per cent laid him off. From this it is very evident that various foremen in the same company see their obligations quite differently when confronted with the same situation.

Even the failure to lay off the guilty worker can get the foreman into trouble because this represents discriminatory treatment, which the union will use in cases of lay-offs made by other foremen. Nevertheless the majority of the foremen did

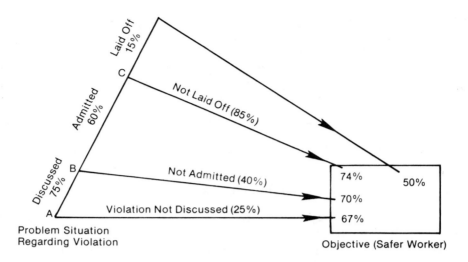

Exhibit 11.6 Decision points a foreman meets when he observes a safety violation. At point A he decides whether or not to discuss the violation. If
he decides to discuss it, he reaches point B. Denial of a violation
will settle the matter since he cannot prove a violation. However, if
the man admits the violation the foreman moves to decision point
C, where he must decide whether or not to lay off the worker. The
outcomes with regard to the objective of future safety are shown
with each of the decisions in the box in the right side of the
diagram.[33]

not comply with the regulation in this specific situation. Although a simulated
situation may not be representative of real life, the foremen regarded the results as
realistic. (Simulated situations are needed to obtain statistical data on the same
situation. Unfortunately real-life instances do not lend themselves to repetitions of
the same situation.)

After the foremen had completed their talks with the linemen the latter were
asked what effect this contact would have on their use of the safety belt in the
future. The percentages of times that the goal of the future safety was accomplished
by the various decisions are very much the same (67 per cent to 74 per cent) for
three of the decisions. But in the case of the lay-off it drops to 50 per cent. Thus as
far as safety is concerned the decision not to discuss the violation was the simplest
way to obtain satisfactory results.

This experiment has been repeated many times. Punishment invariably produces
the least future safety. In addition, the linemen report that their production will
suffer and that they will try to avoid being caught in the future. Subsequent use of
this case indicates that foremen are less punitive than higher management personnel
and that staff workers (including safety engineers) are the most punitive. It is one
thing to have the task of administering punishment and facing the punished worker,
but quite a different thing to have the impersonal task of setting up the regulations.
Punishment takes on a different meaning in a general context or abstract sense than
when applied to a particular individual.

In the foregoing experiment the violation was in doubt. The foremen used this
doubt as an excuse for not laying off the worker. They also said that the punishment (three-week lay-off) was too strict.

In another simulated experiment dealing with a violation of a no-smoking regulation, the violation was made clear cut, the penalty was only three-day lay-off, and the foreman already had laid the man off. What would he do when the union steward attempted to get him to change his decision?

The results are shown in Exhibit 11.7. Only 34.9 per cent of the foremen failed to alter their decision, with another 13.4 per cent unable to settle the matter in the allotted time. Both outcomes resulted in a good percentage of grievances. Foremen who take a less rigid stand and do some problem solving retain the union steward's good will. In these instances grievances are rare.

The legalistic stand of determining guilt and involving punishment, even if accepted by the company philosophy, is not supported by foremen in general. They are inclined to recognize the feelings of the workers and to think in terms of future safety and the effects on morale.

The training of foremen should be in terms of positive motivation. Obviously men do not want to have accidents, so attempts to make accidents less attractive by punishment misses the point. If people behave in unsafe ways it is due to the presence of conflicting goals. Punishment is frequently seen as the price for getting caught and therefore often motivates men to find ways to avoid detection or to reduce the pain of the penalty. The Federal Aviation Agency, which can ground pilots for violating safety regulations, discovered that the pilots' union pays the grounded pilot his salary during the lay-off, thus neutralizing the punishment.

One cannot conclude that all forms of punishment can or should be abolished. Instead, the need is to find better ways to accomplish objectives, by studying the motivational alternatives.

It is fairly obvious that the relationship between punishment and production shown in Exhibit 11.5 does exist between discipline and safety. Many safety professionals believe the relationships shown in the other exhibits between elements of management style and production will also be true with safety. John P. Gausch[40] puts it into this perspective:

If the opinions of most authorities are even close to being true most of our loss problems, at least in the accident field, center around insufficient motivation. Generally programs for loss prevention fail to develop an environment in which managers and employees alike will be motivated to perform their jobs in a safe manner. Specialists declare that from 80 to 90 percent of all accidental loss is caused by people, as evidenced by such remarks as, "The most dangerous part of a

Decision	Frequency, percent	Result in grievance, percent
Full three-day layoff	34.9	45
Reduced layoff	4.6	
Warning	22.7	
Forgiven	7.5	2
Consult higher management	8.7	
Consult workers	3.5	
Other	4.6	
No decision	13.4	43

Exhibit 11.7 Decision regarding punishment for violations challenged by the union steward.[105]

vehicle is the nut behind the wheel" and "We can make a nuclear reactor fool-proof, but we will never make it 'damn-fool-proof.' "

What is missing in the development of the typical motivational program? Evidence indicates that motivation is pursued vigorously but practiced with almost total disregard for the professional "know how" and theories of behavioral science. The result is glamorous programming—contests, posters, and rallies. Statistics appraising performance paint a picture of success for these programs, but the serious satistician shudders at their validity. The treasurer must be quite confused when he has seen the company receive an Award-of-Honor for safety on the same day he's signing an increased premium check for fire protection.

Since one must deal with the "people problem," the results of management science and behavioral studies can perhaps offer some clues to guiding efforts in loss control. Consider, as only one small example, the now classic Michigan studies (Likert's work). In the studies, research was conducted to define the characteristics of high- and low-producing supervisors in many industries. (The results of the studies should be known by every management employee even if he will never be involved in a loss prevention effort.)

The research work compared the amount of production with supervisor characteristics. To assume that characteristics of high or low production are related to loss experience, one must contend that production and safety go hand-in-hand or that enlightened management will be an influential factor in the reduction of loss. Few safety personnel would doubt a direct relationship.

This assumption is supported in the studies done in the mining industry in Great Britain. Two different locations were used in a British study to show the relative effects of various management "styles" on production. The significant aspect for our concern here is that the time lost due to accidents was remarkably less in the high-producing mine.

Let us look at the following elementary example, typical of findings in the Michigan study, which puts a finger on characteristics of high and low producers.

CHARACTERISTICS OF SUPERVISORS

Low Producer

Imposes rigid goals on his employees.

Rigidly sets standards, minutely organizes each job, closely directs employee behavior.

Tries to "con" employees into hard work with artificial human relations techniques. Turns work into a game with contests and other manipulative devices or uses threats of dismissal as his main motivating tool.

Places prime importance on controlling the employee's conduct rather than the results achieved.

High Producer

Supports the "psychological advantage" of letting his workers do their own work their own way.

Sets general goals and standards. Uses these measurements mostly to find out how he can be more helpful.

Expects employees to do their work properly.

Creates loyalty by speaking up for his employees to higher management.

Is more interested in controlling results achieved rather than controlling conduct.

How do these characteristics help to guide a loss prevention effort? Let's take a few problems as examples.

Trying to motivate employees to work safety by using contests. That's asking a supervisor to use a characteristic of a low producer!

Using rigid standards. Perhaps this low-producing characteristic explains why rules-of-the-road safety regulations have been unsuccessful in reducing auto accidents?

These are but small examples of why some sound insight into behavioral science might help, and how it can cast serious doubts on the values and design of today's efforts.

As one pursues other studies in depth, present-day programs become increasingly shallow. An everwaiting danger for loss prevention staffs is self-satisfaction. It is no longer wise to be happy with yourself! Perhaps words by T.V. Learson, former president of IBM, contained a message for loss prevention directors. He described a good executive as "a man with a sense of urgency, a demand for excellence and a healthy discontent with the way things are." There are some good safety programs, but generally, across the broad complex of operations in the United States, as well as abroad, here is the way many of them are.

Programs for loss control drastically lack engineering competence.

New processes—and old ones—that are not evaluated realistically go on stream and operate dangerously close to critical loss.

Controls are developed with little consideration for measurement criteria.

Measurement is based on invalid statistics, and feedback loops are poorly planned.

Goals are established for managers who have no opportunity to participate in setting them.

Activities are forced on management with little or no consideration for "job enrichment" or personal commitment.

Motivation does not take advantage of the strength of intrinsic need.

Present efforts are not entirely wasted; things could be worse. Tough-minded management theory and rigid standards get results over the short haul. Even a "police" safety program with nit-picking inspections gives the impression that it will help to reduce waste. An executive can use many short-term motivators to demonstrate his effectiveness. He can get quick results with discipline. With a little luck he can get promoted before the disease of bad morale sets in and all the beautiful curves and graphs start down the backslope. It's evident that safety and loss control programs in many corporations are now on the back-slope; the short haul programs just won't suffice.

Efforts must now be made to formulate long-range programs including improved engineering disciplines and known theories of management. Activities must be integrated with existing management responsibilities. We should review our effort not from the standpoint of condemning present programs but in order to see how we can develop a comprehensive plan to improve experience in the future. The tendency to forge ahead and increase present earnest but weakly conceived efforts is not likely to produce the desired results.

An evaluation must be made as a grass-roots study and a completely open-minded appraisal of the value and extent of total loss prevention programming for the future. There are many areas of concern.

In many operations it is obvious that the loss prevention staff is too large. The do-it-yourself philosophy which prevails, frequently with poor results, can build a small empire in a short time. Too much control might be centered in the corporate staff office. Motivation might be poorly constructed and could continually reinforce an already serious problem. There are many ills which plague today's programs.

THE CONTINGENCY SCHOOL OF MANAGEMENT

The above discussion on style and effectiveness summarizes one of the more recent schools of management thinking, the contingency school. As you trace the changes in management thinking, at least three distinct schools emerge. This book attempts primarily to fit safety thinking into the latter two. The first school of management, the oldest, is the classical school. It describes management as it must be carried out in the typical bureaucracy. It describes the manager pretty much like McGregor's Theory X. It describes organizations in the classical sense, with organization charts, jobs having job descriptions, everyone fitting nicely into a little box, and communication flow described by solid or dotted lines, as are responsibilities and authorities.

Until the mid 1950s, management thought favored the *Classical School of Management*. This was governed by the rules of a bureaucracy, and favored job descriptions, standards of performance, writing up corporate manuals, and determining rules and regulations to govern worker behavior. This school of thought was based on the assumption that "all people are alike"—they will behave as we dictate and will react to management's manipulative schemes, wage incentives, etc.

The *Classical* school was replaced by the *Human Relations* school which said that behavior is dependent upon happiness. A happy worker will be a productive worker and will do what management wants. In this school of thought, the management task was perceived as concentrating on ensuring that the worker was psychologically comfortable on the job. The thought was also based on the assumption that "all people are alike"—happy folks are productive folks.

In the seventies the *Contingency School of Management Thinking* emerged, and for the first time, management's underlying assumptions about people were in tune with psychological reality. The assumption behind the contingency theory of management is that "everybody's different," therefore our management style and how we deal with a worker must be contingent upon the situation, on the worker, and on his needs. How a manager manages must be appropriate to the situation.

As our individual organizations begin to shift from one school of thought to another, we must make sure that our safety programs shift in accordance with the current management approaches and styles. Safety programs that "fit" the culture and climate of the organization must shift with the times. Too often our safety programs have remained firmly rooted in the classical school of thought, depending on standards of managerial performance, corporate safety manuals, rules and regulations to get the job done. While this is perfectly appropriate to 1950 style management, it is most inappropriate to 1980 style thinking in management. Each of us must assess our organization's management school of thought and build the safety program to be in tune with that style of managing.

Part V

Building
The Safety Program

Chapter Twelve

Why Safety Programs Don't Fly

Somewhere, something has gone wrong. Not many people ever suspected that the conscientious effort that has been applied to control losses would have been unsatisfactory. There is little doubt that losses have been reduced to some extent by present programs; the question is, are present programs the most effective way to gain maximum control, and if not, what went wrong?

John P. Gausch[40]

At the end of Part IV, we used a rather extensive quote from John P. Gausch[40] which seemed to set the stage for what will be said in this part. Earlier in Mr. Gausch's article he stated:

> Many people refuse to believe that the problem is very serious. They think that conditions will improve if we continue our present efforts.
>
> This appears doubtful. Plans for preventing losses have failed to clamp the lid on the mushrooming traffic death toll. As a matter of fact, there is some indication that losses increase as our programs of prevention are intensified.
>
> Is increasing the size of loss prevention staffs the answer? The effect of efforts to control industrial theft seldom can be judged by the number of guards at a plant location. In many heavily staffed plants, property loss from stealing and fraud are such an economic drain that accidents and fire losses take a back seat.
>
> The problem is complex. The first thing that "went wrong" was probably that technology of operations out-ran loss prevention skills and disciplines. Efforts have leaned heavily on a cops-and-robbers game. Staff safety audits, fire inspections, and security surveys are the cops. Management and hourly workers are the suspects being policed. This is the approach followed by the majority of insurance engineering groups, industrial staff auditors, and government inspectors. Seldom is operating management credited with enough initiative or integrity to control their own loss exposures.
>
> Many specialists in loss control also failed to keep pace with the growing complexity of exposure. They did not develop techniques to analyze potentials, and they seldom use scientific problem-solving to study operations or evaluate processes. . . .
>
> The second thing that "went wrong" was that programs were developed in an isolated environment. Companies felt that the staff specialist was more effective if he was located in an untouchable world, so he could shoot without being shot back at. Because of his isolation, the specialist was not aware of the growing knowledge in the field of management science or behavioral research.

Our intent in this part is to look specifically at this growing knowledge and to adapt it to safety management. Here are some of the areas we should consider if we are to answer the question, "Why don't safety programs fly?"

PRIORITES

Many safety programs start out with a "Safety First" approach, with an orientation of building a safety organization separate from the production organization, rather than building safety into regular work systems and activities. This immediately puts top management, middle management, and the first-line supervisor in the situation of making choices. Daily, perhaps hourly, a choice must be made between safety and production. This leads the organization as a whole to place priorities on production and safety, and which will come first? Obviously if a choice must be made, safety cannot win. Profit (defined as production) is the name of the game; it is the reason the manager exists; it is the job. So now we have placed ourselves in a win/lose situation. Proper programming, however, can overcome this. By building in safety we do not force the line to make a choice, we help obtain safe production and we place ourselves in a win/win situation.

COMMUNICATING AND MOTIVATING

Our safety programs attempt to communicate with our people and motivate them; attempt to change their attitudes, or better yet, their behavior. We utilize posters, booklets, slides, tapes, movies, bulletin boards, displays, signs, green stamps, awards, dinners, contests, ad infinitum. Are they effective? Do they change attitudes or behavior? Do they motivate or even communicate? We do not know. We have never proven whether any of these have any effect whatsoever. We use them because it is traditional for safety programs to use them. As we will see later, very little real research has been done in this area, and what little has been done is conflicting and surprising.

In short, there is no real foundation for using these elements in our safety program. This is not necessarily to say that they should not be used. But let us not delude ourselves into believing these traditional safety program components are controlling losses– unless we have facts that substantiate our claim. And we have not even utilized the facts that are available as to what constitutes effective communication.

OTHER COMPONENTS

Training

Our safety programs depend so much on training that they would collapse without it. We spend a great deal of time, money, and effort on safety training for the new workers, the experienced workers, and the supervisors. Laws require it. We believe in it. But, how much of our training is aimed directly at the needs of the individual trainee? How much of it stems from an analysis of job needs, corporate needs, or human needs? How much of our training is based on meeting behavioral objectives and leads toward making the individual trainee able to perform differently as a result of the time, money, and effort spent? And how much of our training is accurately evaluated to determine whether or not the time, money, and effort spent was worth it?

Usually we put together (or purchase) a safety training program that consists of x subjects, we present these subjects to the employees (or supervisors) that the line (or personnel) sends to us and we make no attempt to evaluate whether or not we should have bothered. Research has shown what really should be included in industrial safety training.

Rules

Our safety work rules are another cornerstone of our program. Rules are invariably written by those of us who "know." They are then posted, put in a rule book, or negotiated, with the end result that they are law. Then we must enforce the law and punish whenever it has been broken. Is this the best, easiest, and most efficient way to effect behavior change? Research has a lot to say on this.

Screening and Placing

Often we do not do this. Or when we do, we try to select the "right" person for the job; the one who fits mentally and physically, to lessen the possibility of accidents. How do we make this prediction? Take for example our attempts to select drivers. We call past employers and check the state MVR (motor vehicle driving record of accidents and violations.) We do this even though research suggests that we could get considerably better, more valid information just be asking the individual.

Built-In Hazards

We routinely place workers in situations that are traps: work situations such that normal physiological and logical psychological actions will likely cause injuries. Then we investigate and cite carelessness as the cause. Human factors research could be of real help to us here.

Inspections

We spend untold time, money, and effort inspecting and correcting physical conditions for OSHA or our own idea of safety with little factual information about whether or not the dollars spent are related to the record. However, inspection is traditional in safety programs.

Records

Our safety programs make much use of records. Investigators search only for one cause when there are many. Injury records are kept along the lines of American National Standards Institute (ANSI) standards Z16.1 and Z16.2 which results in a great deal of more or less unusable stored data because the input has little, if anything, to do with cause. (Obviously, however, some records are necessary for legal and other purposes.)

These are some of the reasons safety programs don't fly. Conceptually they are based on tradition, not on facts. They are not based on modern management knowledge or behavioral research. Let us look briefly at each of these.

TRADITION

As indicated above, much of what we do in safety programming is based on tradition. We do it because "that's the way safety programs are done." Attempts to solve new problems with old methods do not solve the problems at all. Dr. Earl Heath[53] has some appropriate comments:

We are told that one of the phrases that Edison lived by was "There must be a better way—find it!"—and this would seem to be an appropriate byword for the accident prevention specialist.

The first step toward achieving a breakthrough in any field of endeavor, including accident prevention, is to recognize the fallacy of the "absolutes" and to stop parroting statements that have been made for centuries.

- A great many people feel that individuals can be "scared" or "frightened" into being "safe." This belief runs counter to everything we know about human behavior.
- Seemingly, many safety people keep searching for the "safe" individual. In the light of current knowledge on the subject, it is improbable that such a person exists.
- We still hear too much about the "accident prone" individual. A statement by Dr. Leon Brody[20] is of interest here:

  We no longer like to speak of the "accident prone," but prefer the more objective term "accident repeaters." Although the psychological concept of accident proneness is still controversial, there are whole constellations of factors making the repeater. Such constellations operate in a high percentage of accidents to more or less average individuals. You might regard these in medical terms as syndromes.

- Too much time is spent on improving our accident reporting procedures and our investigative techniques, and too little on studying error-free performance. In other words, we go about our tasks backwards. If the Yankees or any other ball club wanted to teach its players to be better hitters, they certainly would not get a .100 hitter and tell their players to watch him carefully, but to make certain that they didn't do what he did!

  In effect, this is the way we go about accident prevention. We study accidents and the people who are involved in them almost to the exclusion of studying error-free performance and the individuals who never or rarely are involved in accidents.

- Another myth we have lived with for more than a half-century is that an accident is "an unplanned, unexpected, unforeseen event—a fortuitous occurrence over which we have no control." It isn't.

Illustrating the tunnel vision of tradition, Charles Kettering used to tell one of his favorite stories which involved the drive between Detroit, where he worked, and Dayton, where he lived. A friend who also drove between Dayton and Detroit approached Kettering one day with the statement, "I understand you drive from here to Dayton in 4½ hours." Kettering replied that he was sometimes able to do this, depending on traffic. "I don't believe it," his friend charged. "I'm a much better driver than you are, and I can't do it." The following Friday, the friend rode along. Here is the account in Kettering's words:

So we rode into Dayton in about 4½ hours, or a little more, and he said, 'No wonder you do it. You don't stay on Route 25.'

Now Route 25 is a red line that is marked on all the maps between Detroit and Dayton. If you are a stranger, that is the road you should take. It never occurred to my colleague that you could take any other road on either side of Route 25. There's a lot of country on either side of it; in fact, half the earth is on each side of it.

Another illustration is provided by the poem in Exhibit 12.1.

One day through the primeval wood
A calf walked home as good calves should;
But made a trail all but bent askew,
A crooked trail as all calves do,
Since then three hundred years have fled,
And I infer the calf is dead.
But still he left behind his trail
And thereby hangs my moral tale.
The trail was taken up next day
By a lone dog that passed that way;
And then a wise bell wether sheep
Pursued the trail o'er vale and steep,
And drew the flock behind him, too,
As good bell wethers always do.
And from that day o'er hill and glade
Through those old woods a path was made.

The years passed on in swiftness fleet,
The road became a village street;
And this, before men were aware,
A city's crowded thoroughfare.
And soon the central street was this
Of a renowned metropolis;
And men two centuries and a half
Trod in the footsteps of that calf.
Each day a hundred thousand rout
Followed this calf about
And o'er his crooked journey went
The traffic of a continent,
A hundred thousand men were led
By one calf near three centuries dead.
They followed still his crooked way,
And lost one hundred years a day'
For thus such reverence is lent
To well established precedent.

For men are prone to go it blind
Along the calf-paths of the mind,
And work away from sun to sun
To do what other men have done.

Exhibit 12.1 Path of the Calf, Samuel Walter Foss, 1985.

One reason safety programs don't fly is because they are based largely on tradition. Another is that they do not often utilize modern management or behavioral techniques and knowledge. We will look in some detail in this part at some of these. First we will examine some of the reasons behind some of our safety and other management problems as explained by Chris Argyris[6] of Yale University. Argyris suggests that the problems of worker apathy and lack of effort are not simply a matter of individual laziness. Rather they are often healthy reactions by normal people to an unhealthy environment created by common management policies. More specifically, Argyris states that most adults are motivated to be responsible, self-reliant, and independent. These motives are acquired during childhood from the educational system, the family, and communications media such as books, television, and radio. But the typical organization confines most of its employees to roles that provide little opportunity for this.

First, Argyris examines these principles of management that direct us:

1. *Work Specialization.* Since we have believed the principle that concentrating effort on a limited field of endeavor increases the quality and quantity of output, it follows that organizational efficiency is increased by the specialization. American management has made a science of specialization. Ideas inherent in this assumption are:

> That the human personality will behave more efficiently if more specialized.
> That one best way can be found to define the job.
> That any individual differences in the human personality may be ignored.

2. *Chain of Command.* Following the logic of specialization, the planners created a new function (leadership), the primary responsibility of which is to control, direct, and co-ordinate the interrelationships of the parts and to make certain that each part performs its objective adequately. They then must be motivated to accept direction, control, and coordination of their behavior. The leader, therefore, is assigned formal power to hire, discharge, reward, and penalize the individuals in order to mold their behavior in the pattern of the organization's objectives. This makes the individuals dependent on, passive toward, and subordinate to the leader. As a result, the individuals have little control over their working environment.

3. *Span of Control.* The principle of span of control states that administrative efficiency is increased by limiting the span of control of a leader to no more than five or six subordinates whose work interlocks. By keeping the number of subordinates to a minimum, great emphasis is placed on close supervision.

THE INCONGRUENCY

Bringing together the evidence regarding the impact of formal organizational principles on the individual, Argyris concludes that there are some basic incongruencies between the growth trends of a healthy personality in our culture and the requirements of formal organization. The individual in our culture tends to develop along specific trends as follows:

1. From a state of passivity as an infant to a state of increasing activity as an adult.
2. From a state of dependence on others as an infant to a state of relative independence as an adult. Relative independence is the ability to "stand on one's own two feet."
3. From being capable of behaving in only a few ways as an infant to being capable of behaving in many different ways as an adult.
4. From having erratic, casual, shallow, quickly dropped interests as an infant to possessing a deepening of interests as an adult.
5. From having a short-time perspective as an infant to having a much longer time perspective as an adult.
6. From being in a subordinate position in the family and society as an infant to aspiring to occupy at least an equal and/or superordinate position relative to his peers.
7. From having a lack of awareness of the self as an infant to having an awareness of the control over the self as an adult.

These characteristics are descriptive of a basic developmental process along which the growth of individuals in our culture may be measured.

In the work environment, because of our management principles, people are: (1) provided minimal control over their workaday world; (2) expected to be passive, dependent, subordinate; (3) expected to have a short-time perspective; (4) induced to perfect and value the frequent use of a few superficial abilities; and (5) expected to produce under conditions leading to psychological failure. All of these characteristics are incongruent to those which healthly human beings desire. They are much more congruent with the needs of infants in our culture.

Argyris then suggests that the conflict leads to these escapes for the workers:

1. Leaving the organization.
2. Climbing the organizational ladder.
3. Manifesting defense reactions such as daydreaming, aggression, ambivalance, regression, projection, and so forth.
4. Becoming apathetic and disinterested toward the organization, its makeup, and its goals. This leads to such phenomena as: (a) Reducing the number and potency of the needs they expect to fulfill while at work, (b) goldbricking, setting rates, restricting quotas, making errors, cheating, slowing down, accidents, and so on.
5. Creating informal groups to sanction the defense reactions and the apathy, disinterest, and lack of self-involvement.
6. Formalizing the informal group.
7. Evolving group norms that perpetuate the above behavior.
8. Evolving a psychological "set" against the company.

Then, because of the employees' actions, many managements tend to respond to this behavior by:

1. Increasing the degree of their pressure-oriented leadership.
2. Increasing the degree of their use of management controls.
3. Increasing the number of "pseudo" participation and communication programs.

These three reactions by management actually compound the dependence and subordination that the employees experience, which, in turn, cause the employees to increase their behavior, the very behavior management desired to curtail in the first place.

Is there a way out of this circular process? Argyris suggests the basic problem is the reduction in the degree of dependency, subordination, and submissiveness experienced by the employee in the work situation. It can be shown that job enrichment and employee-centered (or democratic or participative) leadership are elements which, if used correctly, can go a long way toward helping the situation. Close supervision leads subordinates to become dependent on, passive toward, and subordinate to the leader. Close supervision also leads to the problems and behavior described. Obviously, close supervision and tightened control is not the answer to the problem.

Certainly Argyris's incongruency theory has a great deal of meaning in safety. As employees adapt to frustration and conflict in their work situation and manifest defense reactions such as daydreaming and aggression, these surely influence the accident record. As employees exhibit the various forms of escape, our safety record is surely influenced to a very marked degree. When we see these things, our reaction is very much as Argyris indicated:

1. We increase the degree of our pressure to enforce safety rules (three-day layoff).
2. We increase the degree of our controls. We watch even closer—step up our activities.
3. We increase the number of "pseudo" participation and communication programs (posters, contests, gimmicks). This is real. We have all seen exactly these reactions toward a work-

ing population that "does not care about their own safety." And we wonder why our programs don't fly!

What should we do in safety programming? As Argyris suggests, we should consider programs that reduce the degree of dependency, subordination, and submissiveness. We should build safety programs that incorporate the concepts of job enrichment, participation, and employee-centered leadership. We will talk about this approach in the next few chapters.

Chapter Thirteen

Dealing with Conflict

The normal reaction to conflict is "fight" or "flight." Either reaction can be healthy, but usually they are unhealthy to both the worker and the organization.

Dan Petersen

In Chapter 12 we discussed the innate conflict between mature human needs and what the organization does to meet those needs. Argyris' conflict theory suggests rather strongly that the more classical the management style, the more conflict there will be. The intent of this chapter is to focus on this conflict (and others), on people's reaction to the felt conflict, and on dealing with the problem.

The innate conflict Argyris discussed is not the only one people in organizations feel. There are normally more. Typically a subordinate will feel a conflict when the boss uses a leadership style inappropriate to that subordinate's needs. There may be a felt conflict when the subordinate's input is not asked for, is not listened to, or is ignored. In short, conflict will be perceived in many situations where human needs are not being met. And the problem will probably continue to worsen as workers become even more educated, get a taste of participation and want more, and experience recognition from one boss and come to expect the same in other job settings. When they do not get it, more conflict will be felt.

Frederick Herzberg[55] (discussed in depth in later chapters) is another of the many behavioral theorists to discuss this problem. He uses the word "hostility" to describe the same problem:

> The managers today and, increasingly in the future, will have to understand the psychological laws of hostility and the appropriate management of hostility.
>
> The simplest although very incomplete statement of the basic law of psychological hostility is called frustration-aggression. Simply put, if someone hurts you, you will get mad at him. (If you don't get mad when someone hurts you, you are what the psychologists technically call a "nut.") Hostility is not an ephemeral or abstract psychological term, rather it is the substance of a biochemical change in your body manifested by the secretion of neuro-hormones at nerve fiber connections. As in any biological imbalance, the surplus biochemicals must be transformed or neutralized. For example, too much acid in the stomach must be removed by some alkaline action (Rolaids). Similarly, at the behavioral level you must energize and transform the hostility that results from frustration. You have to walk it off, shout, throw something but you must in some way express hostility.
>
> What is the mentally healthy thing to do when your boss frustrates you? TELL HIM TO GO TO HELL. Seriously, if you tell your boss to go to hell when he

frustrates you, you will be mentally healthy—and fired. And what does firing do to you? It just increases your frustration. You see how hard it is to be mentally healthy?

What can you do? If you express your hostility to the frustrating object, your boss in this instance, it only results in increased frustrations. The psychological law, however, states that you must express hostility. One solution is to go home and beat your wife, a technical psychologists call displacement. This basic mechanism operates, very obviously, in prejudice. The bigoted person is frustrated, but should he express his hostility to the frustrating object he would be clobbered, so he therefore finds a convenient scapegoat to release the hostility with relative impunity. But returning to our example, if you go home and beat your wife you will be mentally healthy—but she won't. Then what happens if she has all the money? She leaves you, taking all the money with her, and furthers your frustration.

If you cannot express your hostility directly at the person who is frustrating you and you cannot displace your hostility to your wife, and all other avenues for expression of your hostility are blocked, where can the hostility go? There is only one direction possible. The hostility must be directed back into oneself. Psychologists call this internalization. When hostility is directed toward oneself, it leads to depression, self-hate, self-degradation and, ultimately, psychological suicide. If we look around us in industry today, we can see employees who have no longer any interest in what they do, who merely respond as automatons on their jobs and who have indeed been forced to commit psychological suicide. In this final solution for the individual lies a great hidden cost to the organization.

Herzberg describes the physiological responses to hostility and how they become psychological responses—displacement or internalization. Perhaps an easy way to picture this is to think in terms of reactions of workers to management's action. In physics we learned a simple, but important, law: for every action there is an opposite reaction. While the law is stated for our physical universe, it has broad implications in our psychological universe.

When there is a psychological action against us we react in some way to that action. In Chapter 12 we looked at what organizations do to people. The action against workers is the process of changing their attitudes and even their values to be in tune with the "acceptable" behaviors, attitudes, and values of the system. This process is hard on most people. It creates stress which results in strain on our physical condition. It forces us into mediocrity and it destroys creativity. If we live through it (that is, don't quit) management succeeds in socializing us until we "fit" and become promotable.

But the above law of reaction suggests that such actions against people usually do not go unanswered. People will react in some fashion and typically the reaction will come in one of two ways. Either they will react against the action (fight back), or they will attempt to escape the action. This chapter attempts to look at the "fight" or "flight" reactions available to the worker in the job setting. Some options will be healthy to the individual, some unhealthy; we'll pay for the latter eventually. Some options will be healthy to the organization, some unhealthy and we'll also pay for these eventually. Exhibit 13.1 describes some of the options. There are no doubt more that the reader can add to each list.

As you look at the lists in Exhibit 13.1, a few things seem obvious:

1. There are many more unhealthy reactions than healthy.
2. Many of those reactions that are healthy to the worker are unhealthy for the organization.
3. It is healthy for the worker to fight (at least it is used more often).
4. Workers seem to have a wealth of options from which to choose.
5. Quite a few of the worker's options are related to the organization's safety record.

REACTION OPTIONS					
HEALTHY REACTIONS (For Persons)		UNHEALTHY REACTIONS			
FIGHT	FLIGHT	FIGHT		FLIGHT	
		TO PERSON	TO ORGAN-IZATION	TO PERSON	TO ORGAN-IZATION
Unionize	Get Promoted	Hostility	Ignore Rules	Ignore Rules	Accept Mediocrity
Participate —Committees —Quality Circles		Frustration	Quit (Turn-over)	Accept Medi-ocrity	Turn Off Creativity
Ignore Rules		Stress & Illness	Subjective Injuries —Fake	Turn Off Creativity	Goal Shifts
Quit				Goal Shifts	
Subjective Injuries —Fake —Use Workers Compensation			Alienation —Sabotage	Substance Abuse	Substance Abuse
			Use Workers Compensa-tion		
Use Sick Leave					
Venting —Scream Room —Hit Room —Kick Dog			Use Sick Leave		

Exhibit 13.1 Reaction Options.

Let's look at a few. Exhibit 13.1 suggests that the easiest place to start is with the column "Healthy Flight" as there is only one reaction: "get promoted." This is normally healthy for both the individual and the organization. In the Argyris list of symptoms of conflict, this was the only positive one (from management's viewpoint): get promoted, work up in the organization, and get to the point where you, too, can inflict pain. Why is promotion perceived as "flight?" Because it is the process of avoiding conflict by joining the other side. If this is the only healthy reaction to both sides, why is it not used more often? Probably it is not used for two reasons:

1. Hourly workers have experienced enough pain and feel they need no more; they have seen what the first level of management is like: overwork, hassle, shifting management goals, more work than time to do it.
2. In many organizations a caste system has been built and it is difficult to come up through the ranks. College graduates are hired as supervisors as the first step in their career path. This is the old military system of officers and enlisted men. The whole option of a "healthy flight" is closed off to the worker, who must either fight or choose an unhealthy reaction.

THE HEALTHY FIGHT REACTIONS

Here we have a number of options, all healthy for the individual worker and most unhealthy to the organization. To unionize is one option—to formally band together to fight, or more accurately to formally choose to be represented in the fight by a different organization, the union. Does this solve the conflict—yes and no—or perhaps "depending upon the situation." In some cases it heightens the conflict felt by the worker and stiffens the battle lines. It needs to be said, however, that a strong and militant union is usually well deserved due to past management actions.

A particularly healthy fight in safety is when workers choose to participate to make the environment better through whatever mechanisms are available to them: safety committees, quality circles, ad hoc committees, or a management that asks for help. This is perhaps the best of all worlds because everybody wins. It is discussed in depth in Chapter 15. Every behavioral theory and every managerial theory say participation ought to work, and it does, unless:

1. Management screws it up by dictating how and when people can participate.
2. Management screws it up by using participation as a new and fancy way of getting the workers to "jump through new hoops."
3. Management screws it up by asking for, and then ignoring, hourly worker input.
4. Management has already screwed it up in the past to the point that workers "know" it really doesn't want help.

Here are some healthy reactions for the worker that are unhealthy for management:

1. Choosing to ignore rules of management because of a perception (often learned) that "they don't know" or "they don't care" or "they haven't earned the right to say that to us." This is a fight reaction as it is a distinct (conscious or unconscious) effort to say "Screw you, we'll do it our way."
2. Choosing to quit. This is usually healthy for the worker as it provides an opportunity to improve the situation in a new job.
3. Choosing (consciously or unconsciously) to use the system through a "pseudo-injury," a fake injury, a "Monday morning" injury, or using sick leave to get time off. This is healthy for the worker as it fulfills the need for time off to escape temporarily the pain felt on the job. It is clearly unhealthy for the organization. This option is disucssed in Chapter 25.
4. Venting the frustration by having a way to "let off steam". As Herzberg stated earlier, this can be unhealthy to others—wives and dogs that get kicked a lot, etc. But it is healthy to the worker—displacing is better than internalizing. Some organizations are dealing with this today by providing opportunities for venting, such as a "scream closet" where an employee can go to vent the rage, the hostility, and the conflict. Some Japanese companies offer a room with punching bags, each bag having the face of a supervisor or manager (this might be a good tool for performance appraisals of managers—the most beat up achieves the lowest rating). A more common approach is counselling assistance through employee assistance programs (EAPs).

THE UNHEALTHY REACTIONS

Here we have a long list. Some have already been mentioned above. The worst might be the fight reaction of hostility and frustration with no outlet which leads to stress induced illness. As of 1985, stress induced illness is the #1 health problem under Workers' Compensation. While stopgapped in some states, it is wide open in others. This may be the biggest single safety problem since the back strain emerged years ago. Chapter 26 attempts to discuss this major problem.

Unhealthy flight reactions to the worker are equally serious. For instance:

1. *The Forcing of Mediocrity.* As a person is socialized through time and the organized efforts of management, both abilities and will to work change. The two things affecting an individual's effectiveness are:

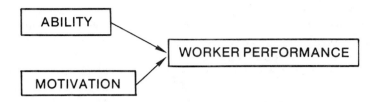

Research in the behavioral sciences in the last 60 years suggests that motivation (the wanting to work) is infinitely more important than ability (knowing how to work). Management initially concentrates on a person's ability. The new employee is taught how to do the job through training and thus over time the worker's ability improves. At this point management invariably makes the fatal mistake that ensures lower worker performance: they assume since training worked before it will work again. The underlying assumption is that performance is a function of training, period. This, of course, is not true; it is a function of both training and motivation, with the heavy emphasis on motivation.

The new employee starts with adequate motivation (will to work) and inadequate ability:

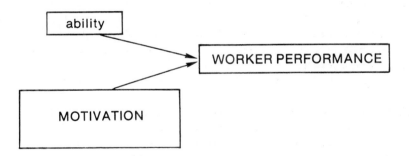

Training does increase ability. However, what typically happens over time is that as management continues to concentrate on ability with more and more training and retraining, ability increases, but motivation decreases, since little attention has been placed on what is motivational to most workers:

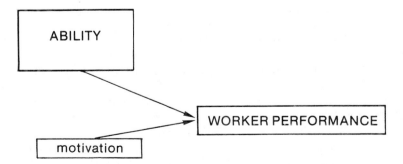

Mediocrity is defined as "less than top performance". When workers produce less than they are able; when managers accepts less from themselves or their subordinates than they could get—all are expressions of mediocrity. Mediocrity was unacceptable in America for many years; today it has become a way of life in many American organizations.

America was the industrial leader of the world, primarily because we were the most productive, most innovative nation. Today we lag behind others in productivity and in innovation. We have accepted mediocrity. Workers know they could produce more and better if they chose to, but most do not want to. And that choice is often a logical choice for a normal, rational person who is simply reacting to the workplace environment.

John Jones was a top name, a top thinker, in his field in 1970. He was 48 years old, published, respected, and known. He had taught in universities and consulted in industry. When people in his profession looked for new answers and new approaches they sought him out. In 1971, John decided to leave academia and consulting and joined a company. He accepted a position to create some new approaches. His motivation for accepting was more money and more security. By 1972, he had created what he had been brought in for—the new ideas and new approaches. His new programs were implemented and were operational. A year later his young boss had been promoted as a result of the new successes.

John Jones continued to create and innovate with even more new and good ideas. However, his new boss was a little less willing to go along with progress, and by 1974, seven new projects had been created, presented, and turned down. From 1974 to 1977 John continued to try, but finally got the message and from 1977 until 1987 not one new idea was presented to anyone. For ten years he merely sat and did nothing but accept the proferred paycheck. Everyone was happy. John got what he wanted, money and security. He channeled his energies elsewhere. The new boss was happy; he didn't have to deal with the new ideas. The old boss was happy; he got his promotion and several others later. Mediocrity was served.

An unusual story? I doubt it—it is perhaps typical of American industry stories. It is much easier for an organization to deal with mediocrity than anything else. It is hard to manage the totally inept and it is hard to manage the brilliant. Mediocrity is easy to manage. In the above story, mediocrity resulted from a systematic (albeit unconscious) management process. It was unhealthy to both the individual and the organization. The individual disappeared from national prominence; he gave up. The organization, once known as a national leader in the field, became one of many and eventually went bankrupt.

What causes mediocrity? A management that fears change followed by an employee that eventually gives up. This is unhealthy flight that hurts both.

2. *The Destruction of Creativity.* The Jones story was the systematic conversion of a highly effective person into a less than mediocre worker. His is also the story of the destruction of innovation. John finally learned and gave up. After four years he did what they wanted— he turned off his brain and never did another day's work.

3. *Goal Shifts.* This describes another unhealthy reaction for the person and the organization. John Jones had goal shifts from needed product development to no development, from creativity to doing nothing. The organization had a goal shift from leader in the industry to "don't rock the boat." In both cases the goals were met. Workers experience the same goal shifts—from buying into the safety program to buying out. Why? Because over time they learn that safety meetings are boring, that safety is negative, and that management does not want their input. To maintain their psychological health they "buy out", but in the long term it is unhealthy to them as well as the organization.

4. *Alienation, Strikes, and Sabotage.* Perhaps the most unhealthy reaction to the organization is alienation so severe it leads to strikes and even sabotage. While uncommon, it does exist as a severe reaction to perceived management abuses.

```
AVOID:        - Ignore
              - Keep from happening

DEFUSE:       - Cool down for rational discussion
              - Resolve minor points/avoid major issues
              - Keep issues unclear
              - Look for conflict of intent, misunderstanding

RECONCILE:    1) Diagnose
                 a. Your own intent
                 b. Type of conflict
                 c. Relation between members

              2) Initiation
                 a. When and how to approach, confront

              3) Listening
                 a. Uncover specific needs, concerns of each
                 b. State what's important to you
                 c. Check what's important to other person

              4) Problem Solving
                 a. Invite/suggest alternatives
                 b. Temporarily alter restrictions
                    - Remove restrictions
                    - Impose restrictions
                 c. Define problem
                 d. Brainstorm

              5) End Discussion
                 a. State your understanding of what has been said and
                    why (if resolved); or
                 b. Make decision, acknowledge right to differ, explain
                    what decided and why (if resolved); or
                 c. Suggest you continue discussion another time (if
                    emotions are intense)
```

Exhibit 13.2 Conflict resolution strategies.

5. *Substance Abuse.* A similar severe unhealthy reaction is drug or alcohol abuse. This is discussed in Chapter 27.

CONFLICT RESOLUTION

In the last chapter, Argyris suggested three specific approaches to conflict reduction: less control, levelling, and job enrichment. These are discussed in some detail in Chapter 15 on participation, Chapter 23 on building worker relationships, and Chapter 14 on job enrichment. These chapters focus on organizational strategies to reduce the causes of conflict.

Some thoughts on individual (one to one) conflict resolution are shown in Exhibit 13.2.

Chapter Fourteen

Enriching with Safety

We will achieve better and healthier work results from making the work itself a rewarding experience. Man is a human being who finds positive satisfaction in the exercise of his human talents. When permitted to do so, he will provide the organization with the kind of extra performance that is essential to success.

Frederick Herzberg[58]

There has been a great deal written in recent years, both pro and con, about job enrichment. The principle of job enrichment was mentioned in Chapter 12 as one of the solutions to the basic incongruency between the nature of mature people and the characteristics of organizations. Job enrichment is in one aspect the opposite of specialization; it is an attempt to give back to the employee a piece of the action that specialization took away. More importantly, it is an attempt to put some meaning into the job, and thus a sense of worth for the person on the job.

DUAL FACTOR THEORY

While Chris Argyris and other behavioral scientists came to the conclusion that job enrichment is a necessity, it was actually Frederick Herzberg who first coined the term and expressed the principle. Herzberg's interest in job enrichment grew out of his discovery of what might be called his *dual factor theory of job satisfaction and motivation.* Herzberg initially reviewed the literature (3,000 books and articles) in the field of job attitudes and motivation in his book *Job Attitudes: Review of Research and Opinion,*[57] a scholarly review of what was known about attitudes from 1900 to 1955. In analyzing the data reviewed, he noticed that the things people said positively about their job experiences were not the opposite of what they said negatively about their job experiences; the reverse of the factors that seemed to make people happy in jobs did not make them unhappy. That was the core insight that led to Herzberg's theory. Job satisfaction and job dissatisfaction are not opposites; they are completely separate continua, like hearing and vision. If this is true, then they must be produced by different factors and have their own causes.

The essence of Herzberg's Motivation-Hygiene theory can be illustrated by comparing the "traditional" way of viewing dissatisfaction and satisfaction as simple opposites on a single continuum with the motivation-hygiene model which places satisfaction on a separate and distinct continuum from dissatisfaction (see Exhibit 14.1). The classical approach to motivation, according to Herzberg, has concerned itself only with the environment in which the employees work, that is, the circumstances that surround them while they work and the things given in exchange

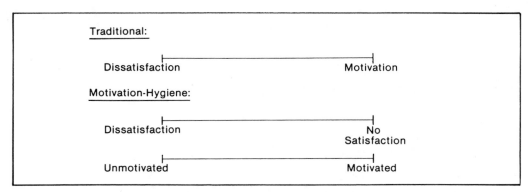

Exhibit 14.1 The dual factor theory.[58]

for work. Herzberg considers this concern with the environment a never-ending necessity for management, but says it is not sufficient in itself for effective motivation. That requires consideration of another set of factors; namely, experiences that are inherent in the work itself.

Herzberg asserts that work itself can be a motivator. Traditionally work has been regarded as an unpleasant necessity, rather than as a potential motivator. For this reason, it has generally been considered necessary for management to either entice people to work by means of various rewards or to coerce them to work by means of various threats, or both. According to Herzberg, the potential motivating power of work was obscured by the fact that most jobs were not stimulating and, therefore, some kind of external pressure, either positive or negative, had to be applied to get people to do them. But when a job provides an opportunity for personal satisfaction or growth, a powerful new motivating force is introduced. Herzberg holds that there is no conflict between the environmental approach to motivation and the approach through work itself. He regards both as important. However, the environmental approach, which he refers to as *hygiene,* is inherently limited in its capacity to influence employee behavior, while the approach through the job, which he refers to as *motivation,* seems to be capable of larger and more lasting effects.

Herzberg uses the term *hygiene* to describe such things as physical working conditions, supervisory policies, the climate of labor-management relations, wages and various "fringe" benefits. In other words, hygiene includes all the various factors through which management has traditionally sought to affect motivation. Herzberg chose the term hygiene to describe these factors because they are essentially preventive actions taken to remove sources of dissatisfaction from the environment, just as sanitation removes from the physical environment the potential threats to health. His research has shown that when any of these factors are deficient, employees are quite likely to be displeased and to express their displeasure in ways that hamper the organization, for example, through grievances, decreased productivity, or even strikes. But when the deficiencies are corrected, even though productivity may return to normal, it is unlikely to rise above that level. In other words, an investment in hygiene may eliminate a deficit, but it does not create a gain. Further, it is inherent in the nature of hygiene needs that satisfactions are not lasting and that with the passage of time a feeling of deficiency recurs. Just as eating a meal does not prevent someone from becoming hungry again in the future, a wage increase will not prevent eventual dissatisfaction with the new wage level. Hygiene, according to Herzberg, is a necessary but thankless task for management. Even a fully effective hygiene program will not motivate employees to *sustain* a higher-than-usual level of efficiency.

Herzberg uses the term *motivation* to describe feelings of accomplishment, professional growth, and professional recognition that are experienced in a job that offers sufficient challenge and scope to the worker. These factors seem capable of producing a lasting increase in satisfaction, and with it, an increase in productivity above "normal" levels. This has been found to be true in a wide variety of jobs and organizational settings.

Herzberg's analysis focuses attention on job design. In most cases, jobs were either not consciously "designed" at all or were designed primarily from the standpoint of efficiency and economy. To the extent that these steps have taken the challenge and opportunity for creativity out of a job, they have contributed to a demotivating effect. That is, apathy and minimum effort are the natural results of jobs that offer the workers no more satisfaction than a paycheck and a decent place to work. These hygiene factors may keep them from complaining, but they will not make them want to work harder or more efficiently. Offering still more hygiene, in the form of prizes or incentive payments, produces only a temporary effect. Herzberg views *job enrichment* as a means of introducing more effective motivation into jobs. He draws a clear distinction between the deliberate enlargement of responsibility, scope, and challenge (job enrichment)—and the movement of an individual from job to job without necessarily increasing responsibility at all (job rotation). Herzberg has found job rotation unsatisfactory as a motivating tool, but has achieved impressive results with job enrichment. After an initial "adjustment" period, during which productivity temporarily declines, efficiency tends to rise well above previous levels.

Exhibit 14.2 lists the main dissatisfiers and motivators. Dissatisfiers seem to be such items as pay, benefits, company policies and administration, behavior of supervision, working conditions, and other factors that are generally peripheral to the task. Though traditionally thought of by management as motivators of people, these factors are actually more potent as dissatisfiers. High motivation does not result from their improvement. Dissatisfaction does result from their deterioration. Motivators are such items as achievement, recognition, responsibility, growth, advancement, and other matters associated with the self-actualization of the individual on the job. Job satisfaction and high production are associated with these motivators, while disappointments and ineffectiveness were usually associated with the dissatisfiers. A challenging job which allows a feeling of achievement, responsibility, growth, advancement, enjoyment of work itself, and earned recognition motivates employees to work effectively. Factors which are peripheral to the job (work rules, lighting, coffee breaks, titles, seniority rights, wages, fringe benefits, and the like) dissatisfy workers. These could be considered tease items which will temporarily please but which will be long-range motivational failures. In other words, the factors in the work situation which motivate employees are different from the factors that dissatisfy employees. Motivation stems from the challenge of the job through such factors as achievement, responsibility, growth, advancement, work itself, and earned recognition. Dissatisfactions more often spring from factors peripheral to the job.

Job dissatisfaction	Job motivation
Company policies and administration	Achievement
Supervision	Recognition
Working conditions	Work itself
Interpersonal relations	Responsibility
Money, status, security	Professional growth

Exhibit 14.2 Motivators and dissatisfiers.[58]

SAFETY USES

Charles Hughes,[63] of Texas Instruments, has related all this to safety management:

> How effective is the safety professional in his job? Safety programs and campaigns may help maintain a level of awareness, but they do not necessarily motivate safe work behavior. We must look at the motivation for safety in relationship to the person's job. To understand the modern theory of employee motivation, we must research the history of the various management strategies that influence work behavior. To understand work behavior, we must understand the nature of work itself, historically.
>
> It all began with Frederick Taylor, the father of industrial engineering. He attempted to establish the relationship between the environment of the work and the means by which the worker performed his assignment. Taylor was concerned with the mechanistic approach to the design of jobs. The basic concept of classical industrial engineering is the subdividing and specializing of work into different functions and tasks to the greatest possible degree. In this way, human beings could be treated in the same way as replaceable parts in a replaceable machine. The thinking parts of people were not utilized, and each employee was quite literally a "hired hand."
>
> Today, many safety professionals still follow Taylor's approach of subdividing the responsibility and planning for safety into several small pieces, with each employee having responsibility for his own safe work behavior only. As with Taylor and many others, I think that many safety professionals still conceive of the employee as being a passive and reactive organism that can be manipulated only if the right combination of pressures can be brought to bear upon him. In short, then, the classical approach to motivation has been to manipulate the work environment.
>
> This is the way the studies at Western Electric's Hawthorne plant started—as an industrial engineering study to discover the relationship between the level of lighting and the degree of efficiency in assembling telephone relays. As the lighting became brighter, production increased. This is where the research ordinarily stopped, but the Harvard professors conducting the study continued to experiment. They reduced the lighting, and the production again increased until it was so dark the employees could not find a screwdriver. This is a gross oversimplification but the message was there—manipulate the environment.
>
> Perhaps many safety professionals believe that if they can arrange things correctly to meet the mechanical requirements of people, these people will perform their work in a safe manner. There were many studies made, and many programs implemented, with "human relations in mind." However, there was no consistent relationship between the "morale" (attitudes about the work environment) and productivity. Sometimes morale would be good and so would production. But, sometimes morale would be poor and production would still be good.
>
> Actually, there is a relationship between morale and production, as Fred Herzberg, author of *The Motivation to Work* and *Work and the Nature of Man,* discovered. He found that a certain factor (hygiene) concerning the work environment (the context in which the job is done) had varying kinds of relationships. No consistent positive relationship between hygiene (or maintenance) needs and job performance was discovered. However, he did find consistent relationships between measures of level of motivation (related to the content of the job) and production. Dr. Scott Myers of Texas Instruments also corroborated Herzberg's concepts with study groups.
>
> In these studies the basic approach is to ask an employee a question: "Think of a time when you felt very satisfied in your work and tell me about it. Describe an actual event. Tell how long the feeling lasted." Then ask him the reverse question:

"Think of a time when you felt very dissatisfied in your work and tell me about that." From these highs and lows in life at work, two factors and a practical theory were derived. When Dr. Myers put these questions to the engineers, their replies showed that the big motivator was achievement. Fifty per cent of the time, engineers mentioned events of achievement. Advancement and recognition were also motivators, but for less than ten per cent of the time. The main motivator was the opportunity to work at a job with meaningful goals to achieve. The opposite, their main source of dissatisfaction most frequently mentioned, was company policy and administration. Competence and friendliness of supervision was never mentioned by engineers as a source of satisfaction and motivation, but only a source of dissatisfaction.

What motivates manufacturing supervisors? Achievement is the most frequent factor again. What dissatisfies them? Company policy and administration again. Supervision, it seems, follows the same pattern wherever it exists.

What about women employees performing assembly work? The traditional supervisor, when asked what motivated these ladies, said: "Oh, the money, pleasant people to work with, and good working conditions and benefits." What do the ladies themselves say? Achievement also. The opportunity to do something meaningful in their work, to have a feeling of accomplishment.

Nearly every organization has spent years teaching supervisors it was not important for women to have a meaningful job. The things supervisors usually think are important (such as pay, peer relations, competence of supervision, policy and administration, benefits, security of employment and friendliness of supervision) hold more potential for dissatisfaction than they do for motivation. These data challenge the traditional safety motivation program.

Compare Civil Service employees. Here research shows that achievement is the most frequent cause of dissatisfaction; many had to prod their memories to discover a time when they felt satisfied and motivated in their work. Policy and administration of the government is 60 per cent of the time given as a source of dissatisfaction, and it does not motivate at all. To quote John Gardner, Secretary of Health, Education and Welfare: "The last act of a dying organization is to issue a new version of the policy manual."

There are maintenance factors that, if properly implemented, can eliminate dissatisfaction caused by employee maintenance needs (physical, social, status, orientation, security and economic). These appeal to the motivation needs (growth, achievement, responsibility and recognition).

Maintenance needs [Exhibit 14.3] are handled on a mass administrative basis. Companies mass-administer benefits, seniority, and facilities. Maintenance is important, and pay is a maintenance factor for many people (particularly automatic pay systems, rather than merit systems). If they are not paid—and if the physical working conditions are bad enough—they will go home.

These factors may be improved without getting a corresponding improvement in productivity or job satisfaction. The motivation needs are satisfied by the job, by work itself. It is not what is around the job but what is in the job. For example, at one company free coffee and donuts were distributed to the employees during the two coffee breaks each day. But no one was really "motivated" by these tiny treats. There are degrees of dissatisfaction, and free coffee and donuts can only eliminate one degree, not all.

Most jobs are very bad; they should be more like sports, such as bowling. In bowling you have both motivation and maintenance. You decide whether or not you will bowl, and you pay to bowl. Physical conditions of bowling alleys are good; the beer is cold, air conditioning is available, and the alleys are standardized. There is a good scoring system and you find out what improvements, if any, you and your team have made. You have status, orientation, and security—therefore opportunity for motivation. You can grow in the sense of improving your bowling abilities. You

Exhibit 14.3 Employee needs. Effective job performance depends on the fulfillment of both motivation and maintenance needs. Motivation needs
include responsibility, achievement, recognition and growth, and
are satisfied through the media grouped in the inner circle. Motivation factors focus on the individual and the achievement of company and personal goals. Maintenance needs are satisfied through
media listed in the outer circle under the headings of physical,
social, status, orientation, security, and economic. Peripheral to the
task and usually group administered, maintenance factors have little motivational value, but their fulfillment is essential to the
avoidance of dissatisfaction. An environment rich in opportunities
for satisfying motivation needs leads to motivation-seeking habits,
and a job situation sparse in motivation opportunities encourages
preoccupation with maintenance factors.[58]

receive immediate feedback, and therefore you get a feeling of achievement, or the
lack of it, without waiting. You have responsibility, because when you are bowling,
you are responsible for what happens. You plan and control your bowling, and you
receive recognition if you perform.

Suppose bowling was installed in the factory. What would management do? Probably hang a curtain half way down the alley and stand a man beside the curtain. He
would look at you, watch the ball, and see what happens to the pins. You would
roll the ball, it would go under the curtain and you would hear a crash. You would
say: "How am I doing?" and he would say: "Never mind. You're doing great. Keep

up the good work." At the next performance review he would say: "Remember last year? You put it in the alley." Worse than that, some jobs do not have the curtain but neither do they have any pins. That is not motivation, that is drudgery.

How can we apply these concepts to making safety a meaningful part of work and increase the level of employee motivation to work safely and to use protective equipment? If we start from the basic premise that the job itself is a source of motivation, then obviously we must build safety into the job. Putting safety into a job might be accomplished through group and individual safety problem-solving. Groups of employees can be brought together as task forces with the responsibility to identify safety problems on the job and obtain solutions to these problems. They could, in addition, set goals for safety improvement, and identify the targets they will shoot for as a group to maintain safe work behavior. They could also establish their criteria for measurement. The individual and the work group as a safety task force of rank and file employees can prescribe their own controls. That is, they can assess and compare the results actually achieved against the plans that they themselves have made. They could identify the successes and failures, and report to management on their own self-designed safety program and its success. Also, they could compare their own safety records with that of other groups in other companies.

In simple terms, it is the job that motivates and the safety professional has no satisfactory alternative except putting safety into the work. This must be accomplished with the co-operation of the employees, and to work most effectively it must have their full involvement in planning and control of safety. If employee task forces plan and assess safe work behavior, someone must still implement the plan. The implementation of the safety improvement plan may be done by the employees themselves, or by line management or the safety professional. The fundamental difference is that the direction comes from the people who have the most to benefit from working safely—the employees themselves.

Here are some of my thoughts about the role of the safety professional [Exhibit 14.4].

The difference between the Don't and the Do is the difference between Theory X and Theory Y. Safety professionals should not make examples of people who violate the safety rules, but should help people to see the need for rules and the consequences of violation. Theory X would prescribe that a public example be made of anyone who breaks management's rules. This is precisely the point. These are management's rules under Theory X rather than the rules of the employee group. Traditional safety management would check to see that people are working safely. If we were truly able to build safety into the job, then we would have a self-

Don't	*Do*
1. Make examples of people who violate safety rules.	1. Help people see the need for rules and the consequences of violations.
2. Impose safety standards and procedures on employees.	2. Involve employees in solving safety problems and setting standards.
3. Check to see if people are working safely.	3. Provide a self-monitoring system.
4. Develop and install new safety procedures.	4. Make improving safety procedures part of each person's job.
5. Ignore safe work practices.	5. Recognize safe work practices.
6. Stimulate safety by emotional appeals and elaborate campaigns.	6. Stimulate safety by making safe work a part of everyone's job goals.
7. Make safety exclusively the safety professional's.	7. Share responsibility for safety with all employees.

Exhibit 14.4 Some suggestions for safety professionals.[63]

monitoring system where people would check themselves and each other to see that they were working safely.

Safety professionals often have the assignment of developing and installing new safety procedures. The concept offered here is that they do not perform that particular function any longer, but now make the improvement of safety procedures a part of each person's job. It is common practice for them to try to stimulate safety by emotional appeals and elaborate campaigns, posters, and prizes. If we are able to accomplish the design of a job in which safety is made a meaningful part, then we can stimulate safety by making safe work a part of everyone's own personal job goals. We should not ignore safe work practices, but should recognize them. Furthermore, safety professionals should not make safety inclusively their own responsibility but should share the responsibility for safety with all employees.

The operating employees should receive the same kind of satisfaction from the management of safety as the safety professional himself does. His own motivation as an individiual comes from the performance of a meaningful job with challenging goals and the opportunity for achievement and responsibility. Safety professionals would not usually be concerned about their own motivation, since they already have a meaningful job in safety. However, some of them keep the real motivating and meaningful parts to themselves rather than telling their employees these basic truths. If achievement and responsibility are motivating factors, then safety professionals must provide the opportunity for employees at all levels to make safety a meaningful part of their work.

Hughes thus gives us some insight into how we can utilize job enrichment principles in safety. Normally in safety programs we do not do these things. We could in most cases. For instance, we could easily resist our impulse to make examples of violators (no three-day layoffs). We could stop checking on people constantly. We could have fewer elaborate campaigns with emotional appeals. And as Hughes suggests, we could involve employees in solving safety problems, in setting standards and in sharing responsibilities. And we could devise a self-monitoring system. All of these would contribute to job enrichment in safety. Hughes does not say it, but it is very possible that our safety program could be the first step toward enriching jobs. We can give employees a piece of the action in safety even if they do not have it in production.

Exhibit 14.5 shows Herzberg's principles of job enrichment. Looking at these principles from our safety management view-point, here are some of the ways our safety programs might utilize them:

A. Throw away the book of "safety rules" for each job, and let each person devise the operational rules for that job with full accountability for developing (and living by) those rules. Job Safety Analysis then becomes a regular part of the job. In those jobs where we have several workers, or quite a few doing the same job, have them devise together the allowable work rules. For instance, rather than publish the driving rules for your fork truck operators, have them decide on the driving rules by which they will live. Give them the Federal Standards or the State Plan as a starting point or as a guideline, and tell them the company will accept whatever they decide on, but make sure they understand they will be expected to enforce their own rules.

B. Use safety sampling and other kinds of measures that give immediate feedback. Then provide the results of samples to the employee or the group.

C. Turn over the complete departmental inspection process to the work group. Let them decide when to inspect, how to inspect, and where to inspect, using OSHA guidelines. Provide them with a priority system so maintenance must fix their safety items. Do not second guess them, but let them know that each month or so top management will tour all departments.

Principle	Motivators involved
A. Removing some controls while retaining accountability	Responsibility and personal achievement
B. Increasing the accountability of individuals for own work	Responsibility and recognition
C. Giving a person a complete natural unit of work (module, division, area, and so on)	Responsibility, achievement, and recognition
D. Granting additional authority to an employee in his activity; job freedom	Responsibility, achievement, and recognition
E. Making periodic reports directly available to the worker himself rather than to the supervisor	Internal recognition
F. Introducing new and more difficult tasks not previously handled	Growth and learning
G. Assigning individuals specific or specialized tasks, enabling them to become experts	Responsibility, growth, and advancement

Exhibit 14.5 Job enrichment principles.[58]

D. Let each department run its own safety campaigns. Give them the National Safety Council catalog and x number of dollars and have them plan the safety programming for their department for the next six months.

E. Redesign the monthly safety reports so that they are interesting enough to be read by each employee. Then let the employees know how they and the company are doing on safety. Write up accidents that have happened (anonymously) and provide each employee with the account.

F. Turn over some of the accident investigation function to employees. Train them in multiple causation. Or ask them to go on a special "Hazard Hunt" (see Exhibit 14.6). This form was used in an effort to get employees involved in preparing for OSHA. Employees were encouraged to submit hazards with the promise that each would be researched and corrected or if not corrected, they would be told why it was not feasible. No reward was offered; it just gave the employees a chance to get in on the act.

G. Assign the inspection function, the investigation function, or any other program component to individual employees on a temporary or rotating basis. Many companies have had employees serve on safety committees on a rotating basis to accomplish a similar purpose.

These safety program suggestions have been proven in both safety and nonsafety situations. For instance, one job enrichment experiment had one group, designated as an achieving unit, have its job enriched by the principles described in Exhibit 14.5. Another group continued to do its job in the same way. Changes for the achieving unit were introduced at an average of one per week of the seven motivators listed in Exhibit 14.5. At the end of six months the members of the achieving unit were found to be out-performing their counterparts in the other group and they also indicated a marked increase in their liking for their jobs. The achieving group had lower absenteeism and, also, a much higher rate of promotion.

This experiment, reported by Herzberg, was scientifically well done, including control for the "Hawthorne effect." Exhibit 14.7 illustrates the changes in performance, measured two months before the study period began and at the end of each month of the study period.

To: _____

From: _____

HAZARD HUNT

I think the following is a hazard: _____

_____

_____

_____

DO NOT WRITE BELOW HERE — TO BE FILLED IN BY SUPERVISOR

Supervisor:

    Agree this is a hazard.

        Corrected by Supervisor on _____ Discussed on _____

        If cannot correct, sent to Personnel on _____

        Job order on _____ Scheduled _____ Discussed on _____

    Do not agree this is a hazard.

        Discussed _____ Conclusion _____

        To Personnel _____ Conclusion _____

DO NOT WRITE BELOW HERE — FOR PERSONNEL USE

Supervisor _____ HH No. _____

Matrix No. _____ (Seriousness)

Exhibit 14.6 Hazard hunt.

(The shareholder service index was a measure representing quality, accuracy, and speed.) The "achievers" were performing less well before the six-month period started, and their performance continued to decline after the introduction of the motivators, perhaps because of uncertainty over their new responsibilities. In the third month, however, performance improved and soon the members of this group had reached a high level of accomplishment. Exhibit 14.8 shows the two groups' attitudes toward their jobs, measured just before the first motivator was introduced and again at the end. They were asked sixteen questions, all involving motivation. A typical one was, "As you see it, how many opportunities do you feel that you have for making worthwhile contributions in your job?" The achievers became much more positive about their jobs, while the attitude of the control unit remained about the same. Many studies, by many people, have borne out these same results.

Herzberg, the father of job enrichment, generally suggests steps such as those listed below to enriching jobs. Some seem applicable to safety management:

1. Select those jobs in which (a) the investment in industrial engineering does not make changes too costly, (b) attitudes are poor, (c) hygiene is becoming very costly, and (d)

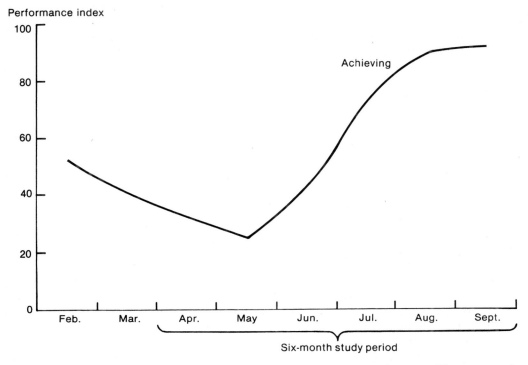

Exhibit 14.7 Shareholder service index in company experiment. (Three-month cumulative average.)[58]

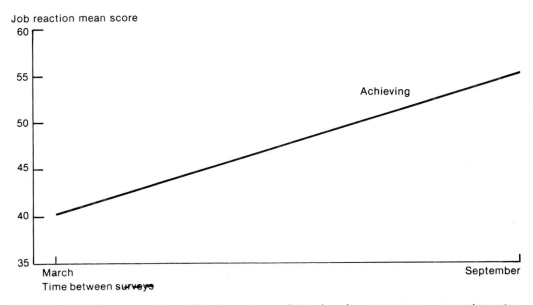

Exhibit 14.8 Changes in attitudes toward tasks in company experiments. (Changes in mean scores over six-month period.)[58]

motivation will make a difference in performance. Many jobs fit these criteria when looked at from a safety viewpoint.

2. Approach these jobs with the conviction that they can be changed. Years of tradition have led (safety) managers to believe that the job content is sacrosanct and that their only scope of action is in motivating the people.

3. Brainstorm a list of changes that may enrich the jobs, without concern for their practicality.

4. Screen the list to eliminate suggestions that involve hygiene rather than actual motivation.

5. Screen the list for generalities, such as "give them more responsibility," that are rarely followed in practice.

6. Screen the list to eliminate any horizontal loading suggestions. (Job enlargement rather than job enrichment.)

7. Avoid direct participation by the employees whose jobs are to be enriched. Herzberg states that, while such participation will result only in short-term movement.

8. In the initial attempts at job enrichment, set up a controlled experiment. Have central groups so you can be sure the results achieved are from the enrichment rather than from some unknown source.

9. Be prepared for a drop in performance in the experimental group the first few weeks. The changeover to a new job may lead to a temporary reduction in efficiency.

10. Expect first-line supervisors to experience some anxiety (even hostility) about the changes made. Anxiety comes from their fear that the changes will result in poorer unit performance. Hostility can arise when the employees start assuming what the supervisors regard as their responsibility. These feelings invariably pass rather quickly.

Chapter Fifteen

Participative Safety

An individual tends to accept his own conclusions. Positive acceptance of safety rules can result if subordinates can suggest and develop their own safe methods and rules. After all, who knows more about the work and its hazards than the people actually doing the work?

Robert L. Burns[25]

Participation is closely related to job enrichment. For example, the principles in Exhibit 14.5 have a lot of participation built into them. In recent years the concept of participative management has become a focal point of interest among both scholars and practitioners of administration. Its acceptance has coincided with the increasing contribution of behavioral science to managerial thought.

THE RESEARCH

The appeal of the participative approach was suggested as early as 1937 by H. Carey[26] in his definition of "consulting supervision" as the procedure whereby supervisors and executives consult with employees or their peers on matters affecting employees' welfare or interest prior to establishing policies or initiating action. From such beginnings, the concept of participative management has grown into a full-fledged approach to administration affecting both patterns of organizational relationships and leadership style. The appeal of this concept is many-sided. Participation viewed in Carey's sense can be construed as a comfortable rationale for the paternalistic manager. Applied in the modern interpretation it complements the political and social philosophies of democracy and individual self-actualization. Today psychologists believe that participation enhances the learning process and that a "democratic environment" may be more conducive to productive effort under certain conditions. These and other advantages enhance the value of the participative concept.

A few of the best studies of participation can be cited to show how the application of this technique can be used to further the adjustment of workers to changing conditions. The best-known studies are those carried on in the Harwood Manufacturing Company. In this company an attempt was made to come to grips with the workers' reactions to retraining. Harwood had had especially severe turnover problems among workers forced to transfer from one job to another because of changing manufacturing needs. Many of these workers actually quit shortly after their retraining had begun. There were also a large number of voluntary quitings during the period just before the workers reached factory standards of competence at their new jobs. In addition, retraining was slow and costly. Transferees took much longer to learn new jobs than did new workers, even those who had been highly skilled at their previous jobs. A controlled experiment was set up in which one group of workers about to transfer was given routine information about the nature of the change and nothing else. Three other groups were asked to plan the way in which the factory would adapt to shifts in the market and in technology. These latter groups arrived at the necessity for their own retraining by means of group discussion. The group which had merely been ordered to retrain showed high turnover and a long training period. The groups which had discussed the shift and arrived at their own decision about its necessity showed virtually no turnover and a phenomenally shortened training period. At a subsequent period, the original "no participation group" was reassembled and again retrained. This time they participated fully in the decision-making process. Exhibit 15.1 shows the differences in response of this group to retraining when they did participate and when they did not participate in planning the shift.

Participation has been found to be successful in many situations. It does not always increase productivity (sometimes it increases, sometimes it stays the same, but it never declines), but it does always improve attitudes, turnover, morale, etc. Workers do want to be involved in

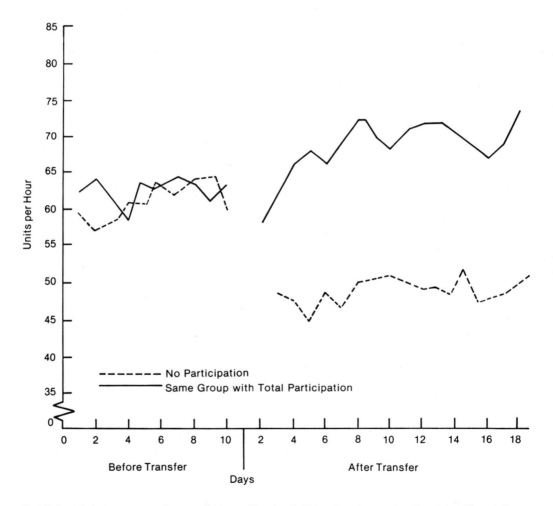

Exhibit 15.1 A comparison of the effect of "No Participation" with "Total Participation" on the same group.[82]

decision-making where they feel they have a "right" to be involved. Workers' attitudes toward both company and union are strongly affected by the degree to which foremen and shop stewards welcome the participation of the rank-and-file workers in arriving at decisions. Involvement of all levels of employees in decision-making is a major factor in morale. In fact, the idea that it is a good thing to get employee participation in making changes has become almost axiomatic in management.

Participation, however, is not something that can be created artificially. We cannot buy it. We cannot hire industrial engineers, accountants, and other staff people who have the ability "to get participation" built into them. We cannot tell our supervisors and staff persons to just start participation. Participation is a feeling the people must have, not just an act of being called in to take part in discussions. People respond to the way they are usually treated, rather than by a strategy of sudden participation. Many supervisors have had unhappy experiences with executives who read about participation and pick it up as a new psychological gimmick for getting other people to do what they want. Participation will never work so long as it is treated as a device to get somebody else to do what you want. Participation must be based on respect and respect is acquired when management faces the reality that it needs the contributions of the operating people.

There is considerable belief among safety professionals that participation improves the safety record. One of the main proponents of participative safety is Robert Burns (General Electric). He states:[25]

> One of the problems inherent in the mtraditional way that safety rules are formulated and administered is that the few at the top of the organization dictate what is proper or improper safety-wise for those at the bottom. The job of the supervisor is to enforce the rules, and his subordinates are to follow them. Since most people resist being told what they can or cannot do, each rule becomes a restriction of one's personal liberty. Acceptance of restriction is difficult for most people, unless they are convinced of the validity of the restriction. For example, most people stop for red traffic lights because they are aware of the validity of a rule which prevents accidents. Without acceptance, any rule or law will either be ineffective, or it will cause problems in enforcement.
>
> If we admit that there must be a system of rules in order that an organization function effectively, then it is necessary for industry to find an imaginative approach to a system which will initiate and develop a solid set of safe-practices specifications which will be accepted and enforced as valid guides to safe behavior by those who must work within their bounds. Acceptance is the key requisite, and acceptance hinges on the obvious validity of any rule. One of the basic principles of nondirective counseling is that an individual tends to accept his own conclusions. A group will tend to accept its own conclusions as well. Positive acceptance of safety rules can result if subordinates suggest and develop their own safe methods and rules. After all, who knows more about the work and its hazards than the people actually doing the work? Motivation to follow the safe practice, then, would come from within each person as a part of the group, and is not the result of outside pressure. Once the people of the department decide what rules of safety should apply to the activity of their department, they will, when necessary, enforce their own rules much more effectively than can the supervisor under the traditional punitive system.
>
> In a positive approach towards safety, the role of the supervisor becomes one of key importance; he becomes the leader in helping the group develop its rules of conduct. The role of the safety director becomes that of the expert to whom the group looks for professional help and advice. The safety director and the supervisor listen and encourage the members of the group to express their ideas. They ask questions in order to be sure that all sides of an issue or method have been considered by the group, and they supply necessary technical information when it is

required. They function as they should—as leaders, not tellers. With the help of the leaders, the departmental group is able to develop standards of behavior that will readily meet the needs of the department. Those who might refuse to do things the group's way, because they think their own way is best, can pretty well be persuaded to conform in the areas of safety under the threat of ostracism. Since the need to belong to the group is a powerful motivator, the potential rule-breaker will probably tend to conform.

Another benefit from this approach to rule-making is that it gives a number of people, perhaps for the first time in their lives, a degree of participation in directing the activities of the department. This can create a sense of involvement and commitment to the goals and the objectives of the department—goals and objectives that they have a part in establishing. From the area of safety, this involvement and commitment can spread into many other facets of the department, permitting each member of the organization not only to contribute to his maximum, but also to achieve greater job satisfaction than ever before.

Participative safety programs have been tried and they work. One Midwestern company[128] had these results:

As this program developed employees recognized that they had a stake in safety, and we—the management—stayed away from telling them how to do it. Then, instead of giving lip service to our efforts to enforce rules, employees reacted positively because the program for safety was theirs.

Another[128] stated:

Physically, little has changed; the machines, the protective devices, the products, the procedures, and most of the employees are the same as they were 18 months ago. What has changed is the attitude of the employees. Today, they are concerned with the safe method of performing the job, because they gain individually and collectively by being safety conscious. They also are aware that only their efforts can prevent accidents.

And the workers[128] agree:

We had a safety committee for six years, but nobody took it to heart. Now we're thinking safety and quietly anticipating hazards before they create accidents. Some of the older employees make a definite effort to get any new fellows in on the program and watch their work on the machines to head off anything before it happens.

ARE COMMITTEES PARTICIPATION?

In the safety program of the past, the closest thing we had to real participation was the safety committee. The pros and cons of safety committees have been aired for years without changing anyone's mind about them. They exist; and in some companies they fulfill a role of communication, quasi-participation, or perhaps in some cases, real participation, for the few workers on the committee. The safety committee, however, is not what we are talking about when we talk about participative safety. We are talking about employees participating in the safety decisions that pertain to their jobs.

PARTICIPATION AND DISCIPLINE

One of the real advantages of participative safety is that it almost eliminates the need for discipline. Robert Burns states[25] :

> One of the main problems inherent in safety is that there is no particular satisfaction gained from working in a safe manner: you may avoid an accident by working in a safe manner, but you never really appreciate an accident that is avoided, because you are usually unaware of it. While the motivation to work safely is vague, the motivation to violate safe working practices is more concrete: the prescribed method is inconvenient; it requires more effort; it takes time when one is in a hurry; there may be a fear that coworkers may think one is timid, or afraid if he follows all the safety rules; perhaps dislike for the boss or the company is enough reason to ignore safety regulations. When there are more motivating factors working against safety than for safety, the tendency of workers will be to follow the unsafe method.
>
> Once workers have been trained in safe working methods, the problem of motivation—usually the responsibility of the supervisor—becomes primary. Most supervisors approach the problem of motivating safe work practices by means of the use of punishment. In punitive systems, the subordinate is presumably motivated to follow the safety rules because he is afraid that punishment will follow if he does not. A rule with a set penalty for infraction is an example of the fear punishment method. While it is true that in many cases this method is effective, in others it has a negative effect.
>
> Punishment has negative implications: it may result in wrong learning, that is, learning how to avoid getting caught, and it can create hostility within the worker when he is caught. Fear can motivate people to avoid unsafe methods, but this avoidance does not necessarily lead to an acceptance of the safe method. It is a negative emphasis rather than positive, and, as a rule, the gains realized by negative motivation are short term. As a negative force, punishment can create hostility: a schoolchild who is punished for fighting may fight even more in the future, because of the hostility which can develop as a result of the punishment, and hence, further infractions may result. In the factory or the office the behavior of a subordinate who is punished for breaking a safety rule can become hostile in the same manner as the behavior of the younster punished in school.
>
> Punishment makes impossible the development of a strong, positive relationship between the subordinate and his supervisor because it is negatively oriented. In fact, it can encourage a deterioration of the company-supervisor-subordinate relationship. In companies where policy calls for a disciplinary layoff for infraction of safety rules, supervisors may take great pains not to see a violation in order to keep themselves from being placed in the position of administering punishment. This "voluntary blindness," in effect, implicitly encourages the unsafe method. Since safety rules cannot be eliminated—since they are mandatory in every facet of industry—the problem is one of how to increase the positive influences of safety rules, while minimizing negative reactions by subordinates.

Burns' views on participative safety and on punishment tie in closely with what the psychologists now know about punishment. Punishment following an act tends to teach a person not to get caught in the act again, rather than not to perform the act at all. In safety this is not good enough. We must teach people not to perform unsafe acts whether or not management, or the supervisor, are observing them. Participating in the decision as to what is an unsafe act is the point where participation leads to the behavior we want. Having participated in that crucial decision, the employee no longer is a discipline problem to us. As Burns[25] summarizes:

If industry cannot eliminate safety rules, then we, as leaders in industry, should abolish the traditional methods of establishing them. Part of the rule-setting responsibility belongs to those people for whom the rules are intended. Under the direction of a safety director and his subordinate, the supervisor-leader, positive motivation towards safety rules and positive enforcement of safety rules can result, first, in sound interorganization, and second, in sound economics: fewer lost manhours, less compensation for injuries sustained on the job and less replacement of skilled help. Each means gained dollars.

IS PARTICIPATION FOR EVERYONE?

Roderick Forsgren[38] (University of Maine) made a detailed study of participative safety programs and identified the conditions needed for participation to work:

There are some very basic conditions which must exist before individual employee participation in safety programs becomes an effective technique. A person must be free to rearrange his particular goals in light of his experiences. Involvement in safety programs must provide experiences which allow a person to alter his beliefs and sentiments concerning safety as it relates to his work.

Individuals must also be willing to accept participational activity before it can become an effective managerial technique. People with low independence needs, people who work under safety programs that are highly authoritative and reflect central control, might well find opportunities for participation as signs of weakness in leadership. Thus, hopes for immediate acceptance and compliance via employee participation may be dysfunctional in light of the safety objectives.

Workers in most organizations are, however, accustomed to some degree of participation, although it may be at a very low level. There is probably far less danger in extending participative techniques in the safety area than is imagined by many safety directors.

Participation in safety programs must also be viewed by the individual employee to be relevant and important. His involvement in safety programs should be perceived by him to increase positive goals and reduce negatives ones. It might well be that these positive goals extend beyond the realm of safety and include such factors as recognition, status, power, and achievement.

Thus, use of employee participation in safety programs carries with it the "risk" of permitting employees to make decisions in areas that management may have traditionally considered its prerogative. Changes in work-flow or work-methods design that may have safety implications would be an example of a "hot" area; whereas extending participation to areas which are judged irrelevant by the employee result in very little or no return to either the individual or the company.

A final criterion for effective participation in safety programs is the belief by employees that the area in which their decision-making is permitted is one which they view as legitimate. By legitimate is meant an area in which the employees feel it is right and proper to engage in the decision process.

Although there may be some exceptions, participation in safety programs generally does not offer situations in which individuals doubt they have a right to make decisions; for very often, unsafe conditions are directly related to work and almost always viewed as affecting a person's safe being.

One must be careful, however, not to confuse legitimacy in safety with importance of safety. These two concepts are not identical. It may well be because almost all safety practices can be interpreted by the individual to have some direct bearing on his well being; and, hence, he may view it as a legitimate area. Whether or not he considers it important is a separate matter.

Thus, in summary, the approach by line personnel in implementing safety programs is dependent upon four basic assumptions which must exist before participative techniques are useful. These are (1) psychological involvement, (2) fulfillment of a need for independence, (3) the relevancy or importance of the area of decision-making, (4) the extent to which the employees view the area as a legitimate one in which to make decisions.

The absence of any one of these four factors makes the implementation of safety programs through the line organization via employee participation an ineffective venture.

In addition, certain organizational requirements are necessary. For example, the use of participative techniques must be economically feasible. Admittedly, it is virtually impossible to determine the exact costs of participation. However, there are times when the costs of the participant's time, for example, would be prohibitive. There are practical restrictions in addition to time and sheer cost, including employee experience and adequate communication structures, which can severely restrict the organization's use of participative techniques.

Participative safety casts each worker into a role quite different from the role in a hierarchical safety program. The participative program changes the worker from an inferior subordinate to an equal, a peer; from a person who receives an order and does the work as ordered to one who helps decide what is to be done and who then acts in accordnace with those decisions. This molds new values, new attitudes, and changes in behavior. In short, under participative safety, the worker becomes a motivated individual because the role is that of a person who wants to accomplish some specific results. The decided goals are group goals, not goals thrust upon them. And since the group has set the goals, achieving those goals is accepted; it is the group norm. Behavior that leads to group goal achievement is normal behavior. Behavior that does not lead to this becomes deviant behavior. A perfect example is in the wearing of personal protective equipment. Thirty years ago, hard hats were virtually unknown in the construction industry. Today, practically every construction worker wears one at all times. The hard hat has actually become a symbol of the industry. Before hard hats had become "accepted" they were considered a taboo item of wear. Anyone wearing a hard hat was considered a "sissy" on most jobs; the wearer was the deviate. Today the individual consturction worker is almost forced into self protection from serious injury. To not wear a hard hat is deviant behavior today.

John P. Gausch,[42] is an advocate of striking a balance between authoritarian safety programs and participative programs. His study of fifteen locations with 10,000 employees resulted in these conclusions, which he labeled as opinions:

- It is *not* supported that an expression of strong participation—self-control, group decision making, and reduced authority—will assure any improvement or be related positively to safety experience.
- It is *not* supported that an expression of strong authority—warnings, threats, reprimands and reduced participation—will assure any improvement or be related positively to safety experiences.
- It *is* supported that an expression of strongly balanced involvement tends to have a respectable validity coefficient with improved safety experience.

He further offers the prediction model shown in Exhibit 15.2.

Dr. M. T. McDowell,[94] President of McTar Petroleum of Vancouver, B.C., Canada, recently described some cases studies of safety programs that utilized participation:

The two cases summarized are from very different industries (construction and manufacturing). The studies exampled are longitudinal in nature (3–5 years). In

Contents of program				
Strong participation	+	Strong control		Result
No	+	No	=	Poor
Yes	+	No	=	Questionable
No	+	Yes	=	Questionable
Yes	+	Yes	=	Good

Exhibit 15.2 Prediction model.[42]

both cases the Traditional Safety Programme was replaced with a programme involving some contemporary behavioural science wisdom (The Modern Safety Approach). For both industries the introductory process as well as content and format of the Modern Safety Approach was identical. It should be noted that Safety and Accident Prevention is used by the author as an entry vehicle to conduct larger Organizational Development change interventions.

How Traditional Safety Programmes Differ from Modern Safety Programmes

Traditional Safety Programmes Typically Have:

1)   One Head Office person responsible for Safety.
2)   An organizational safety specialist usually referred to as a Safety Director (Sometimes Safety Directors have Departmental support staff).
3)   Safety as a staff function with director and department as staff positions.
4)   A set of specific Do and Don't Company Safety regulations set down by management.
5)   Marginal line commitment to safety and moderate involvement in accident prevention.

The Modern Safety Approach Has:

1)   On the job worker and work group responsibility for accident prevention and safety goals.
2)   Organizational safety specialist usually referred to as a Safety Coordinator.
3)   Safety and accident prevention as a line responsibility of line organization.
4)   Emphasis on upward communication from work group to supervisors to middle management to top management. Also lateral communication to the Worker's Union and the Worker's Compensation Board as well as between workgroups.
5)   Umbrella policies from management protecting worker input as well as specifying structure of committees, nature of communication flow and reward system. These policies focus on the *process* of the Safety Programme. The exact content of the rules and regulations are left to the particular work groups to formulate and enforce (The Buddy System).
6)   Management and work group involvement in and commitment to accident prevention goals and safety programmes.

Safety has a rather unique quality relative to all the other organizational dynamics. It is pervasive in its importance throughout the organization. This catholic aspect of of safety affords the consultant instant organization-wide visibility while at the same time allowing the O.D. change agent to enter the system on an important non-threatening issue. A successful impact on safety performance and worker well being can have a lasting and profound impact on other improvement efforts during the O.D. intervention.

The consultant is as legitimate in the executive office discussing safety with the president as he/she is on the floor or in the trenches with the workgroup.

The central thrust of the Modern Approach to Safety and Accident Prevention is the committed involvement of intact work groups. Workers in traditional settings often report a sense of powerlessness in their jobs. That is a feeling of being unable to influence policies, procedures and conditions. The resultant attitude is often verbalized by a conclusive comment like, "that's the way it is and will always be". This attitude often comes from years of confirmed experience. By placing Safety and Accident Prevention directly in the hands of the workers and work groups the "ever thus" attitude can become a "yes we can" commitment. The involvement in Safety can overflow into decision making and problem solving around other aspects of the work environment.

Dr. McDowell then describes two case studies:

Case 1   *Manufacturing Company*

Semi- to highly skilled organization employing approximately 60 people. Company was characterized by poor productivity, absenteeism, high costs, grievances and excessive accidents. The statistics listed below focus upon the safety record only, however, similar data exist for production, absenteeism, waste and profits.

The results achieved for Case 1 are shown in Exhibit 15.3.

Case 2   *Construction Company*

Semi-skilled organization employing approximately 300 people.

Results are shown in Exhibit 15.4. Case 2 is diagrammed in Exhibit 15.5. He suggests that the modern approach seems to yield:

Year	No. of M/hrs (Approx.)	Total No. Accds.	Time Loss	No. of Man Days Lost	Actual Comp. $ Costs	Comp. Costs Adj. for Inflation to Show $ Comparison with Equitable 1975 Values
1975	62,726	41	24	257	11,360	
1976	69,598	34	7	125	5,661	12,543
1977	58,563	25	9	95	6,069	13,923
1978	61,965	30	5	68	4,069	15,454
1979	104,897	42	24	422	19,350	17,154
1980	122,000	40	15	218	11,500	19,040
1981	107,000	44	17	259	20,399	21,135

Exhibit 15.3 Accident Statistics Showing Manufacturing Company's Safety Record Since the Inauguration of the Modern Safety Programme February 1976 to 1981.[94]

Year	No. of Man Hrs. Worked	Total No. of Accds. Reported	No. of Time Loss Accds.	No. of Man Days Lost	Act. Comp. Cost	Comp. Cost Adj. for Inflation to Show $ Comparison with Equitable 1975 Values
1974	649,000	65	37	1,123	110,088	
1975	479,000	31	14	1,090	68,899	122,198
1976	571,000	37	15	816	42,945	135,640
1977	535,000	32	7	885	43,225	150,560
1978	504,235	36	5	727	50,189	167,122
1979	510,356	33	4	487	33,817	185,505
1980	472,000	30	8	530	33,690	205,910
1981	570,000	42	11	482	29,999	228,560

Exhibit 15.4 Accident statistics showing construction company's safety record since the inauguration of the Modern Safety Programme introduced in February 1976 to 1981.[94]

1. Fewer time loss accidents
2. Lower accident cost
3. Less severe accidents
4. Fewer worker days lost

In both the cases exampled above the companies enjoyed improved profits, fewer grievances and less absenteeism. Additionally both companies enjoyed an assessment rate reduction from the Worker's Compensation Board.

QUALITY CIRCLES IN SAFETY

One of the popular participation techniques is the use of Quality Circles. Quality Circles started in Japan in the early 1960s and some ten years later found their way to the United States. A Circle consists of a small group of workers doing the same type of work forming together under management's auspices to solve problems they face daily on the job. They tend to meet on a regular basis, select their own group members, select their own problems to work on, and are purely voluntary. Often they use sophisticated statistical tools to help solve these problems. They then submit their analysis and suggested solutions to management for decision and implementation.

Since a number of behavioral theories and theorists are discussed throughout this book, Exhibit 15.6 is included to show how the Quality Circle (QC) concept fits with these theories. Usually an organization will first form some type of a structure to allow the participation to occur. As shown in Exhibit 15.7, this structure often mirrors the organizational structure. Often the Quality Circle concept is installed slowly, in perhaps a pilot plant or pilot section of the plant, and allowed to grow on its own. Typically the growth is fast, as it is popular in most organizations.

Exhibit 15.8 diagrams the installation process. Training in the problem solving process (see Exhibit 15.9), and often training in the use of statistical methods in problem solving, precedes group installation. While you don't need Quality Circles to get participation, they have been found to be highly successful.

Obviously, not everyone agrees whole heartedly with participative safety. Safety professionals are showing cautious interest conceptually, but few are doing much yet with the concepts. There is no proof that participation will definitely yield a lower accident rate than nonpartici-pation, so few try it. There is no proof that authoritarian safety programs will yield a lower accident rate than participative programs either, but it is here and we are comfortable with it, so mostly we stay with it.

Following is a final word on participative safety program implementation from H. Harvey Cohen,[29] writing in Professional Safety magazine:

> Considerable preparation is necessary for all taking part in implementing such programs. This includes not only technical personnel handling the implementation, such as facilitators and higher management, but all individuals who participate.
>
> Lower and middle management need to be prepared in several ways. First, they must be trained in their new and different roles, especially vis-a-vis subordinates

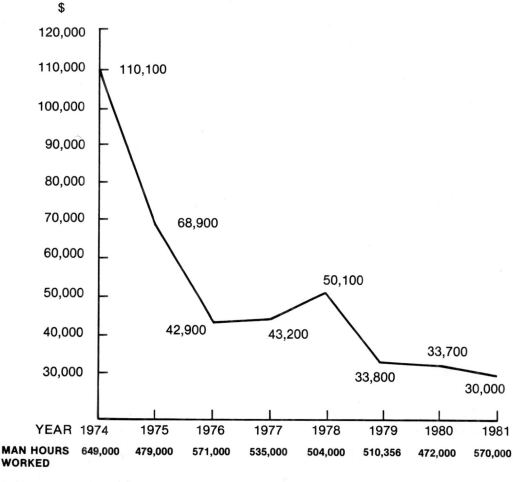

Exhibit 15.5 Case 2 — construction company results.[94]

Theory or Principle	Compatible QC Concepts
"Scientific Management" *(Taylor)*	Quality circles place an emphasis on statistical tools and work analysis.
"Hawthorne Effect" *(Mayo)*	Special attention is available to circles via presentations to management and communication of results.
"Hierarchy of Needs" *(Maslow)*	The structure and processes of QCs help employees to meet higher-order needs, particularly those dealing with social belonging, esteem, and self-actualization.
"Theory of Personality and Organization" *(Argyris)*	Mature, responsible behavior is the essence of circle membership.
"System 4" *(Likert)*	Quality circles are a tool for achieving a participative group approach to management.
Power, achievement, and affiliation motives *(McClelland)*	Quality circles provide opportunities for satisfying all three motives through autonomy, accomplishment, and social interaction.
"Motivation/Hygiene Concept" *(Herzberg)*	Quality circles supply the motivational factors of growth and learning, recognition, achievement, the chance for advancement, and an improved perception of the work itself.
"Theory X" and "Theory Y" *(McGregor)*	The managerial attitude on which QCs are based consists of respect, trust, and belief in the capability of workers to solve problems, create ideas, and take responsibility for their own behavior.

Exhibit 15.6 Management theories and compatible Quality Control concepts.

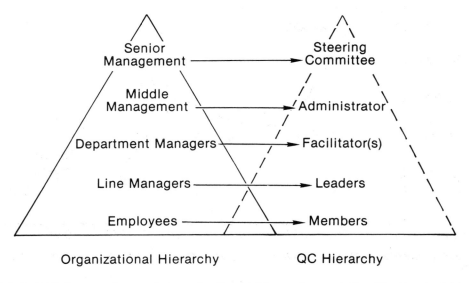

Organizational Hierarchy   QC Hierarchy

Exhibit 15.7 Comparison of organizational hierarchy and Quality Control hierarchy.

who have increased roles in decision-making. This will make their adjustment easier, leaving them with less of a feeling of lost power and importance, and will help them to adapt to their new roles and responsibilities.

Second, managers at all levels must really believe that their subordinates have something to contribute; otherwise, employee attempts to provide input will not meet with much acceptance or interest.

Employees, too, must be prepared. They need training in new techniques, new work patterns, and the like. Less obviously, they need guidelines in areas of participation. For example, if problem-solving is an important new part of employees' jobs, then training in problem-solving methods is a must. If this will be done by committee or group, training in effective interpersonal communications is essential.

Implementation of an employee involvement program (or, for that matter, virtually any organizational restructuring) takes time. Immediate positive results should not be expected, and in fact, some grumbling and resistance is inevitable. (One goal of the preparation discussed above is to reduce such resistance by increasing understanding of the new methods.) It takes time for workers and management to become accustomed to new ways of doing things and to take maximum advantage of them.

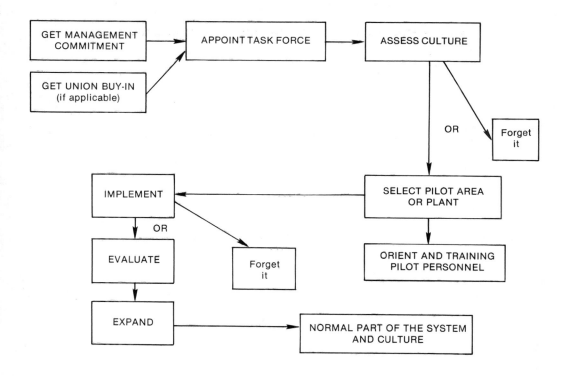

Exhibit 15.8 Quality Circle installation process.

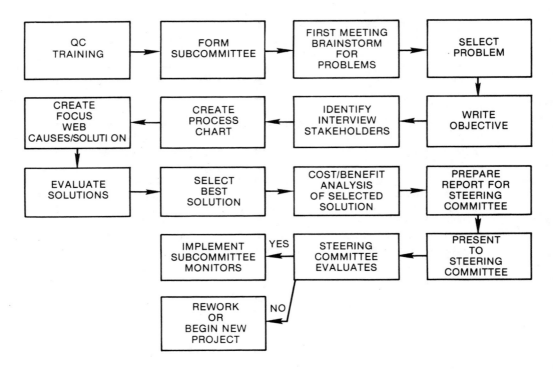

Exhibit 15.9 The Quality Circle problem solving process.

Chapter Sixteen

Coping with the Group

Each group makes its own decisions. It sets its own work goals, which may be identical with management's goals or different. Each group sets its own moral standards. The group also sets its own safety standards, which it lives by, regardless of what management's standards are.

D. C. Petersen[127]

If our programs are to be successful in accident prevention, we must motivate our employees: we must make them want the same goals that we want. It becomes increasingly obvious that when we are talking about motivating people—about changing, influencing, or controlling the behavior of people—that none of us really has the key. It is becoming increasingly apparent that even as we talk about motivation, we really cannot actually motivate. Motivation by dictionary definition means "to incite to action." It seems that, at least in safety, we cannot incite action; we can only influence attitudes and hope actions results from that "improved" attitude.

Management, through its definition of policy, makes the decision that it *wants* safe performance from its people. Management, however, cannot *force* safe performance, even with the most sophisticated procedures of fixing accountability that are now finally coming on the safety scene. Employees decide for themselves whether or not they will work, how hard they will work, and how safely they will work. They decide this based on their attitudes—attitudes towards themselves, the environment, the boss, the company, the entire situation. They decide based on knowledge, skills and the group's attitude toward the problem. All we can do is to create influences to help them decide, to extend some influence where we cannot fully control, and to recognize and understand those influences over which we have no control.

Perhaps this sounds weak, as if we have little real power. This is not true. Even though we can only influence, some of the influences that we can bring to bear are powerful indeed. It is true that in the final last second of decision making before the accident (or nonaccident) we have no power; but if we have used our influence well, we have done a great deal to determine whether or not the accident will occur. What are some of our influences? Here are some of them:

1. The influence of training.
2. The influence of supervision.
3. The influence of special emphasis programs.
4. The influence of media.
5. The influence of the attitudes of the group towards safety.
6. The influence of power.
7. The influence of authority.
8. The influence of leadership.

A great deal has been written about the importance of training in molding individual safety consciousness and attitude. We all know the crucial role of supervision as an influence over employees. We have all read about and used some form of special emphasis program to mold an attitude. (The "Knowing's Not Enough" program is perhaps the best example of a special emphasis program aimed directly at attitudes.). We all use safety media of all varieties to try to shape the thoughts and attitudes of our people toward safety. But how much have we really thought about the influence of the work group on the individual in that group who must make the daily decisions that can result in injury? How much does the group attitude toward safety mold or affect the individual's attitude? If safety is "sissy stuff" to the work group, how will the individual look at safety? If the group has decided hard hats are not to be worn, how will the individual react?

Each employee is an individual, and also an integral part or a member of a group. Each manager must manage the crew as different individuals, but also as a group. As in chemistry, elements combine together to make other substances with entirely different properties; individuals combine together to produce a group which has entirely different properties, a distinct personality of its own. We have to recognize the group properties as well as the individual's properties. Each group makes its own decisions and sets its own work goals. These may be identical with management's goals or they may be different. The group also sets its own safety standards, and it lives by *its* standards, regardless of management's standards, and regardless of the OSHA standards.

WHAT IS A GROUP?

A group is a number of people who interact or communicate regularly and who see themselves as a unit distinct from other collections of people. Also, the members of a group are drawn, or perhaps even bound, to one another in one state or another of interdependence. In other words, there is something at stake, and the group members share in or will share in that something. This interdependence may have nothing to do with the task that the group performs, but has to do with the group itself or with the relationships within it. Thus, each member may depend on the others for the satisfaction of needs for affection or affiliation or on the existence of the group and membership in it for security. All parts of the definition must be true: The people in question must communicate or interact regularly. There must be some sort of interdependence. Members must think of themselves as a group.

The formation and development of a group depends on two things. One of these is a collection of individuals with needs and desires and the second is a task. The most important of the needs or desires of the individuals is the social need to belong to a group. Industrial studies have shown that those workers who work alone or with but one or two people are not as happy in their work as are those workers who work with several people. Related to this social need for affiliation is the need to give and to receive affection. This need is an extension of the need for affiliation. The need for affiliation causes one to want to be with other people. The need for affection is what happens when they are together. They tend to conduct themselves so that they please those with whom they interact regularly.

An additional characteristic of individuals who enter into groups is the desire to further their self-interest. The individual feels that self-interest will be best furthered by acting with, or by being associated with, other people. The idea is sometimes strictly a case of strength from unity, but it is more often a case of the individual feeling that if several people want and work for the same goal, the chances of success are greatly enhanced. Thus, individuals with their varied needs and desires are important in the formation of a group.

The second indispensable input is a purpose, a goal, a task to be done. The purpose is its reason for being. In business, it is usually definable in terms of some sort of production. In education, it would be in terms of some kind of learning. In social terms, it would be definable as an event, function, or experience. Whatever the situation, the formation of a group requires both a collection of individuals with certain needs and desires and a purpose. Neither is sufficient by itself. A collection of individuals without a common goal is still a collection of individuals, not a group.

A factor that often influences the safety of a person, and yet which is not always understood, is the problem sociologists call *group norms*. Group norms are really the informal laws that govern the way the people that belong to a group should and should not behave. Very often when members of a group are asked what their norms are, they cannot identify them and yet unconsciously their behavior is strangely influenced by them.

Groups norms are the accepted attitudes about various things in the group situation. These include attitudes about how workers behave toward their boss, how they react to safety regulations, and how they react to production quotas. It "codifies" their attitude about the company, manner of dress, and merit systems into recognized, accepted, and enforceable behavioral patterns. If a member of a group takes on a pattern of behavior or expresses an attitude that is in violation of that commonly accepted by the group, there are methods of punishment to encourage conformance.

In an industrial organization, if the norms developed within the work group are favorable to safety, the group itself will encourage and even enforce safe practices much better than can supervision (hard hats in construction, for instance). Group norms often develop which are against our safety rules. A group might have an attitude that safety is for sissies. We do not really know why this type of philosophy becomes entrenched, but we know it often does. Often management's first response to such a situation is to pass a regulation which will force the person to violate the norms of peers and follow management. If the group is a strong group (with a high degree of cohesiveness), the member will violate management's direction rather than run the risk of being shut out of the work group. We ought to understand this phenomenon, and our objective should be to find some way to get this phenomenon working for safety, rather than against it.

GROUP PRESSURE

The power of group pressure has been shown many times in controlled experiments. Exhibit 16.1 illustrates a test on high school graduates, all men, with good eyesight. They were to tell whether two lines were the same length or not. Most of the pairs of lines were noticeably different in length.

When these men were alone, they were almost 100 percent correct in judging which line was longer. But then they became the unsuspecting victims of group pressure. They sat in with a group and all judged the lines. All the other people in the group were conspirators who had been told to call out wrong answers. Each conspirator called out the wrong answer before the turn came for the innocent victim to report how the line lengths looked to him. The first two columns on the left show that the victims were only slightly misled when one or two others preceded them with the wrong answer. But when three or more others gave wrong answers before them, one-third of the victims gave in. They took the group's word, rather than what their eyes told them—the majority effect. It is significant that a group of three others was as powerful as a group of sixteen for misleading the victims.

Some of the victims said later, after being told of the practical joke played on them, that the line lengths honestly seemed to change as they heard each of the others in the group call out

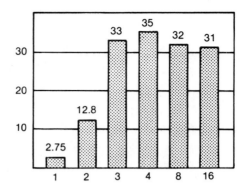

Percent of Times
Persons Adopted
the Wrong Estimate
the Group Had Made

Exhibit 16.1 The power of the group in keeping people from trusting their own judgement.

the wrong answer. A few of the victims deliberately gave answers they felt were wrong; they did not want the group to think them queer. But most of the victims felt the group must be right, because it was unanimous. Somewhat similar tests have been made when judging the size of a rectangle. In these tests people tended to change their esitmates to be more like the one they were told a group of twenty to thirty people had made.

The way a change of one's group affiliations shifts attitudes was demonstrated in an analysis of 2,500 blue-collar workers in a Midwest factory. The psychologists had records of the attitude of each of these men toward the union and toward the management. A year later the men who had been promoted to foremen or elected stewards were followed up. The men elected as stewards had been no more pro-union than the rest, and the men promoted to foremen had not been any more pro-management than the others. Within a short time after the promotion, or election, the men's attitudes had "spontaneously" shifted. The newly elected stewards became much more pro-union than they formerly had been. The new foremen became much more pro-management than they had been a few months previously. The shifts in attitudes were larger, and more widespread, among the new foremen than among the new stewards.

These group pressures put more pressure on workers than do the standard procedures which are written in the manuals. In safety, work-group pressures and group norms are perhaps the single most important determinant of worker behavior. To restate our original thesis: the group sets its own safety rules and they live by their rules, not management's.

Safety programs then must not only speak to the individual; it must also attempt to understand the group's safety norms and to influence those group norms so that they are safety-oriented norms. Safety programs must help to build strong work groups with goals that coincide with management safety goals.

BUILDING STRONG GROUPS

We can determine the strength of groups by observing some characteristic symptoms of strong and weak groups. In a strong group, the members voluntarily:

1. Try to deserve praise from the rest of the group.
2. Seek recognition from the group leaders (not management).
3. Exert pressure on weak group members.
4. Put special efforts into achieving the group goals.

The key to identifying strong groups seems to lie in the word "voluntarily." In a strong group the members seem to want to achieve all of the above. It is important to them as individuals to conform to the goals and mores of the group.

How about in weak groups? Here are some characteristic symptoms of the weak group:

1. Form cliques or subgroups.
2. Exhibit little cooperation.
3. Are unfriendly.
4. Use no initiative.
5. Avoid responsibility.
6. Have no respect for company policies.

Some causes of weak-group cohesiveness are too much pressure from management, lack of support, and job inadequacy.

Each manager might do well to stand back and look for these identifying characteristics. To build a strong group with goals identical to ours, there are four things that are essential:

1. *Individual Competence.* Each member of the group must have the ability to perform competently.
2. *Individual Maturity.* Each member must be mature. This means the group dislikes the "prima donna" who can but who will not do a fair share. They dislike the "yes man" who exerts more effort pleasing the boss than the group. They dislike the "let-George-do-it type."
3. *Individual Strength.* Each member must have not only the ability and the maturity, but also have the strength to do the job, to earn group respect. This means no "weak sisters," no "loners."
4. *Common Objectives.* These are essential to a strong group.

We can readily see that the first three items deal with qualities or characteristics of individuals that cannot be changed readily. We can extend some influences, of course, but basically in all three the strength of the group depends more on whom we place in the groups initially. We should then examine several areas:

1. How do we select and place employees initially? Are the three characteristics (individual competence, maturity, and strength) considered in finding people and are they most important in placing them into the work group? This does not mean we cannot hire and use the loner or the prima donna. It does mean we should carefully consider the placement so as to not destroy a strong group or a strong group essential to the defined goals.
2. We might consider shuffling individuals into other groups in some cases. At times, transfer of only a few individuals makes a major difference in several groups.

The fourth essential to a strong group is a common objective. In this area, management can do quite a bit. In safety it means doing a much better job of goal setting, motivating and communicating so that our safety goal becomes their chosen group goal.

STEPS TO BUILDING STRONGER TEAMS

In the literature, there is a classic demonstration of the social and psychological forces which can build group cohesiveness. A group that lacked cohesiveness was transformed into a highly cooperative team within a few days. This quick building of cohesiveness was not done by asking for teamwork. No platitudes about cooperation, or slogans, or posters picturing the one weak link in the chain were used. The scientists just touched off the natural forces which are available when people are in groups.

This demonstration was made with twelve-year-old boys in a secluded camp provided by the Yale University department of psychology. The boys did not realize they were part of an experiment and the man they thought was the caretaker was Dr. Muzafer Sherif,[148] the chief psychologist. Here is an outline of the procedures by which the random collection of boys was built into highly cohesive groups almost overnight.

1. *Physical Proximity.* The boys were strangers to each other at the start. After they had been in camp a few days, they were divided into two groups. Each group had separate living quarters. This is similar to most business situations, where strangers are put together in a room separated from other work groups. Being thrown together physically gave a chance for interactions which would not otherwise have taken place, thus providing an entering wedge for building team spirit.

2. *Sharing Common Goals.* These separated groups of boys proceeded to set goals for their respective groups. They decided on decorations and the arrangement of their quarters and on other activities which appealed to their own group. The members participated fully in deciding their shared activities, which ties in with the participation aspect previously discussed. Sharing in making decisions and then working together to reach these shared goals are prime factors in building cohesiveness.

3. *Setting Up an Organization and Accepting Leadership.* These boys had not worked together more than a few hours before they began to pool their efforts. They spontaneously organized duties within the groups. They noticed that some members were adept at special activities, so they made niches in their organization charts for these experts. The quickly divided the work and defined the responsibilities of different members. Each member soon understood what role he was expected to play in the group life. These groups also quickly came to look to a few members to play "higher roles" in coordinating the others. Captains and lieutenants emerged and group activities began to center around them. The groups set up their own social levels, or hierarchies of power, within the group. But their accepted leaders were from within the group, not from outside it. In the same manner, the person the company designates as boss may not be the one the group would have designated; thus we usually have our appointed leaders and the group has its informal leaders.

4. *Developing Group Symbols.* The boys had scarcely agreed on their accepted leaders before the members were clamoring for symbols to identify themselves as distinct groups. They invented nicknames and some jargon for their activities. Industrial groups do this if they are cohesive and the vocabulary of one department may sound like Greek to the department down the line. The boys also developed some secrets as offices do through the grapevines, and as families do in family jokes.

The boys' groups bought caps and T-shirts in the colors they chose and distinctive symbols. Adults seem to have much of this same kid stuff in them. Railroaders favor a certain style of work clothes which are trademarks of their group. A house painter feels disloyal to the occupational group unless working in painter's whites. The blue shirt and the white shirt are group symbols in the business hierarchy. The work clothes are part of the role that members are expected to play. When groups want such distinctive symbols, it is evidence of cohesiveness. But it does not necessarily follow that wearing a work uniform designed by the general manager's wife will build cohesiveness.

5. *Competing with Natural Enemies.* The situation with these boys was such that each group quickly looked upon the other as a natural enemy. Groups tend to hold together more firmly when threatened by some enemy, or when some stress makes the members realize they are dependent on cooperation for security or perhaps for survival. Rivalry and stress situations are not rare within a business. One department often looks on another as a natural enemy. One clique considers the other clique a rival, each clique then holds together more strongly than before, and cooperates less and feuds more with the rival clique. If the appointed leader is dogmatic and self-centered, the workers may become more cohesive, but will rally around the goal of frustrating the boss rather than cooperating with him.

In the case of these boys' groups, the natural rivalry was exploited by egging the two groups into competitive contests. Little encouragement was needed; each group was itching to prove its superiority. To intensify this rivalry, the experimenters rigged some of the contests. This made the losers furious at their opponents. Each group held closer together than ever and engaged in open as well as secret warfare. There were pitched battles in which the boy who had previously been the crybaby overnight became hero of his group, but a despised villain to the other group. To protect life and limb, it became necessary for the experimenters to order the hostilities stopped. Merely giving the order and policing the groups was not even adequate at this point.

6. *Liking the People in the Group.* Many social psychologists put this item near the top of the list for usefulness in building cohesiveness. You must like the people to become a part of the group. There is, however, a reciprocal relationship here, because cohesiveness seems to develop most easily when the people are mutually attracted initially. But as cohesiveness does develop, people who have not previously seemed attractive become so, if they are in our group. The people in "our group" usually seem to us to be a little more capable as well as more attractive than their counterparts in competing groups.

Research and experimentation then suggest some ways of building strong work groups. Safety programs could utilize these ways by:

1. Placing people in physical proximity. If the workplaces do not automatically allow this (as in a shipping department), bring them together periodically, as in our 5-minute safety talks, or in meetings called to allow them to participate as a group in the decision-making process.
2. Allowing the group to set their own goals in safety.
3. Allowing the group to organize itself in safety. Let them select their own "enforcers," inspectors, talk-givers, committee representatives, departmental monthly safety director, etc.
4. Allowing the group to develop symbols and safety nicknames for themselves.
5. Setting up competitions with rival departments in the safety record, in inspection results, and in sampling results.
6. Allowing departments, or groups, some latitude in selecting their own members or allow transferring to occur where department members do not seem to get along well.

Chapter Seventeen

Communications

Some clown with a college degree made this up. A lot of them actually don't come in contact with the type of work they are trying to preach safety in. They should approach it from the worker's point of view.

A welder, quoted by Levy and Greene in their safety media study[86]

The final area we will look at in building a safety program that "sells" to employees is the area of communications. After selecting the components of the safety program the plan must then be communicated to someone or to many. In this particular chapter we will look primarily at the communication that must take place between management and the employee. There is a wealth of research and information on communications, so much that it is difficult to sort it all out into usable chunks. There is even some research on safety communications to give some direction.

The logical starting point would be to define the term *communications:* however, this is not easy. Hundreds of articles have been written on defining communications by hundreds of scientists and all are different. There are many communication models in existence today: systems models, information theory and cybernetics models, transactions models, helical models. While all of these attempt to explain, they also confuse. A simple definition which suggests a simple model is that communication is a process by which senders and receivers of messages interact in a given social context. For our purposes here the senders are management (including staff safety and line supervision), the receivers are workers, and the social context is the industrial organization. The model might look something like Exhibit 17.1. With this simplistic model an assessment about what is known about each component from research, perhaps they can be used to improve communication effectiveness in safety.

MANAGEMENT VARIABLES

Research tells us quite specifically that there are three components of source (sender) that influence the effectiveness of a communication: its credibility, its attractiveness, and its power (or control).

Credibility

There is considerable literature showing that the amount of attitude change produced by a given message can be varied by ascribing the messages to sources that differ on such things as

173

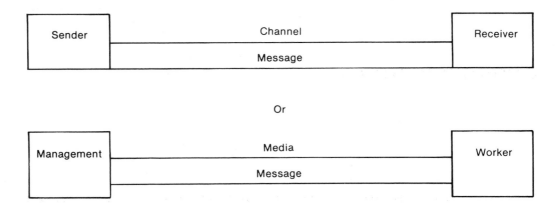

Exhibit 17.1 Our communication model.

knowledge, education, intelligence, social status, professional attainment and age. Researchers have also tested the influence of the objectivity of the sender.

It has been found that the less objective the source was felt to be (that is, the more the receiver suspected that the source was really out to persuade), the less opinion was changed. Sources who stood to profit from the subject's being persuaded were judged to be less fair and tended to produce significantly less opinion change. Several researchers have varied objectivity by forewarning some subjects, and not forewarning others, of the source's persuasive intent. They found that the warning had no impact on the amount of attitude change in their samples as a whole, and failed to find a lessening of opinion change after a warning of the source's intent to persuade.

One safety example of source (management) credibility as an influence on the effectiveness of the communication is in the Levy and Green study[86] referred to earlier. That study found that employees have clearcut ideas about their company's safety activities. The methods, goals, and the amount of vigor and sincerity put into them, and the way a worker views company safety efforts, strongly influences behavior on the job and ability to learn from and respond to safety media. In short, the effectiveness of the safety message presented by safety media (posters, films, booklets) is directly dependent on the employee's perception of management's interest.

Attractiveness

There is considerable evidence that a person is more influenced by a persuasive message when it is perceived as coming from a similar type person. The receiver then concludes that what the source is urging is good for "our kind of people," and thus attitude changes accordingly. The closer the worker identifies with management, the more likely the safety message will be accepted.

Power

One condition to be satisfied if management is to effectively produce opinion change is that it must be concerned about the employee's compliance. It has been shown that when the group's task is depicted to the members as involving higher need for unanimity, the group's members exert more influence over one another. It also has been demonstrated that power to punish will shape behavior only when its possessor can indicate to the person the validity of the power for imposing this punishment. Another condition contributing to the source's influence over the person by virtue of this power is that the source can discern whether or not the person has complied.

These then are some of the things research has discovered about the sender in communication effectiveness. Factors to be considered in safety messages are: (1) What is management's credibility? Is management *really* as interested in my safety as it is trying to tell me? How has management demonstrated that interest in the past? (2) Is my supervisor really "like me" enough to be believable or so "different" as not to be trusted? and (3) Is my supervisor's power over me real? Must I really comply with the safety rules?

Research also speaks to us about what happens when workers receive conflicting messages from two different sources: management and their work group. Almost invariably, we find that people are more influenced by their peers than by their superiors.

ONE–TO–ONE COMMUNICATIONS

In an earlier chapter the Argyris conflict model was discussed. He suggested the answer to the problem fell into three areas. One of those was "levelling." In the one-to-one relationship between two people, a boss and a subordinate, the closer it is to a levelled relationship (that is, a relationship of equality rather than superior-subordinate), the better off both parties will be. You can have that levelled relationship one to one; you cannot in a group or meeting format. Earlier, motivation was defined as a person attempting to satisfy his or her current needs. A boss can only find out what those needs are in as one-to-one format—not in a group or meeting format. In the last chapter we looked at the power of the informal group, it being the largest single determinant of individual worker behavior. Management can only break through that power one-on-one; in a meeting or group format, that power is only strengthened.

One-on-one communication by a supervisor is absolutely crucial to safety success.

Since that supervisor's credibility is essential to the success of the communication process one-on-one, supervisors must be deemed credible by their subordinates. Several studies have found that supervisory credibility is a function of the following behaviors:

1. Delegating responsibility in decision-making to subordinates.
2. Asking subordinates' opinions concerning upcoming decisions.
3. Giving subordinates opportunities to give additional ideas or information over and above what the supervisor has asked for.
4. Giving prompt answers to questions and suggestions.
5. Making sure that subordinates find it easy to get help with their problems and complaints.
6. Being aware of and responsive to subordinates' feelings and needs.
7. Being "frank" and "open" with subordinates.
8. Being supportive of subordinates concerning the complaints to upper management.
9. Expressing a sincere concern for the welfare of subordinates by:

   a. maintaining reciprocal relationships by exchanging ideas with subordinates;
   b. showing interest in the personal lives of subordinates;
   c. being helpful when help is needed;
   d. being concerned about subordinates getting ahead in the organization;
   e. being supportive with upper management;
   f. and by complimenting subordinates.[52]

It is important to note that these behaviors must be sincere or they may be counterproductive. For example, complimenting subordinates (or anyone for that matter) should be sincere or it will probably be viewed with suspicion as an obvious attempt at false flattery.

All communications between supervisors and subordinates may not be pleasant and enjoyable.

If a subordinate must be reprimanded, there are certain things that should be avoided so that the situation is not worsened. These are referred to as the seven deadly sins of reprimanding.

1. Failing to get all relevant facts and acting on hearsay evidence.
2. "Flying off the handle" or losing one's temper and acting while one is not in control of oneself.
3. Failing to let the other person know the precise offense for which punishment is being invoked.
4. Failing to get the other person's point of view, side of the story, or interpretation of the facts.
5. Letting the subordinate talk the supervisor out of the punishment which should be invoked.
6. Failing to maintain some record of what has gone on, why the punishment was given the subordinate, and what the next step will be if the matter is not corrected, and
7. Nursing a grudge or "holding it against the employee" for having had to be disciplined.[54]

MESSAGE VARIABLES

More attitude-change research has focused on message variables than on any of the other communication variables. Some of the knowns are that the general skill of the sender has not proved to be a very powerful determinant of persuasive effectiveness. While well-organized messages have been found to be more effective than poorly organized ones in affecting comprehension, they have no special effect on opinion change. A second factor is the comparison of dynamic with subdued styles of presentation. A dynamic presentation is less effective in producing attitude change and more likely to be labeled "propaganda" than is a conversational delivery. Very little research has been done on the use of humor, but it has been found that the use of humor in a speech on a serious topic makes it neither more persuasive nor more interesting.

Fear Appeals

One of the most interesting lines of research on the message is the use of fear in persuasion. Some studies have shown that minimal fear arousal in connection with one's recommendations rather than high fear arousal is more effective in producing attitude change and resistance to subsequent propaganda. Other studies have only occasionally supported these findings. A considerable number of studies have found predominantly positive relationships: the higher the fear arousal, the greater the opinion change. On the whole, this literature can be summed up by saying that there have been a few findings of a negative relationship, but many more of a positive relationship, between intensity of fear arousal and amount of attitude change. Certain situational and individual difference variables affect the optimal level of fear arousal. For example, Leventhal and Watts[84] found that high fear arousal is better than moderate in getting subjects to reduce the amount of smoking, but that low fear arousal is better for inducing smokers to take x-rays to check for the possibility of lung cancer. The work on fear appeals and attitude change suggests that simple relationships are not likely to be found in this attitude-change area.

The effects of fear-arousing communications seem to depend on two factors: the recipient's personality and predisposition toward the recommended action. The more susceptible the individual is to the message, the more sensitized he or she is and, consequently, the less resistant to the message. Secondly, where action can be taken immediately, a fear-arousing communication should be more effective than a more bland appeal.

If a fear appeal is to be maximally effective, it should be directed at an audience already predisposed toward the recommended practices (low resistance) and the recommended action should be of such a nature that the audience can engage in it immediately. So, fear-producing posters should not be used for innovative or new behavior nor should they be used in situations where immediate action is not possible. Fear-producing posters should *not* be used in a plant to promote off-the-job safety, especially if the worker will not be engaging in the depicted act in the immediate future. However, they could be used in the punch press department.

Campaigns

Communication campaigns are prevalent in safety programming. Should they be? Are they effective? Apparently we really do not know. The fact that seems to stand out the strongest in the research is that we do not have any facts. Hawkins[52] reviewed the literature on campaigns and came up with these results:

1. There has been relatively little research on the numerous campaigns that have been conducted. It is probably safe to say that the majority of safety communication campaigns get little or none, with the major decisions as to campaign contents and the evaluation of campaign effects being carried out purely subjectively.
2. The research design on most of the studies has been inadequate. Particularly in the realm of campaign evaluation studies, the principal conclusion has been that due to other factors operating, no conclusions were possible. In many cases where conclusions are drawn, they are unjustified.
3. In every case uncovered by item two, no one knows whether previous safety communications campaigns have had any effect in the real world or not. The exceptions are the studies by Laner and Sells[80] and by Kaestner, Warmoth and Syring.[68]
4. A great many safety campaigns utilize negative and/or threatening approaches, containing fearsome, scary, gory, anxiety-producing themes or treatments. This, despite overwhelming evidence that such approaches are most likely to have either no effect (due to avoidance of such messages) or even negative effects. It appears that health and safety campaigns particularly lend themselves to a "negative" approach in which the disastrous effects of not following good practices are stressed. Available evidence would overwhelmingly favor the general principle of a "positive" approach, emphasizing the good effects of actively following approved practices.
5. Despite the paucity of adequately designed research, there is evidence that mass communication campaigns can have a positive effect on safety behavior. This does not mean that just any campaign will have beneficial effects on any sort of safety practice. The greatest likelihood of success is in those campaigns where systematic use is made of known communications principles based on existing research, accompanied by appropriate pre-testing research at various stages of the development process.

The Insurance Institute for Highway Safety[65] conducted a study of a television campaign on seat-belt usage. This campaign was a nine-month saturation campaign of television commercials urging seat belt use. It had no effect whatsoever, according to their research project for the evaluation of the effectiveness of such public-service television announcements.

The campaigns promoting the use of seat belts were based on inadequate knowledge of the factors contributing to lack of use. Slogans such as "Buckle up for safety," "Lock it to me," "What's your excuse," and the like, were the hallmarks of the campaigns. Even with public-service time and space contributed by television, radio, and newspapers, the costs of the campaigns were high and the results inconclusive, according to the researchers. They reported that in 1968 alone a campaign by the National Safety Council, which has traditionally sponsored such "buckle up" drives, used the equivalent of $51,509,034 in public-service time and space in various media.

During a nine-month period in 1971 and early 1972, the Institute sponsored a television advertising campaign consisting of six professionally produced commercials. During the same period, observations were made of cars at numerous locations throughout the city in which the campaign was conducted to determine whether seat-belt use levels changed from previously observed levels and whether any changes that might occur could be attributed to the television campaign. At the end of the nine-month period the researchers concluded that the television campaign did not affect the use of seat belts. In fact, they pointed out, during the time of the campaign the observed level of belt use actually declined. However, they speculated that the decline was not related to the campaign but rather may have been due to seasonal conditions that caused added inconvenience in "having to adjust them (belts) to fit over bulky winter clothing."

Each message employed a theme emphasizing possible injuries that could result from nonuse of belts. Some included such product advertising devices as family responsibility and physician endorsement. Each commercial was directed toward one or more specific population segment, including young married adults, parents, small children, teenagers, and sports fans. The messages were broadcast over television in a medium-sized American city that is frequently used for test marketing. The system serves about 13,000 households, of which approximately 6,500 were shown the seat-belt commercials. The other half of the cable subscribers served as a control group and did not receive them. The messages were given a level of exposure equivalent to that of major advertising efforts used by companies to promote new products, and they were broadcast during periods that are considered prime time. The researchers estimated that the average viewer on the cable which broadcast the messages saw one or more of the messages two to three times per week over the nine-month period.

While the effectiveness of campaigns are in doubt, we really are not sure of their impact. The seat-belt campaign failure described above should not be interpreted as a failure for all motivated campaigns. An in--plant attempt, for instance, operates under totally different circumstances than a television campaign.

MEDIA VARIABLES

The major conclusion of the Levy and Greene study[86] was that safety education materials and programs seem to be tremendously effective, to the point that they are deeply incorporated into the working person's way of life and outlook. They concluded that different materials perform different functions. Posters are quick, amusing capsule reminders that form part of the decor of the plant, providing both impact and background; leaflets teach and explain, feeding thoughtfulness; films and meetings are apt to stimulate motivation and emotional involvement; and so on. All are part of the reinforcement process.

They also offered some suggestions about the development and use of safety media:

1. Employees like those posters and leaflets which communicate quickly and do not use a lot of words. But they do appear to want to know about some more complex and more technical things, about what the best way is to get things done and *why* this is the case.
2. There seems little question but that safety education materials are most useful when they are tied into company safety programs, meetings, clubs, interactions, demonstrations, eating places, and film events.
3. Films are felt to be especially informative. Workers appear to like them very much. There is a feeling among the workers in most companies that films are reserved for special people in the company and that fewer are being shown to the lower echelon employees.
4. There seems to be room for a manual or yearbook which will instruct but which will also include a verbal and pictorial safety record for a given company or industry during that year. Such a yearbook might include pictures of accidents, and how they could have been prevented; demonstrations of damaged safety equipment which spared lives; and some of the proud safety records and events of the year and what went into achieving them.

There is other research which tells us about media variables. While some of this will be covered later under training, one piece of information is particularly pertinent here. Learning psychologists have found that one of the prime factors in learning is the meaningfulness of the material. If workers see usefulness and meaning in the material being presented, they will learn it much better and much more quickly than it they do not. Trying to teach them something they already know, or something they feel they do not need, turns them off very quickly.

Researchers have also found:

1. In general, when a persuasive communication is presented in both spoken and written form, the spoken word has more persuasive impact.
2. It is generally found that comprehension is greater with reading than with listening.
3. The sources of communication are perceived as more credible with oral than with written messages.
4. It is clear that any impact that the mass media have on opinion change is less than that produced by informal face-to-face communication.

RECEIVER VARIABLES

The final area that communication researchers have looked at is the receiver, in our case the worker. What are some of the variables here? The following are a few of the findings that might give us some direction:

1. Learning is enhanced with participation. For instance, calling on a subject to improvise a speech produces more attitude change than passively reading, an already prepared speech.
2. The relationship between individual susceptibility and persuasion is very complex. For instance, there are relationships between sex and influenceability, age and influenceability, background and influenceability, and education and influenceability, but the research results are confusing and, at times, conflicting. While we know they exist, right now it does not appear that we can successfully use these relationships.
3. The amount of attitude and behavior change depends to a great extent on the group and group norms with which the person is involved.

One additional study was made in 1972 which attempted to find out how employees view attempts at communicating a safety message to them. In this study, through an anonymous questionnaire, we asked the employees of a medium-sized manufacturer in the Midwest to give us their opinions on the effectiveness of the usual safety program components. We then asked them to rate the amount of interest in their safety that they perceived their supervisors, the management, their peers, and they themselves had. Our intent in this study was to obtain some insights into such questions as: Does the employee's perception of management interest in their safety affect the effectiveness of all components of a safety program? Does it, for instance, influence the effectiveness of discipline or of training? Does the employee's perception of the supervisors' interest in safety affect the effectiveness of safety program components? Does, for instance, the fact that George knows his supervisor continually forces him to work on an unguarded press influence the effectiveness of his supervisor's weekly 5-minute safety talk or of the safety poster in the lunchroom or of the three-day layoff given for a safety rule violation?

Most research and data we have relates to media and the bulk of that research relates to safety poster effectiveness. Relatively little insight has been provided by these studies into the relationship of program component effectiveness to the employee's perception of management, supervisor or peer interest. The Levy and Greene study[86] looked at this relationship but only the employee's perception of management was considered.

Industrial safety programs consist of management and supervisors carrying out certain activities which locate and define hazards and operational errors, control the behavior of people, motivate people, and record results. These various activities, systems, or components are present in all safety programs, with varying structure and varying degrees of success. Most of what we do is based on theories from the past, or we have stolen ideas from management theory and adapted them for our use. Safety managers have assumed that discipline is an effective supervisory tool in safety. This commonly used tool comes from management theory. It is common standard safety practice to lay off a worker for three days for continued violation of a safety rule. Similarly we use safety posters in abundance, safety films, safety literature of all types, 5-minute safety talks, inspections, competitions, and many other components as integral parts of our safety programs. All of these are based on old concepts or theories borrowed from other fields which say that employees will be influenced by them and change their behavior accordingly. Are they really effective? Do the employees believe they are effective? We really do not know.

Most of what we do in safety has really never been tested. We do not actually know except by the "test of time" and the improvements achieved whether or not these components so often included in programs have a real effect. And if we have been successful in our plant program, we really do not know which component or which interaction of components achieved the success. In short, we really do not know what works in safety programming. We do not know what is effective, or even what employees and supervisors think is effective.

COMMUNICATIONS STUDY

In this 1972 study we merely asked the employees what they thought was effective by using the survey form shown in Exhibit 17.2. We then correlated the results of their answers, department by department, to determine if the effectiveness of program components as judged by employees co-varied with their perception of interest by the various people and groups. Exhibit 17.3 shows an 8 x 5 correlational matrix which summarizes the results obtained and Exhibit 17.4 shows the overall scores. The survey was administered to 316 employees in 28 departments. Each employee completed the form and sealed it in a return envelope to assure anonymity. We expected several results based on the Levy and Greene study:

Department _____

The Effectiveness Of Safety Programs

In order to put together a better safety program for XYZ Corporation, we would like your help in the form of your personal opinion on what kind of things are the most effective. Listed below are some of the common things usually thought to be important in safety programs. We'd like your opinion on the effectiveness of each. Please place a check mark somewhere on the scale. This is anonymous so *do not* sign your name.

Please send this directly back to Personnel in the attached envelope.

I. Following are standard communication methods in safety. How good are they in reminding you or motivating you to work safely?

	Most Effective	Very Effective	Somewhat Effective	I Pay Little Attention	Not at all
a. Safety Posters	Most	Very	Somewhat	Little	None
b. Safety Films	Most	Very	Somewhat	Little	None
c. Safety Booklets, Brochures	Most	Very	Somewhat	Little	None
d. Short Talks From Your Foreman	Most	Very	Somewhat	Little	None
e. On The Job Training From Your Foreman	Most	Very	Somewhat	Little	None
f. Inspections by your Foreman & Management	Most	Very	Somewhat	Little	None
g. Competition Between Departments	Most	Very	Somewhat	Little	None
h. Reprimands, Discipline Layoffs, Etc.	Most	Very	Somewhat	Little	None

II. Please rate how much you think the following people are interested in your safety.

	Most Interested	Int.	Average	So-So	No Interest
a. The people you work with	Most Interested	Int.	Average	So-So	No Interest
b. Yourself	Most	Very	Average	So-So	None
c. Your Lead man	Most	Very	Average	So-So	
d. Your Foreman	Most	Very	Average	So-So	None
e. Management	Most	Very	Average	So-So	None

III. What would you like to see included in the XYZ Safety Program?
     (Use the back)

Exhibit 17.2 Survey form.

1. The employees' judgment of the effectiveness of safety media would co-vary with their perception of management interest in their safety.
2. The employees' judgment of effectiveness of *all* safety program components would co-vary with their perception of management interest.
3. The employees' judgment of the effectiveness of all program components would also co-vary with their perception of the foreman's and leadman's interest.

In some of these hypotheses we were somewhat surprised. We found some correlation between poster and film effectiveness and the employees' judgment of management interest, but it was not very significant. There was no significance in the relationship between literature effectiveness and management interest. This does not support the findings by Levy and Greene.

CORRELATIONS OBTAINED

	Posters	Films	Booklets	Talks	Training	Inspection	Competi-tion	Discipline
Manage-ment	+0.302 P<.10	+0.501 P<.10	-0.283 NS	+0.057 NS	+0.516 P<.01	+0.342 P<.10	+0.065 NS	+0.329 P<.10
Foreman	+0.236 NS	+0.506 P<.10	+0.060 NS	+0.517 P<.01	+0.502 P<.01	+0.652 P<.001	+0.367 P<.10	+0.260 NS
Lead Man	+0.554 P<.01	+0.394 P<.05	-0.153 NS	+0.725 P<.001	+0.292 NS	+0.314 NS	-0.832 P<.001	+0.533 P<.01
Peers	+0.354 P<.10	+0.494 P<.10	-0.426 P<.05	+0.881 P<.001	-0.289 NS	+0.475 P<.02	+0.678 P<.001	+0.740 P<.001
Self	+0.754 P<.001	-0.752 P<.001	+0.575 P<.01	+1.314 NS	+0.559 P<.01	+0.231 NS	+0.114 NS	+0.226 NS

Top figure is the correlation. Bottom figure shows significance.
NS means "not significant."

Exhibit 17.3 Correlations obtained. Top figure is the correlation. Bottom figure shows significance. NS means not significant.

The Importance of Management Interest

This seems to mean that most media we use can be used regardless of managment credibility. Whereas the Levy and Greene study suggested *not* to use media where employees do not believe management is interested (for it can actually do more harm than good), this study disagrees. We found a significant correlation between the effectiveness of training and the employees' perception of management interest, but no significance and only small correlations between management interest and 5-minute safety talks, inspections, and discipline. Employees tend to view training as growing out of management interest; less so with inspections and disciplines. They do not connect the foreman's talks and departmental competitions with management interest.

Item	Score
I. How effective are:	
a. Safety posters	3.0
b. Safety films	3.5
c. Safety booklets, brochures	2.6
d. Short talks by the foreman	3.4
e. On-the-job training	3.6
f. Inspections	3.4
g. Competitions	2.7
h. Reprimands, discipline	2.8
II. How interested in your safety is (are):	
a. The people you work with	3.4
b. Yourself	4.3
c. Your leadman	3.6
d. Your foreman	4.0
e. Management	3.7

Exhibit 17.4 Overall scores.

The Supervisor's Interest in Them

At the outset we hypothesized that the perception of foreman interest would co-vary with safety-component effectiveness. We assumed this would hold true in the use of safety media. It did not. Our results showed a correlation (but not a significant one) between the effectiveness of films and foreman interest, but no correlation between foreman interest and posters or other media. Employees do not perceive most safety media as having anything to do with what their foreman wants from them.

However, the relationship between foreman interest and effectiveness of talk, training, and inspections are all significant. If employees think the foreman is interested in their safety, they think safety talks, on-the-job training and inspections are all effective. Conversely, if the foreman does not succeed in convincing his people that he is interested in their safety, they do not think any of these activities are effective.

Competitions do not co-vary with supervisory interest and, surprisingly, we found no correlation of any significance between the use of discipline and foreman interest. Employees think discipline is effective (or ineffective) regardless of how interested they believe their foreman is.

The correlations obtained indicate that employees see their leadman more as a peer than as a representative of management. Talks are highly correlated with leadman interest; discipline, films, and posters less so; booklets, training, and inspections not at all. Competition is highly correlated with leadman interest in safety for some unknown reason.

Peer Safety Interest

Some rather interesting relationships turned up with peer interest correlations with the various components. Posters, films, and booklets all showed a correlation, but it was not significantly high. Still there is some indication that if the work group is interested in safety (safety is a group norm), these tools are thought of as effective.

Safety talks showed a highly significant correlation. There are more than 999 chances in 1,000 that employees who work in safety-conscious groups believe 5-minute safety talks to be effective, and that employees who work in groups where safety is not a group norm believe safety talks to be ineffective. Since this is obviously true (according to the statistics), we were surprised to note that no correlation exists between peer interest and training. The informal group opinions have no influence on whether or not training is found to be effective in safety.

There is a significant relationship between peer interest and inspection effectiveness. Also, highly significant relationships exist between peer interest in safety and the effective of discipline and competitions. Discipline is effective as a safety program tool only when the informal group is interested in safety. If the group is not interested, discipline will not work. The same holds true with competitions.

The Workers' Interest in Their Own Safety

Some interesting relationships also were found in the workers' self interest in safety. The most surprising was the highly significant correlation between self-rating of safety interest and the effectiveness of posters and films. Those employees who are interested in safety believe posters are effective; those who are not safety-minded do not believe in poster effectiveness. The opposite was found in the relationship between films and self-safety interest. Those employees who are interested in safety believe films are *ineffective*. Those employees who are not interested in safety believe films *are* effective. On these relationships there is less than 1 chance in 1,000 that they do not exist. They are highly significant.

While less significant, there is a relationship (1 chance in 100 to be untrue) between booklet effectiveness and training effectiveness and self-rating. Employees who rate themselves as interested in their own safety believe both these to be effective tools.

We found no significance between self-rating and 5-minute safety talks, inspections, competitions, and discipline.

Summary and Conclusion of the Study

Here are the major conclusions reached:

1. Posters are thought to be effective by employees who rate themselves interested in safety.
2. Films are thought to be effective by employees who rate themselves with little interest in safety, and to a lesser degree by employees who think their leadman is interested in their safety.
3. Booklets are thought to be effective by employees who rate themselves as safety conscious and by those who are in interested informal groups.
4. Five-minute safety talks are thought to be effective when the informal group norm is for safety, when the leadman is interested, and when the foreman is interested.
5. On-the-job training is thought to be effective when the employees themselves are interested, or when the foreman and management are interested.
6. Inspections are thought to be effective when the foreman is interested and when the informal group is interested in safety.
7. Competitions are thought to be effective when the informal group is interested in safety and when the leadman is disinterested.
8. Discipline is thought to be effective when the informal group is interested and to a lesser degree when the leadman is interested.
9. Other correlations not listed above were found not to be significantly greater than chance using a level of 1 chance in 50 as the cutoff point for significance.

In structuring future safety programs, these results indicate that we should:

1. Hold off on the use of media initially until we have some information about employee perceptions.
2. Use training to build a belief in management interest (improve credibility).
3. Use inspections, training, and 5-minute safety talks to build supervisory credibility in safety.
4. Use talks, competitions, and discipline where we are relatively sure the informal group norms include safety. Do not use them where there is evidence that the group's goals do not include safety.
5. Use posters to appeal to the safe individual.
6. Use films to appeal to the unsafe individual.

UPWARD COMMUNICATION

Most of the above discussion pertains to downward communication—the communication process from management to the worker. What about the opposite, upward communication, the communication process from the worker to management.

Several years ago Opinion Research Corporation[163] conducted a study among employees and their immediate bosses to explore the climate for upward communications in industry. Included in the research were 265 engineers and scientists, 499 white-collar employees, 1,049 hourly employees, and 334 of their immediate supervisors in eight companies—a metropolitan bank, a manufacturer of household products, a producer of paper products, the branch store of a major retailer, two chemical companies and two public utilities. At three of the companies, hourly employees were represented by unions.

The desire to be heard is, as one would expect, widespread among employees. In quantified terms, about 90 percent of the employees say that it is "very" or "fairly" important to them to be able to discuss their ideas about work problems with higher management (people above their immediate boss). Job occupation or sex makes little difference: the desire is as strong in the production worker as in the engineer or clerical employee, and it is as strong among women as among men.

There are important differences, however, in the character of this desire. Engineers, for example, have high interest in helping to make important work decisions. Hourly employees see communications with management mainly as a way of facilitating the work; they want to tell management about obstacles to performance which, in their view, management should do something to remove. Differences in such matters as educational achievement, social background, career aspirations, and concepts of the work role probably account for those differences in expectations regarding the function of communications.

Many employees find it difficult to get the ear of management. Although the results differ from company to company, over half the employees studied complained of lack of opportunity to make contact with those above their immediate supervisor.

In these studies a number of obstacles to upward communication were found:

Obstacle Number One: Many employees fear that expressing their true feelings about the company to their boss could be dangerous. The boss is often seen as untrustworthy, a person to whom it is dangerous to talk with full candor.

> "An employee who told his immediate supervisor everything he felt about the company would probably get into a lot of trouble."

	Per Cent Agreeing With This Statement
All employees	54%
Engineers and scientists	50%
White-collar employees	50%
Hourly employees	59%

Although important differences exist from company to company, as they do on all the issues in the study, what's interesting is that many scientists and engineers, on whom management depends for innovation and who are usually thought to be "close" to management, don't trust their bosses.

Obstacle Number Two: The fairly widespread belief that disagreeing with the boss will block promotion. In the eyes of employees, management retaliation can probably take many forms. A major one is fear that speaking frankly will kill one's chances of advancement. Many employees in some companies do in fact believe that the way to get ahead is to be a "yes-man."

> "The best way to get ahead in this company is not to disagree very much with your boss."

Nonsupervisory employees . . .

	True	Hard To Decide	Untrue
Household products company	54%	14%	27%
Retail store	43%	14%	36%
Chemical company	42%	17%	38%
Utiltiy company	32%	14%	54%

("No opinion" omitted)

Obstacle Number Three: The widespread conviction that management is not interested in employee problems.

For many employees, management is remote, shut off, preoccupied with its own concerns and therefore out of touch with employee values and worries.

"Many problems important to an employee are not considered important by management."

	Per Cent Agreeing With This Statement
All employees	72%
Engineers and scientists	61%
White-collar employees	70%
Hourly employees	80%

Obstacle Number Four: The feeling that employees are not rewarded for good ideas. Some companies have a lot of work to do convincing employees that they will be rewarded for their ideas.

Note, however, the contrast below between the retail firm and the chemical company, a firm with a long history of trying to improve communications, on the issue of whether good ideas are rewarded.

"It's easy for people with good ideas to get ahead in the company."

Nonsupervisory employees. . .

	True	Hard To Decide	Untrue
Retail firm	26%	30%	44%
Chemical company	53%	26%	21%

Obstacle Number Five: Lack of supervisory accessibility and responsiveness.

About a third of the employees rate their boss as "good" on being easy to see with a problem and only about a quarter rate him "good" on such matters as ability to handle complaints and encouraging suggestions. On the other hand, majorities of employees are apt to rate their boss "good" on matters having to do with job knowledge and operating problems.

Supervisors in part disagree. The vast majority believe they are readily accessible to employees. At the same time, however, many do admit to lack of skill in performing the upward communications task and ask for more training in specifics.

Percent of supervisors saying they need more training in. . .

Communications — particularly in how to listen	72%
How to handle employee problems and complaints	67%
How to lead group discussions	57%

Obstacle Number Six: The conviction that higher management doesn't take prompt action on problems.

Prompt management responsiveness to employee problems and concerns is unquestionably a key element in successful upward communications because it completes or closes the cycle of interchange.

Many employees do not have a high opinion of management performance in response to employee demands for action on their problems.

"How would you rate higher management in the company on taking action on employee ideas, complaints, and problems?"

Say management "usually" or "always takes prompt action"

	Per Cent Agreeing With This Statement
All employees	18%
Hourly	10%
Engineers & Scientists	15%
White-collar	27%

SOME GUIDES TO EFFECTIVE COMMUNICATIONS

To summarize the thoughts expressed here on communication, here are some guidelines from *Organizational Behavior,*[73] by Joe Kelly (Sir George Williams University of Montreal):

1. Before communicating, a person should analyze his problem in as great detail as possible in order to determine exactly what he wishes to communicate.

2. The purpose of the communication should be defined. Specify your intentions, then define your aim.

3. The physical and human environment should be considered, i.e., timing, location, social setting, and previous experience.

4. Communication is not exclusively verbal. Consult with others if this is thought to be necessary.

5. Objectivity is not always necessarily a criterion of good communication in every circumstance; sometimes two-sided messages are useful.

6. Try to influence the person with whom you are communicating: try to see things from his or her point of view. Remember, different people have different perceptual slants.

7. Assess the effectiveness of the communication if this is possible; this is usually done by encouraging feedback.

8. Choose carefully the type of communication process best-suited to your purpose.

9. Many executives seem to believe that it is possible to manage by exclusively vertical forms of communication, but research reveals that a great deal of communication is lateral.

From research in upward communication, several conditions emerge that appear to encourage increased upward communication:

1. *Frankness within management*—Establish genuine two-way communications between all levels of management. When critical discussion is choked off at higher levels of the company, it ceases to flow at lower levels.

2. *Supervisor accessibility*—Develop an awareness among managers that the keys to better listening are accessibility and responsiveness. Employees don't want to be heard all the time. But when they do have a problem, they need the assurance that their boss will learn and act.

3. *Welcome for the new and different*—Tolerate all kinds of ideas, those that are foreign, silly or hostile as well as those that management considers constructive, i.e., those that it is willing to accept. Looking with disfavor on employees for thinking differently leads to closed minds.

4. *Visible benefits*—Visibly reward those who have creative new ideas. This is the strongest encouragement management can give.

5. *Acceptance of criticism*—Regard criticism as healthy and normal and lack of criticism as dangerous and undesirable, an indication that employees have given up trying to get through to management.

6. *Sensitivity to employees*—Be willing to wrestle with the problem of interpreting what an employee is really trying to say. A gripe about working conditions may mask a belief that the boss doesn't appreciate the individual's job performance.

Checklist for Communications

For the purpose of analysis, it is useful to look first at:

The Message—What is this supposed to be? What language is it to be put in? What information does it contain?

Communicators—Who are they? What are their roles? Where do they stand on the status scale? Is there a status gradient? Does either or both have a vested interest in communicating the message? Are their personalities likely to interfere with the communication process?

Media—What form should be used? What are the mechanics of the information handling? What is the density of the communication system? What is the time pattern?

Environment—What are the circumstances of the communication? Who must know and who must not? Is there a protocol which is appropriate on this occasion? What about situational factors, especially the social setting?

Effects—How effective is the system? How capable is it of adaptation? What are its aims—are they being achieved?

The study of communications is already a major focus of research, and there is likely to be further extensive research. In this chapter we have attempted to give a brief overview and to suggest application of present knowledge to safety.

Part VI

Building Safe Behavior

Chapter Eighteen

Identifying Safe Behavior

Accidents are costly. From the organization's point of view, they result in lost time, in treatment and insurance costs, in morale problems, and in poor public relations. Applicants who are likely to have accidents should be detected and either rejected or used in jobs where the chances of accidents are small.

Robert Guion[47]

Lest the reader be misled by the opening quotation, I will continue the rest of Guion's statement[47] about predicting who will have accidents:

> Accidents are costly. From the organization's point of view, they result in lost time, in treatment and insurance costs, in morale problems, and in poor public relations. Applicants who are likely to have accidents should be detected and either rejected or used in jobs where the chances of accidents are small.
>
> Unfortunately, such a glib statement implies a big assumption. It assumes that accidents are largely a function of individual accident proneness. If accident proneness refers to any characteristic of an individual that is permanent enough to allow prediction to be based upon it, Mintz and Blum (1949)[112] have shown clearly that this assumption simply will not hold. Accident proneness, as a sufficiently stable trait to allow for measurement in the employment office, accounts for very little of the total variance in accident statistics. This means, at best, that one should not expect to obtain very high correlations between tests and measures of accident rates.
>
> Within these limitations, however, it is still sound policy to reduce accidents by selection as much as possible.

Our purpose here is to look at what possibilities there are to reduce accidents by selection.

CAN WE PREDICT WHO WILL HAVE ACCIDENTS?

Reducing accidents by selection and placement assumes we can predict who will have accidents. We stated earlier that severe injuries are fairly predictable in certain situations and we listed some of those situations. However, predicting who will have accidents is quite a different thing, for this assumes that those who have accidents are different in some identifiable way from those who do not have accidents. We can examine what such identifiable ways might be by examining briefly the various accident causation theories, and then looking at whatever research there is on the relationship between accidents and personal measurable factors.

PRONENESS THEORIES

The psychological causation theories at which we might logically look from a predictability viewpoint center around the various proneness theories.

We will start out by looking at the terms we are using. First, there is a difference between an accident repeater and accident proneness. An *accident repeater* is an individual who has more than one accident of the same type. *Accident proneness* describes a person who has significantly more accidents than others. Research indicates that there is no such thing as one type of accident prone person. Rather, for each individual some behavior is safe and some unsafe, depending on many things, including the environmental hazards to which that person is exposed. It seems that in present-day American thinking accident proneness does exist in some people for at least short periods of time, exists in others for relatively long periods of time, and in both instances is predictable if properly measured at the right time. If an individual has one or more accidents, it does not mean accident proneness. Accident proneness refers to relatively consistent characteristics which make the person more susceptible to accidents. There are such people, but their number is small, and their contribution to the total accident problem is slight. Almost all people have accidents. When a person has difficulty adjusting to the environment, it is referred to as *temporarily accident prone* or *accident susceptible.*

The more recent studies have tended to deemphasize the concept of accident proneness as a major cause of accidents. A survey of 27,000 industrial and 8,000 nonindustrial accidents indicated that the accident repeater contributed only 0.5 percent of them, whereas 75 percent were due to relatively infrequent experiences of a large number of persons. This analysis by Dr. Morris Schulzinger in his book, *Accident Syndrome,*[145] came to these conclusions:

1. The tendency to have accidents is a phenomenon that passes with age, decreasing steadily after reaching a peak at the age of twenty-one. The accident rate between the ages of twenty and twenty-four, in both our industrial and nonindustrial series, is 2½ times higher than between the ages of forty and forty-four, four times higher than between the ages of fifty and fifty-four, and nine times higher than between the ages of sixty and sixty-five.
2. Most accidents are due to the increased liability to accidents of youth. Seventy percent of the nonindustrial accidents occurred before the age of thirty-five and nearly 50 percent before the age of twenty-four.
3. Men are significantly more liable to accidents than women; the ratio of male-to-female accidents being 2 to 1 in the nonindustrial series and, apparently, even higher in the industrial series.
4. Most accidents (74 percent) are due to relatively infrequent solitary experiences of large numbers of individuals (86 percent). These figures were identical for the industrial as well as for the nonindustrial series, and remained constant for nearly every year of a twenty-year period.
5. Those who suffer injuries year after year, over a period of three years (3 to 5 percent), account for a relatively small percentage of all the accidents (0.5 percent).
6. Irresponsible and maladjusted individuals are significantly more liable to have accidents than responsible and normally adjusted individuals.

Simple chance is, of course, also a factor in the unequal distribution of accidents. A "normal" distribution of accidents (all random happenings) would be to have just about what we have, that is, a few people who have many more accidents than most other people.

Schulzinger's studies indicated that, when the period of observation is sufficiently long, most accidents occur in individuals with a low degree of proneness, and that the relatively small percentage of the population that contributes a disproportionate number of accidents is essentially a shifting group, with new persons constantly falling in and out of the group. His experience suggests that in the course of a life span almost any normal individual under emotional strain or conflict may become temporarily accident prone and suffer a series of accidents in fairly rapid succession. Most persons, however, find solutions to their problems, develop defenses against their emotional conflicts, and drop out of the highly accident-prone group after a few hours, days, weeks, or months. Some persons remain highly accident-prone throughout life, with or without lapses of years of freedom from the accident habit. The latter are the truly accident-prone individuals. They contribute, however, only a relatively small percentage of all the accidents.

Thus, the concept of accident proneness, which on first thought would seem to lead us to an improved safety record through selection, on close examination seems to pose some real difficulties. We could, perhaps, screen out the tiny percent of irresponsible and maladjusted individuals truly accident prone, but the cost would undoubtedly not be worth it. We could hire only older workers or only females, but these approaches, obviously, are not feasible either. On having successfully identified this year's crop of accident-prone workers, we would find that they are not next year's crop. Therefore, the proneness causation theories do not lead us very far.

CAUSATION THEORIES

Related to the proneness theories are two other theories which state that a worker is more prone to accidents in certain work situations.

Goals-Freedom-Alertness Theory

This theory simply states that great freedom to set reasonably attainable goals is accompanied typically by high-quality work performance. This theory regards an accident merely as low-quality work behavior that happens to a person instead of some other error in performance. Raising the level of quality involves raising the level of alertness. Considerable evidence supports this theory. Factory departments with more movement of personnel among departments, that is, intracompany transfer mobility, have fewer accidents. The same is true of departments with greater promotion probability for the typical employee. Departments with the best suggestion records tend to have fewer accidents. There are fewer accidents in individual-type jobs than there are in crew-type jobs. Accidents are more frequent in jobs of lower-rated prestige.

This individual goals-freedom-alertness theory suggests the need to provide emotional reward opportunities for alertness such as special economic incentives, prestige-building honors, extra privileges, machine and work-space-decoration contest participation, and representation on special committees and councils. These rewards held as attainable goals by workers in relatively "dead-end" jobs should operate to raise the average level of alertness not just to hazards but to everything.

Adjustment Stress Theory

This theory states that unusual, negative, distracting stress on a person increases liability to accident or to other low-quality behavior. This theory refers to distractive negative stresses imposed upon the individual either by internal environment (such as disease, alcohol, or toxic items) or by external environment (such as temperature, poor lighting, high noise level, or physical strain). Its stresses are different from those experienced by the accident prone; their stresses result from a constitutional inadequacy. Ordinary adjustment stress is not the result of constitutional inadequacy but of temporary conditions.

Temporary stress factors which have been found to correlate significantly with accident rates include employee age, workplace temperature, illumination, mean rated comfort of the shop, degree of operational congestion, obvious danger factor threateningly present, manual effort involved in job, weight of parts handled, frequency of parts handled, alcohol consumption, and influence of disease.

These two theories related to proneness lead us more to prevention of accidents through participation, enrichment, and environmental control than they do to prediction and selection. It seems that what we know about the concept of accident proneness, or even accident susceptibility, does not lead us very far toward an ability to predict who will have accidents. Let us then examine the relationship between certain psychological traits and the occurrence of accidents of various kinds or in various situations. This is a more specific approach to the study of the variability of the individual's contribution to accidents.

HUMAN TRAITS AND ACCIDENTS

Considerable research has been done on personal factors of all kinds that are related to accidents. Depending on the situation and the needs of the job, selection tests could be devised for many of these. The following is just a listing of the areas of research on this subject.

1. The senses—their functions and their limits.
    A. *Vision—blindness, adaptation, color blindness, depth perception, vision, illusions.* The eye, the most important human sensor, has been neglected in accident research. There is some indication that there is a relation between poor visual function and accidents, but better studies are needed.

    Only two industrial studies specifically investigate the relationship between accidents and visual performance of workers. One studied 1,384 miners with 1,265 accidents over a period of two-and-a-half years. No strong relation between poor acuity and accident repeaters was found. This might be expected when all types of accidents are investigated.

    Another study attempted to select accidents "that could possibly have been caused by low visual performance," and found the accident free did in fact have a "superior visual performance."
    B. *Audition—intensity, frequency, deafness.* Much has been written on hearing loss due to noise, but few studies have been made to see if hearing defects had any importance in industrial accidents. One study was "slightly in favor of the deaf," and another reported that sixty deaf persons employed in a variety of jobs had no lost-time accidents over a period of three years.
    C. *Motion sensors—joint and head.* While there is no research here, we do know that in most industrial situations unusual problems with balance organs will seldom arise since the person is usually standing on the ground; but if any rapid or unusual acceleration of vehicles, conveyers or self-propelled head movements occur, it is preferable not to require close or precise work for a few seconds afterwards.

    These stability sensors are intimately related to the whole process of self-propelled body movement. Thus, disorders and misinterpretations can lead to falls.

D. *Skin sensors.* Under the surface of the skin there are special sensors for touch, temperature, pain, and pressure. There has been no research to correlate these sensors with accidents.

E. *Taste and smell.* These are considered to be unimportant with respect to accidents, except to those working with toxic chemicals. There have been no studies in this area. Generally, very little research has been attempted to search for sensory errors in accident causation.

2. *Responses, including alertness, vigilance, memory, decision making, and risk taking.* While there is a vast amount of research in these areas, we cannot yet bring this research all together into simple sets of rules, nor can we yet utilize what is known in predicting which applicant is better for which job. Some of the things that are known in these areas are listed below.

A. *Alertness.* It is hypothesized that accidents are more liable to occur to a person when experiencing a very high fear arousal level because of inability to coordinate all the necessary information and correctly act on it. Also, if the arousal level is low, accidents could occur because the person is not observing the environmental clues and thus cannot act on them. Optimum alertness is thus vital for efficiency of output and safety.

B. *Risk taking.* Tom Rockwell[153] attempted to study risk-taking behavior in a controlled laboratory setting. He found that a person's judgment of risk is not directly related to the hazard as measured by performance capabilities; skilled workers take less risk than unskilled; younger persons take higher risks; and females took considerably less risk than males. These laboratory tests could not find any correlation between risk taking and various psychological or biographical measures, though the high risk taker tends to be a person with a high anxiety level, high sociability, and low emotional stability.

C. *Intelligence.* Several researchers have found that accident frequency is unrelated to intelligence, when intelligence is adequate for the situation. Obviously there is little to direct us on the research.

How about the relationships that have been found between various types of common tests and accident frequencies?

1. *Sensory Motor Tests.* Measures of muscular coordination have been shown to be related to certain types of accident situations. It seems reasonable to suppose that clumsiness, inadequate skill, slowness of response, and defective sense organs contribute to accidents. Individuals low in motor ability are unable to get out of accident situations without sustaining an injury; also, in many cases, they do not have the skill or sensory acuity to avoid certain kinds of accident situations. Here are some research findings on the subject.

   Reaction time. Reaction time seems to be an important consideration in some types of work and less important in others. Consequently, it is necessary to analyze the requirements of certain jobs in order to determine the importance of reaction time to safe and efficient performance in those jobs.

   According to John Larsen,[81] tests can show whether a worker falls above or below the minimum safety requirements. But reaction time seems to involve other factors, such as manual or manipulative dexterity, rhythm (or timing), and visual acuity. Reacting quickly by itself will not necessarily prevent a worker from injury, but also there must be reaction in the right way and at the proper moment. To do these things, of course, the worker must foresee an imminently dangerous situation. Where it is purely a matter of vision, reaction time and dexterity seem to be the most important qualities. But when anticipation rather than visual acuity alone is involved, fast reaction time appears to be less significant. The importance of reaction time per se in accident causation depends largely on the nature of the work involved. Moreover, anticipation and, therefore, attention seem to reduce the need for a worker to have quick reaction time.

Visual Acuity.   In one study, 828 workers in twelve job categories were studied and divided into two groups; those whose visual skills met the requirements of the visual pattern for the job and those whose visual skills did not. From the accident data, workers in each job category were split into those who had two or less accidents in six to eight months, and those who had three or more accidents in the same period.

After comparing groups, the conclusion was that vision is one of the factors related to accidents but many other factors are involved. The relationship shown in the study was not sufficiently high to account for all industrial accidents, but that among workers whose visual skills are adequate for the task they are performing, there is a higher percentage of workers whose frequency of accidents is low. Here again, visual acuity is perhaps one factor, but by itself it is not enough to predict accidents.

Another study used forty operators who represented a cross section of a department of seventy-nine female workers. After administering three motor manipulation tests and two visual perception tests, twenty-three "accident-prone" workers were contrasted with seventeen accident-free workers. No significant relationship existed between an accident index and the several sets of test scores. However, the accident index figures tended to be high when the scores on the three motor tests were higher than the scores on the two perception tests, and vice versa. The conclusion was that individuals whose levels of muscular reaction are above their levels of perception are prone to more frequent and more severe accidents than those individuals whose muscular reactions are below their perceptual levels. This conclusion was qualified by the suggestion that "accident proneness" probably arises from several factors and not from a single factor.

2. *Psychological Tests.*   Here are some research findings on the value of psychological tests as predictive devices of accidents.

Intelligence.   As noted above, research shows that intelligence bears little relationship to accidents, except where accidents are due to a failure in judgment. In these cases, it is possible that variations in intelligence may play a significant part in accident susceptibility. Many studies bear out these conclusions. Except at the extremes, intelligence is not associated with accidents to any significant degree, therefore intelligence tests will not help us predict accidents.

Personality.   Some instruments which measure the degree of emotional reactions have been found effective in showing a relationship between certain aspects of emotionality and accident frequency. Even closer relationships are found when an individual's responses are measured under disturbing and distracting conditions.

Moods also seem to be important. In one study it was found that half of 400 minor accidents occurred while the employees were emotionally "low," although this emotional condition existed only 20 percent of the time. Production was 8 percent higher during the happy moods, showing that emotional conditions favorable to accident prevention are also favorable to production. Also, emotional adjustment as such is a factor in accidents. Although personality questionnaires have been found inadequate for detecting accident susceptibility, some believe that accidents and poor adjustment are related.

In one study a high-accident group was compared with a low-accident group matched for age, education, intelligence, sex, socioeconomic background, and exposure to hazard. The responses of the high-accident group to a sentence completion test indicated that they were significantly different from the low-accident group. The low group was higher on responses indicating optimism, trust, and concern for others.

Poor adjustment is related to accident causation. However, even knowing this we cannot yet use this knowledge well to predict or select since, in most cases, it is impractical to spend the kind of time and money needed to obtain a psychoanalytic analysis for each applicant.

Mechanical Ability.   Several studies made of psychological tests and safety have found no correlation, except between the mechanical ability interest tests and the accident record. Generally, the results indicate that the individuals who have had fewer accidents possess a higher degree of mechanical interest than those having a record of frequent accidents. This does not necessarily suggest that these tests should be used as predictors. There is just not enough research on this.

OTHER SELECTION METHODS

Exhibit 18.1 gives a broad outline of the process available to management in selecting employees. Ideally the process is based on job standards which state that for a particular job a particular type of person is required. This is utopian, and the "exactly right" person for the job is seldom (if ever) found, but such a goal is still worth trying to attain. As the exhibit shows, we have two basic sources of information about an applicant: biographical data and test results. As we have discussed above, the various psychological and sensorimotor tests available are not too useful for predicting accident-producing behavior. Depending on the job for which we are hiring, the job knowledge tools and skill tests might be of value to us. For instance, if we are hiring a driver, we want to know how skillful the applicant is at driving. Similarly, a physical examination is most important in some situations. The importance of the examination and its scope should be based on the job for which the applicant is being considered; the determination of what is included should be made by the physician.

On the biographical data side of Exhibit 18.1, there is some research on the effectiveness of these standard selection tools as predictors of future job behavior, particularly in interviewing and reference checking.

Interviewing

The most used selection device in industry today as in all past years is the interview. A 1957 survey by Spriegel and James[174] revealed that 99 percent of 852 firms surveyed used the interview. Few people are, in fact, ever hired without going through an interview. For years the interview has been challenged only by the application blank in terms of frequency of use by employers. It is almost certainly the most important selection decision tool. Surprisingly, however, there is a preponderance of evidence indicating a general lack of validity for the interview as a selection device. One researcher (Schwab[146]), while admitting the interview is extremely popular, questions its value in total. He asserts that twenty-year-old studies, as well as more recent works, conclude that the interview as normally conducted in a selction situation is of little value.

Research has left us with some facts about the interview process as a selective device:

THE PROCESS OF SELECTION

Biographical data	*Tests*
Application	Physical examination
Interviews	Qualifications
References	Defects
Other sources: credit, bureaus, schools,	Skill tests
schools, driving record, agencies	Job knowledge tests
	Psychological tests

Exhibit 18.1 The selection process.[127]

1. Interviewers develop a stereotype of a good candidate and seek to match interviewers with stereotypes.
2. Biases are established by interviewers early in the interview and tend to be followed by favorable and unfavorable decisions.
3. Unfavorable information is most influential on interviewers.
4. Interviewers seek data to support or deny hypotheses and, when satisfied, turn their attention elsewhere.

In other words, we say we interview to get factual data on the person, but in actuality we form a very early opinion, make up our minds, and then spend the remainder of the interview time ferreting our those facts which will substantiate that early opinion. Thus, even our trusted interview may not give us the information we need to hire safe employees.

Reference Checks

The previous accident record of a person can be checked through the previous employer and, in the case of the driving record, through motor vehicle reports. Several researchers question the real value of the information obtained and most agree that if previous employers are checked, the check should be made by telephone not by letter. The validity of information here can be vastly improved when the two people know each other.

Thus, even our old standbys of interviewing and checking references do not give us the kind of information we need to predict who will be safe and who will not. We can get more information from the applicant than from references, but also, we will not believe him or her much anyway because we have already made up our minds at the start of the interview. At least this is what research tells us.

WHAT'S ALLOWABLE IN SELECTION

Another reason that selection is less a tool of safety today than ever before is that much of what used to be possible is no longer allowable under Equal Employment Opportunity Commission (EEOC) guidelines. Exhibit 18.2, compiled from the EEOC Compliance Manual, outlines what currently is lawful and unlawful.

CAN WE PREDICT SAFE BEHAVIOR?

Where does all this lead us? Not far, apparently. Research tells us that

PRE-EMPLOYMENT INQUIRY GUIDE		
Subject	*Lawful Inquiries*	*Unlawful Inquiries*
1. Name	"Have you worked for this company under a different name?" Maiden name.	Inquiries about the name which would indicate applicant's lineage, ancestry, national origin or descent. Inquiry into previous name of applicant where it has been changed by court order or otherwise.
2. Address or Duration of Residence	Applicant's address. Inquiry into place and length of current and previous addresses. "How long a resident of this state or city?"	Specific inquiry into foreign addresses which would indicate national origin.
3. Birthplace	"Can you after employment submit a birth certificate or other proof of U.S. citizenship?"	Birthplace of applicant. Birthplace of applicant's parents, spouse or other relatives. Requirement that applicant submit a birth certificate, naturalization or baptismal record. Any other inquiry into national origin.
4. Age	If a minor, require proof of age in the form of a work permit or a certificate of age. Require proof of age by birth certificate after being hired. Inquiry as to whether or not the applicant meets the minimum age requirement as set by law and requirement that upon hire proof of age must be submitted in the form of a birth certificate or other forms of proof of age. If age is a legal requirement: "If hired, can you furnish proof of age?" or statement that hire is subject to verification of age.	Requirement that applicant state age or date of birth. Reqirement that applicant produce proof of age in the form of a birth certificate or baptismal record.
5. Religion	An applicant may be advised concerning normal hours and days of work required by the job to avoid possible conflict with religious or other personal conviction.	Applicant's religious denomination or affiliation, church, parish, pastor, or religious holidays observed. "Do you attend religious services or a house of worship?" Applicant may not be told: "This is a Catholic/Protestant/Jewish/atheist/etc. organization. Request pastor's recommendation or reference or any other religious references. Applicants may not be told that employees are required to work

Exhibit 18.2

PRE-EMPLOYMENT INQUIRY GUIDE *(Continued)*		
Subject	*Lawful Inquiries*	*Unlawful Inquiries*
5. Religion (continued)		on religious holidays which are observed as days of complete prayer by members of their specific faith.
		Any inquiry to indicate or identify religious denomination or customs.
6. Race or Color	General distinguishing physical characteristics such as scars, etc.	Applicant's race.
		Color of applicant's skin, eyes, hair, etc., or other questions directly or indirectly indicating race or color.
		Applicant's height where it is not relative to job.
7. Photograph	May be required after hiring for identification.	Request photograph before hiring.
		Requirement that applicant affix a photograph to his application.
		Request that applicant, at his option, submit photograph.
		Requirement of photograph after interview but before hiring.
8. Citizenship	"Are you a citizen of the United States?"	"Of what country are you a citizen?"
	"If you are not a U.S. citizen, have you the legal right to remain permanently in the U.S.?"	Whether applicant or his parents or spouse are naturalized or native-born U.S. citizens.
	Statement that if hired, applicant may be required to submit proof of citizenship.	Date when applicant or parents or spouse acquired U.S. citizenship.
		Requirement that applicant produce his naturalization papers or first papers.
		Whether applicant's parents or spouse are citizens of the U.S.
9. Ancestry or National Origin	Languages applicant reads, speaks, or writes fluently.	Applicant's nationality, lineage, ancestry, national origin, descent or parentage.
		Date of arrival in the U.S. or port of entry; how long a resident.
		Nationality of applicant's parents or spouse; maiden name of applicant's wife or mother.
		Language commonly used by applicant.
		"What is your mother tongue?"
		How applicant acquired ability to read, write, or speak a foreign language.

Exhibit 18.2 Continued.

PRE-EMPLOYMENT INQUIRY GUIDE *(Continued)*		
Subject	*Lawful Inquiries*	*Unlawful Inquiries*
10. Education	Applicant's academic, vocational, or professional education; school attended. Inquiry into language skills such as reading, speaking, and writing foreign languages.	Any inquiry asking specifically the nationality, racial or religious affiliation of a school. Inquiry as to what is mother tongue or how foreign language ability was acquired.
11. Experience	Applicant's work experience. Other countries visited.	
12. Conviction, Arrest , and Court Record		Ask or check into a person's arrest, court, or conviction record if not substantially related to functions and responsibilities of the prospective employment.
13. Relatives	Names of applicant's husband or wife, and dependent children. Names of applicant's relatives already employed by this company. "Do you live with your parents?" Names and addresses of parents or guardian of minor applicant.	Name of address of any relative of adult applicant other than applicant's spouse or children.
14. Organization	Inquiry into the organization of which an applicant is a member providing the name or character of the organization does not reveal the race, religion, color, or ancestry of the membership. "What offices are held, if any?"	"List all organizations, clubs, societies, and lodges to which you belong." The names of organizations to which the applicant belongs if such information would indicate through character or name the race, religion, color, or ancestry of the membership.
15. Notice in Case of Emergency	Name and address of person to be notified in case of accident or emergency.	Name and address of relatives to be notified in case of accident or emergency.
16. References	"By whom were you referred for a position here?" Names of persons willing to provide professional and/or character references for applicant. "Who suggested that you apply for a position here?"	Require the submission of a religious reference. Request reference from applicant's pastor.
17. Sex	Only if nature of the work or working conditions provide valid reasons.	Sex of the applicant. Any other inquiry which would indicate sex.

Exhibit 18.2 Continued.

PRE-EMPLOYMENT INQUIRY GUIDE *(Continued)*		
Subject	*Lawful Inquiries*	*Unlawful Inquiries*
17. Sex (continued)	Where a bona fide occupational qualification is reasonably necessary to the normal operation, motif, culture, or atmosphere of that particular business or enterprise.	"Are you expecting?" or "Are you pregnant?"
18. Work Schedule	Inquiry into willigness to work required work schedule.	Any inquiry into willingness to work any particular religious holiday.
19. Miscellaneous	Notice to applicants that any mistatements or omissions of material facts in the application may be cause for dismissal.	

Any inquiry is forbidden which, although not specifically listed among the above, is designed to elicit information as to race, color, ancestry, sex, religion, or arrest and court record unless based upon a bona fide occupational qualification.

NOTE: The above Figure may not reflect the most recently legal situation, either nationally or in your state. Consult your lawyer for current advice.

Exhibit 18.2 Continued.

1. Accident proneness is real, but only in a tiny percentage of people, and is probably a system of maladjustment. We probably cannot afford to find those people.
2. Accident susceptibles are an ever-shifting group. Finding them will not help much because next year the susceptibles will be a different group.
3. Susceptibility comes from situations more than from permanent traits.
4. For some jobs requiring certain psychomotor traits, we can measure to see the quality of those traits. Similarly, if we could identify job psychological needs, we could measure for those. (Unlikely.)
5. Job knowledge and skill tests can be administered effectively, provided the job criteria can be established in these areas.
6. Similarly, physical exams can be helpful, provided physical job criteria are available.
7. Interviews are generally invalid, not because the applicant is lying, but rather because our biases and stereotypes invalidate them.
8. Checks with previous employers (if we have established a good working relationship) will tell us an applicant's past history (assuming this is indicative of the future, which it is not and probably cannot be proven).

In short, when we consider the present state of the art in the light of economics and practicality, we really must conclude that we cannot use selection as an effective loss control method. When it comes right down to what we know today, we end up forced to suggest that we continue to do what we have been doing; that is, to use only the most obvious concepts of selecting and placing based on physical traits and other traditional guidelines and trust our safety training and programs do the job.

Chapter Nineteen

Behavior Change Through Training

There once was a teacher
Whose principal feature
Was hidden in quite an odd way.
   Students by millions
   Or possibly zillions
Surrounded him all of the day.

When finally seen
By his scholarly dean
And asked how he managed the deed,
   He lifted three fingers
   And said, "All you swingers
   Need only to follow my lead.

To rise from a zero
To Big Campus Hero,
To answer these questions you'll strive:
   Where am I going,
   How shall I get there, and
   How will I know I've arrived?

                      Robert Mager[104]

In several earlier chapters, we have discussed training with an emphasis on supervisory training as one of the determinants of the supervisor's performance. Our emphasis there was on analyzing the performance problem to determine if training were the answer, or if something other than training and the setting of behavioral objective was required. We also briefly discussed method. In this chapter we will discuss building safe employee behavior. We must again look at training, for this is (rightly or wrongly) by far the most used behavior change method in industrial safety.

In Chapter 18, we examined the likelihood of helping to solve our safety problem by selecting the "safe" employee and found some real difficulties with this idea. While economically practical, it just does not seem feasible to use selection as our primary control strategy. Perhaps someday, but not now. Where does this leave us? It leaves us with hiring all types of people—some susceptible, some not, perhaps some actually "prone"; some knowledgable about safety from previous jobs, some not; some eager to work safely, some indifferent; etc., etc. It leaves us with a total work force of people who may or may not have to be trained on safety. Where do we go from here? To answer these questions we will paraphrase Robert Mager's[104] ditty:

> To answer these questions we'll strive:
>   Where are we going?
>   How shall we get there?, and
>   How will we know we've arrived?

WHERE ARE WE GOING?

Putting the training process in its simplest form, we can say that it has three steps:

1. Finding out where we are.
2. Finding out where we want to be.
3. Providing the difference.

Unfortunately, in safety training we usually spend almost all our time in determining content and method. Theorists tell us to spend the bulk of our effort and concentration on 1 and 2 above. If we do a good job in analyzing these, 3 almost naturally falls into place. Theorists tell us that content is no more than the difference between 1 and 2 and that method of presentation is almost immaterial to the learning process.

There is a considerable amount of literature on defining training needs. While oversimplifying, it might clarify to say that the definition of training needs encompasses 1 and 2 above; 2 being a function of policy and 1 being the major planning job of the trainer or safety person. Management tells us, through its policy statements, where we want to be and what each person's function is once there. We must then find out how able our people are to fulfill these duties and responsibilities: "Can they do it if their lives depend on it?" Defining needs is, of course, the key point in the process. It provides the objective and sets the criteria for measurement. We would expect that any function this vital to the success of the training would be in widespread use among industry. In one study it was found that a careful and systematic investigation was conducted in only one out of ten companies. Very little training effort is based on any systematic appraisal of the training and development needs. Instead, we choose or prescribe training according to past rituals, current politics, and faddish misconceptions, according to Dr. Donald Newport,[118] who describes four of industry's most popular approaches. First, there is the "Bandwagon" approach where a company tries to keep up with the incorporated Joneses. The "Smorgasboard" approach allows the favorite sons to pick and choose which courses offer the newest buzz-words. Training takes on hindsight in the "Crisis" approach. It is not needed until the crisis arrives and remedial rather than preventative action is required. Finally, the "Excursion" principle is used to measure course value. A program held 1,000 miles from home is twice as good as one only 500 miles distant.

One of the better texts on industrial training is *Training in Business and Industry* by McGehee and Thayer.[96] To these authors the assessment of training needs involves a three-part analysis. An analysis of the organization should uncover the resources and objectives of the company which relate to training. An analysis of the operations should define corporate jobs and tasks. The manpower analysis explores the human dimensions of attitudes, skills, and knowledge as they relate to the company and the employee's job. To quote from their book.[96]

> The determination of training needs requires something more than armchair cerebration. It requires careful digging for facts, frequently with inadequate tools.

It is a grimy business, frustrating, and often carried on under increasing pressure to get something, just anything, going. Consequently, many training efforts are begun without any reason, continued with no purpose, and end with no results—at least in terms of the only legitimate reason for training in industry.

There are probably as many rationales for approaching the problem of determining training needs as there are persons who are concerned with planning and directing training. The approach which we present is no easy side-step of the basic investigations which are necessary in determining training needs. We hold no particular belief for it except that it has and does assist in our ordering of the problems in this complex area. It is a three-fold but closely interrelated approach to thinking about the training requirements of an organization or a component of an organization. It consists of the following:

1. Organization analysis—determining where within the organization training emphasis can and should be placed.
2. Operations analysis—determining what should be the contents of training in terms of what an employee must do to perform a task, job, or assignment in an effective way.
3. Man analysis—determining what skills, knowledge, or attitudes an individual employee must develop if he is to perform the tasks which constitute his job in the organization.

Organization analysis involves a study of the entire organization: its objectives, its resources, the allocation of these resources in meeting its objectives, and the total environment of the organization. These things largely determine the training philosophy for the entire organization. The first step in organization analysis is to obtain a clear understanding of both short- and long-term goals. What is the company trying to achieve in safety in general and specifically by department. A second step in organization analysis is inventory of the company's attempts to meet goals through its human and physical resources. The final step is an analysis of the climate of the organization.

The climate of an organization is, in essence, a reflection of its members' attitudes toward various aspects of work, supervision, company procedures, goals and objectives, and membership in the organization. These attitudes are learned; they are a product of the members' experiences both within and outside the work environment. A training program may be designed to effect certain changes in the organizational climate, for instance, our safety training certainly hopes to influence the employees' attitudes toward safety.

Job analysis for training purposes involves a careful study of jobs within an organization in a further effort to define the specific content of training. It requires an orderly, systematic collection of data about the job. We are familiar with this through our Job Safety Analysis procedures. Other ways are also available for job analysis.

1. Observations. (Is there obvious evidence of unsafe acts or poor methods? Are there incidences on the part of individuals or groups that reveal poor personnel relationships, emotionally charged attitudes, frustrations, lack of understanding, or personal limitations? Do these situations imply training need?)
2. Management requests for training of employees.
3. Interviews with supervisors and top management personnel to accumulate information about safety problems, as well as interviews with employees concerning safety.
4. Group conferences with interdepartmental groups and safety advisory committees to discuss organizational objectives, major operational problems, plans for meeting objectives, and areas in which training could be of value.

5. Comparative studies of safe versus unsafe employees to underline the bases for differentiating successful from unsuccessful performance.
6. Questionnaire surveys.
7. Tests or examinations of safety knowledge of current employees; analyses of safety sampling.
8. Supervisors' reports on the safety performance of employees.
9. Accident records.
10. Actually performing the job.

Manpower analysis focuses on the individual and job performance as it relates to safety. The performance analysis chart in Exhibit 6.3 intends to help with this analysis.

SUPERVISORY NEEDS ANALYSIS

One important element of safety training is supervisory training. Here again a needs analysis is the starting point. An important study in this area was made by Lester Bittel and Jackson Ramsey of James Madison University as reported recently by Karl J. Trommler:[156]

> The survey data used came from a sample of 8,000 supervisors, the largest ever collected. Data were gathered by written questionnaires from 116 participating businesses, non-profit institutions and public agencies, and from 43 chapters of the International Management Council representing 448 companies, for a total representation of 564 organizations in 37 states. The survey was analyzed under the Statistical Program for the Social Sciences, and all figures were found to have a 99-percent chance of being representative with the population represented by the sample. The sample was representative (in geographic, industry, sex and racial distribution) to the total supervisory population of the United States as inferred from data published by the Bureau of Labor Statistics.
>
> A major conclusion of this study is that analysis of the segments of the supervising population can be more meaningful than that of the whole. For example, under the Job function and setting subheading in Exhibit 19.1, the key distinguishing

Column A Job function and setting	Column B
Blue-collar employees	White-collar employees
Unionized employees	Non-unionized employees
Production-oriented functions	Clerical-oriented functions
Manufacturing, construction and/or mining industries	Banking, insurance, service, institutional and/or government/military establishments
Level of supervision	
First-level: supervise only rank-and-file hourly or salaried employees	*Second-level:* supervise other supervisors as well as rank-and-file employees
Educational level attained	
No higher than a high school diploma	College degree or higher

Exhibit 19.1 Who do supervisors supervise?[155]

factor was found to be whether or not the employees supervised were blue-collar or white-collar. White-collar supervisors were found to give more time and attention than their blue-collar counterparts to interviewing and placing employees, consulting with staff, attending departmental meetings, and improving methods and procedures. However, blue-collar supervisors spend significantly more time than do white-collar ones in planning production schedules, making job assignments, and coping with safety, production, and quality control problems. The other items listed under Job function and setting in Exhibit 19.1 reinforce even more the blue-collar/white-collar distinctions.

These are important items to know in designing a flexible supervisory training program and, later, in the selection of who will attend which sessions in the interest of easier personalization of the program. Significant items are also to be considered under the Level of supervision in Exhibit 19.1. For example, second-level supervisors, which I will hereafter refer to as managers, were found to be more concerned than first-level supervisors with production planning and control, discussing employee personal and performance problems, devising improved methods and procedures, and solving cost-related problems.

Exhibit 19.2, also derived from the Bittel and Ramsey survey, shows how various paired categories of supervisors stacked up against their counterparts in rating their own confidence in their abilities to handle various important job activities. However, Saul Gellerman warns that "supervisors have a predilection for confusing confidence or intent with actual performance." Some conclusions that can be drawn from Exhibit 19.2 are:

Supervisor's greatest confidence lies in general human relations skills. This includes items such as communications, conference leadership, discipline, performance appraisal and motivation.

Supervisors show less confidence in specific administrative skills. Here again, they are more confident in their ability in generalized areas such as problem solving and the ability to attain goals. This drops off when specific policies and procedures are involved. Notice how respondents rated their ability to explain OSHA and employee benefit plans and in counseling drug or alcohol abusers and employees about to retire.

HOW SHALL WE GET THERE?

We discussed training methods earlier. We did not, however, look at what we know about human learning behavior. There is more empirical data here than perhaps anywhere else in psychology. Here are some of the "knowns."

Motivation and Learning

Learning theorists generally agree that an individual will learn most efficiently when motivated toward some goal which is attainable by learning the subject matter presented. It is necessary that the outcome must be desired and the behavior must appear to the learner to have some relation to achieving that outcome. If the behavior-outcome relationship is obscure and the learner is striving toward a desired goal, attempts to teach new behavior may be ignored and other kinds of behavior which appear to be relevant to the goal will be tried.

In conducting a supervisors' safety training course, for example, some people may feel that they have more important production problems to worry about and will spend training time thinking about them and complaining about being taken away from the job to learn a lot of

Important Job activities	Average Rating (Percent with high* degree of confidence)	Ranking (From most to least confidence)	Blue-collar	White-collar	1st-Level	2nd-Level	Least educated 7-12 years	Most educated 16-20 years	Male	Female	White	Non-white	Under 50 years	Over 50 years	Non-union employees	Union employees
Talk to employees on a one-on-one basis	88	1	88	90	87	92	86	90	89	86	89	86	88	88	89	88
Maintain harmony within your department	80	2	79	80	78	82	79	79	80	77	80	80	79	78	78	80
Solve departmental problems as they arise	78	3	77	81	75	85	74	84	81	75	79	71	80	78	76	80
Conduct a group meeting in your department	76	4	72	83	71	85	65	86	77	74	74	72	76	74	70	79
Attain departmental goals set by the company	75	5	73	78	72	81	72	79	76	74	76	71	75	76	74	77
Plan and control use of personal time	72	6	66	75	71	75	70	76	73	73	73	69	72	75	72	74
Enforce disciplinary rules	72	7	76	69	67	79	74	68	69	68	73	70	72	72	74	72
Conduct an effective performance appraisal	72	8	71	74	70	76	70	74	73	70	73	72	73	71	69	73
Sell your ideas to your boss	69	9	67	72	68	75	66	74	70	68	70	65	71	65	69	71
Motivate employees	68	10	69	68	64	75	66	70	70	64	68	70	68	67	67	69
Write clear memos, letters and reports	68	11	61	76	66	72	54	82	67	70	68	65	69	67	64	70
Develop new ideas for improving productivity	67	12	66	67	62	74	61	71	68	61	66	67	67	66	66	67
Explain benefit program to employees	56	13	58	54	51	66	57	52	56	53	55	61	54	62	50	58
Use statistical techniques	51	14	53	51	49	57	49	54	54	44	52	58	52	52	51	51
Explain computer inputs and outputs for your department	47	15	39	54	45	51	38	54	47	47	47	48	49	40	39	51
Enforce OSHA rules and regulations	45	16	54	35	43	50	52	35	47	41	44	51	43	54	52	43
Counsel an employee who abuses alcohol or drugs	39	17	44	35	36	45	40	37	42	32	39	54	39	39	41	39
Counsel an employee who will retire next year	36	18	39	32	32	42	37	33	36	31	35	41	33	45	38	34

*"High" (4) + (5) percent rating, where scale is:
(1) not confident
(2)
(3) somewhat confident
(4)
(5) very confident

Exhibit 19.2 Supervisors' Confidence in Their Abilities to Handle Important Job Activities.[155]

nonsense. Or they may enjoy the opportunity to get together with "the boys" and swap stories. Still others may see the training class as an opportunity to show how much they already know and to strive for greater recogniton in the eyes of the trainer and their fellow trainees. A few may see that training could aid them in their job. The behavior of people is oriented toward relevant goals, whether these goals are safety, increased recognition, production, or simply socialization. People attempt to achieve those goals which are salient at the moment, regardless of the trainer's intent.

Reinforcement and Learning

Any event which occurs so as to change the probability of a given response is said to be reinforcing. Positive rewards for certain behavior increase the probability that this behavior will occur again. Negative rewards decrease the probability. Actually, punishment seems to inhibit the occurrence of response rather than eliminate or extinguish it. Instead, the response which leads to the avoidance of punishment is reinforced. However, it should be emphasized that the role of punishment in learning is not completely clear as yet. Often, failure to reinforce a response may have a better long-range effect than a reprimand from the trainer.

Whether or not an event is reinforcing will depend on the perceptions of the individual who is learning. What one person regards as a rewarding experience may be regarded by another as neutral or nonrewarding or even punishing. In general, however, there are various classes of reinforcers—food, status recognition, money, companionship—which are reinforcing to almost everyone at one time or another.

Practice and Learning

An individual learns that which is practiced. In practicing a skill, those kinds of behavior which are performed and which are reinforced will be acquired and maintained. Without practice, learned skills are lost quickly. In safety training, then, the followup and practice is as important as the initial learning. Also, spaced practice (a little at a time) seems to be more effective than massed practice. This seems to be true both for learning rates and for retention. It is better to have a number of short sessions than one long one.

Feedback and Learning

Most experts state that giving a trainee feedback is essential for good training. It is difficult for the trainee to improve performance unless given some knowledge about that performance. If not performing correctly what is the nature of the errors? How can they be corrected? This is essential. Some theorists strongly emphasize immediate reinforcement to each bit learned. The success of programmed instruction is based primarily on this concept.

Meaningfulness

In general, meaningful material is learned and remembered better than material which is not meaningful. It is difficult to state just what is meant by the term *meaningfulness* because no universal measure of this important concept has been developed. For instance, new information in your specialty is more easily learned than material in another specialty. In your own specialty, you are familiar with the meaning of many technical terms and have used these concepts in many situations; they are "your own." Delving into a new area requires the acquisition of new concepts and principles as well as the acquisition of the information being presented.

The implications of the rather consistent experimental findings on meaningfulness both with respect to original learning and to transfer to other situations is quite clear. The material must be made as meaningful as possible to the trainee. In reading through training courses or watching safety films, we often wonder whether the course or film was being aimed at the trainee or at impressing peers and management.

The concept of meaningfulness has implications not only for the way in which material is presented to the trainee, but also for the preparation of the material which the trainee must carry to others. The trainee must try to think in the trainee's terms, to put the material across with examples and language familiar to the worker. It is important to supply as many associations for new ideas and concepts as possible so that they become more meaningful.

Climate and Learning

The classroom environment makes a difference. Research has concluded that to encourage high rates of achievement in highly technical subjects, the environment must be challenging. To encourage achievement in nontechnical areas, classes should be socially cohesive and satisfying.

Research where anxiety and stress are built into the training situation shows that quantity of ideas increased; increases did not transfer to other tasks; there is a curvilinear relationship between anxiety and performance, with poor performance being associated with intermediate anxiety; and anxiety interferes with verbal ability.

These then are some of the insights into the behavior of people involved in learning. Whichever methods are chosen, they should take into account the things we do know from the research.

HOW WILL WE KNOW WHEN WE'VE ARRIVED?

The evaluation of training is not simple. We try to determine what changes in skill, knowledge, and attitude have taken place as a result of training. We also try to determine how these skills, knowledge, and attitudes contribute to organizational objectives. McGehee and Thayer[96] suggest evaluation measures in four broad categories: (1) objective-subjective, (2) direct-indirect, (3) intermediate-ultimate, and (4) specific-summary.

Objective-Subjective—The major distinction between an objective and a subjective measure is its source. A measure is objective if it is derived from overt behavior. If the measure represents an opinion, a belief, or a judgment, it is subjective.

Direct-Indirect—A measure is classified as direct if it measures the behavior or the results of the behavior. An indirect measure assesses the action of an individual whose behavior can be measured only by its influence on the actions of others. For instance, supervisory training effectiveness is usually measured indirectly as we read the results of the participant's (supervisor's) efforts for our determination of training effectiveness. As a result of our supervisory training program on safety, did the department have a better safety record?

Intermediate-Ultimate—An intermediate measure might be the supervisor's test grade—an ultimate measure of the results back on the job.

Specific-Summary—Somewhat related to the problem of intermediate and ultimate measures of training outcome is the problem of measures which are used as an index of successful performance of a specific phase of a job or as an index of the degree of performance of the total job against its potential contribution to organizational goals. Should we measure the effectiveness of our supervisory inspection training program by the number of code violations or by improvement in the department's accident record?

These four categories of measures are not mutually exclusive. A rating scale could be a subjective, intermediate, direct, and specific measure. It also could be a subjective, direct, ultimate, and summary measure. Regardless of their type, measures must have certain characteristics if they are to be used in studying training outcomes. These characteristics are relevant to job criteria and corporate goals, reliability (consistency), and freedom from bias.

Measurement Techniques

The basic procedures for measuring which allow causal inferences are the well-established procedures for the investigation of the relationship between phenomena in both the physical and the social sciences. Stated without elaboration, they consist in showing that condition X is the direct result of condition Y and not of conditions a, b, c to n. In training, this means that what occurs as a result of training is attributable to training and not to chance or systematic changes in the circumstance under which the training takes place.

The procedures used can be classified into certain broad categories according to when the measures are secured and whether or not a control group is used. By control groups, we mean employees who did not receive the training or who received training different from that received by the group for whom we are trying to determine the results. These procedures, using this system of classification, fall into the following types:

1. Measures after training without a control group.
2. Measures before and after training without a control group.
3. Measures after training with a control group.
4. Measures before and after training with a control group.

After Training, No Control Group is the method most frequently used for evaluating training. This is the least meaningful measure for we end up with no solid factual evidence that the training caused the change.

Each of the other three mentioned above are better, with 4, of course, being the best. While best on obtaining factual information on training effectiveness, objections may arise to the use of control and experimental groups. This is particularly true if the control group is simply pre- and post-tested and given no training at all.

HOW MUCH TRAINING IS ENOUGH?

Training Magazine[155] made a survey of the number of hours of training that is given for typical employees at different levels (see Exhibit 19.3). They also presented the information in Exhibit 19.4 on whether that training was provided in-house or by outsiders, and on the specific types of training (see Exhibit 19.5).

	Mean No. of Hours	Median No. of Hours	Mode	% Offering No Training
Executives	28.3	19.6	0	26.0%
Senior Managers	22.7	16.4	0	28.7%
Middle Mangers	32.5	23.5	0	15.9%
First-line Supervisors/ Foremen	32.5	23.6	0	17.8%
Sales Representatives	19.5	0.4	0	54.8%
Professionals	27.2	15.8	0	33.2%
Administrative Employees	11.9	5.3	0	42.1%
Office/Secretarial	10.8	5.2	0	38.1%
Production Workers	12.3	0.3	0	60.6%
Customer Service Employees	13.4	0.4	0	53.1%

Exhibit 19.3 Hours of training by level.[155]

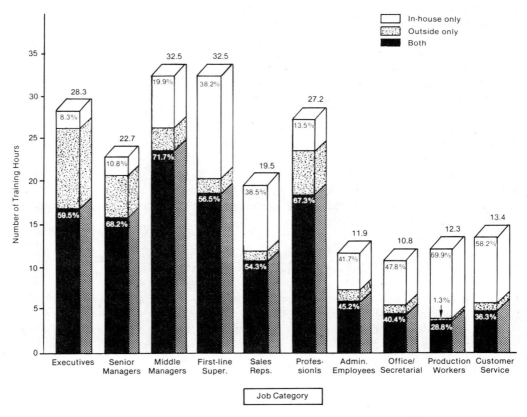

Exhibit 19.4 Training hours by job category.[155]

Types of Training	% of Total Providing	% In-house Only[2]	% Outside Only[3]	% Both In-house & Outside[4]
New Employee Orientation	87.0%	79.1%	0.8%	7.1%
Performance Appraisals	76.7%	48.3%	3.5%	25.0%
Time Management	73.9%	22.8%	18.6%	37.6%
Leadership	73.3%	20.8%	9.8%	42.7%
New Equipment Operation	68.9%	24.6%	7.0%	37.3%
Hiring/Selection Processes	66.9%	34.9%	7.6%	24.4%
Product Knowledge	64.3%	39.6%	2.3%	22.4%
Train-the-trainer	63.6%	24.3%	13.2%	26.1%
Interpersonal Skills	63.5%	21.2%	6.0%	36.2%
Word Processing	62.4%	17.8%	19.4%	25.3%
Problem Solving	61.4%	22.5%	7.2%	31.7%
Goal Setting	60.0%	27.3%	4.0%	28.6%
Listening Skills	59.9%	21.8%	10.1%	29.0%
Stress Management	58.5%	17.2%	17.1%	24.1%
Decision Making	57.2%	17.9%	7.5%	31.9%
Safety	56.9%	27.2%	2.4%	27.3%
Team Building	56.8%	22.4%	5.6%	28.8%
Delegation	56.8%	22.7%	6.5%	27.6%
Planning	55.2%	20.6%	4.3%	30.3%
Public Speaking/ Presenting	55.0%	17.0%	15.9%	22.1%
Customer Relations	53.1%	24.4%	3.5%	25.2%
Computer Programming	53.0%	8.7%	13.3%	31.0%
Management Information Systems	52.5%	11.3%	12.2%	29.0%
Conducting Meetings	50.9%	23.8%	4.8%	22.3%
Writing Skills	50.5%	15.4%	15.7%	19.4%
Data Processing	49.8%	9.9%	8.2%	31.6%
Beginning Sales Skills	48.2%	23.8%	4.4%	20.0%
Negotiating Skills	45.6%	10.8%	17.6%	17.1%
Advanced Sales Skills	44.2%	14.9%	6.2%	23.1%
Strategic Planning	40.1%	11.3%	11.5%	17.4%
Outplacement/ Retirement	32.6%	16.7%	5.9%	9.9%
Reading Skills	26.8%	6.4%	11.6%	8.8%
Nutrition	21.1%	8.1%	5.4%	7.6%
Foreign Language	12.6%	1.1%	9.8%	1.7%

[1] % of sample who offer this type of formal training.

[2] Among respondents providing, % who offer only training designed, developed and delivered by in-house staff.

[3] Among respondents providing, % who offer only training designed, developed or delivered by outside consultants or vendors.

[4] Among respondents providing, % who offer training designed, developed and delivered by both.

Exhibit 19.5 Specific types of training.[155]

Chapter Twenty

Attitudes

"Things surrounded by unpleasantness are seldom surrounded by people"

—Robert Mager

Attitudes are, of course, not crisp and easily measurable. Even so, we should attempt to recognize and assess employee attitudes toward safety. Let us look briefly at what we know about attitudes and how they develop.

DEVELOPING AN ATTITUDE TOWARD SAFETY

Robert Mager's book, *Developing Attitude Toward Learning,*[104] gives us a lot of insight into attitude development which can be applied directly to developing safety attitudes. He reports a study he made several years ago which was designed to determine student's attitudes toward different academic subjects and what formed the attitude in each case. To summarize his results, he states that a favorite subject area tends to become favorite because the person seems to do well at it, because the subject was associated with liked or admired friends, relatives or instructors, and because the person was relatively comfortable when dealing with the subject. Conversely, a least-favored subject seems to become so because of a low aptitude for it, because it was associated with disliked individuals, and because being in the presence of the subject matter is associated with unpleasant conditions.

Things that help to mold an attitude toward a subject are:

1. The conditions that surround a subject.
2. The consequences of coming into contact with a subject.
3. The way that others react toward a subject (modeling).

Conditions

When a person is in the presence of subject matter, it should be accompanied by positive conditions. Conversely, when in the presence of subject matter, there should not at the same time be aversive conditions. If a subject that initially has no special significance is presented to someone on several occasions while experiencing an unpleasant condition, that subject may become a signal that triggers an avoidance response (keep away). If a person is presented with a neutral subject and at the same time is in the presence of pleasant conditions, that subject may become a signal for an approach response (likes it). For instance, Mager cites the reaction of most people when a physician moves a hypodermic needle toward their arm. Many tend to back away, if not physically, they may turn their heads to avoid seeing this signal of forthcoming pain. There is nothing aversive about the sight of the hypodermic needle—the first time we see one, that is. It is a neutral subject. But after experiencing aversive conditions (pain) while in the presence of a hypodermic needle, it becomes a signal for an avoidance response. It is as though the mere sight of a hypodermic needle becomes a condition to be avoided.

How can we use this information in our safety programming? First of all, if we do not already know what our employees consider to be positive and aversive conditions, we certainly ought to find out. To begin with, any form of punishment is obviously aversive; most forms of social interaction are positive; competition and game-type situations are positive; participation is positive; being "told" what to do is aversive.

We should arrange our safety instruction in the presence of positive conditions and in the presence of as few aversive conditions as possible.

Consequences

Whenever contact with subject matter is followed by positive consequences, the subject will tend to become a stimulus for approach responses. Conversely, whenever contact with subject matter is followed by aversive consequences, the subject may become a stimulus for avoidance responses. This statement can be found in any freshman psychology text. It has been documented experimentally and practically perhaps more than any other psychological principle.

Mager translates the principle into a classroom situation by asking us to picture ourselves as students in a situation where when we correctly answer a question, the instructor, smiles and says something like, "Good." When we answer a question incorrectly, the instructor makes a comment such as, "Well, let's look at the question again." In this situation Mager suggests we would no doubt be willing to answer questions and come into contact with the subject matter again. In any case, this kind of interaction would not adversely influence our response toward the subject. Conversely, suppose each time we answer a question the instructor says, "Well I see old Dumbo is at it again." How long would it be before we stopped raising our hand? How long do you think it would be before we began to think of excuses for not attending class?

When experience with a subject is followed by a positive (pleasant) consequence, the probability is increased that the subject will be approached again in the future. When experience with a subject is followed by aversive (unpleasant) consequences, the probability is reduced that the subject will be approached in the future. If you want to increase the probability that a response will be repeated, follow it immediately by a positive consequence. If you want to reduce the probability that the behavior will occur again, follow it immediately by an aversive consequence.

This all seems obvious, but perhaps one pont is a little confusing. In one case we seem to say punishment is bad, in the next breath we say it is good. Which is it? In the first case, we were talking about conditions. Here punishment is not recommended. A safety program built on a punishment atmosphere will not succeed. It is aversive. Seeing the other guy get punished creates an aversive atmosphere for you. Also, a punishment program means that safety rules must be tightly set to make punishment fair. This is aversive because it is "telling them what to do." On the other hand, punishment following an unsafe act might work for the individual along, but here the problem is whether we can be sure that worker will no longer perform the act or just be sure not to get caught the next time.

Positive and Aversives

(handwritten: FAVORABLE   UNFAVORABLE)

What in a work situation is aversive and what is positive? [Although it is not always possible to know whether an event is positive or aversive for a given individual, some conditions and consequences are universal enough to provide us some direction.]

Mager suggests we define an aversive as any condition or consequence that causes a person to feel smaller or diminishes his/her importance. Here are some common aversives, adapted from Mager, that might apply to safety and safety training:

1. *Pain.*   Not too applicable to training, but this is learning by experience (the hard way) on the job.
2. *Fear and anxiety.*   Things that threaten various forms of unpleasantness, such as
   — Telling the worker by word or deed that no accomplishment will being success.
   — Telling the worker, "You won't understand this, but . . . ."
   — Telling the worker, "It ought to be perfectly obvious . . . ."
   — Threatening the exposure of "ignorance" by forcing the worker to do something embarrassing in front of the peer group.
   — Basing an attrition rate on administrative fiat rather than on worker performance. ("Half of you won't be here a month from now.")
   — Being unpredictable about the standard of acceptable performance.
3. Frustration creators, such as
   — Presenting information in larger units, or at a faster pace, than the student can assimilate. (The more motivated the worker, the greater the frustration when efforts are blocked.)
   — Speaking too softly to be heard easily (blocking the worker's effort to come into contact with the subject).
   — Keeping secret the intent of the instruction or the way in which performance will be evaluated.
   — Providing unreadable print; type too small or too ornate, or reading level too high.
   — Providing obscure text or implying more profundity than actually exists, as in OSHA standards.
   — Teaching one set of skills and then testing for another.
   — Testing for skills other than those stated in announced objectives.
   — Refusing to answer questions.
   — Using test items with obscure meanings.
   — Forcing all to proceed at the same pace, thus frustrating the slow and boring the quick.
   — Calling a halt when the worker is absorbed with the instruction or attempting to complete a project.
4. Humiliation and embarrassment, for instance.
   — Publicly comparing a worker unfavorably with others.
   — Laughing at a worker's efforts.
   — Spotlighting a worker's weaknesses by bringing them to the attention of the group.
   — Belittling a worker's attempt to approach the subject by such replies to his questions as, "Stop trying to show off," or "You wouldn't understand the answer to that question."
   — Repeated failure.
   — Special classes for accident repeaters.

5. Boredom, caused by
   — Presenting information in a monotone.
   — Insisting the worker sit through repeated sessions covering the same topics.
   — Using impersonal, passive language.
   — Providing information in increments so small that they provide no challenge or require no effort.
   — Using only a single mode of presentation (no variety).
   — Reading the safety rules aloud.
6. Physical discomfort, such as
   — Allowing excessive noise or other distractions.
   — Insisting the worker be physically passive for long periods of time.

Here are some positive conditions or consequences:
   — Acknowledging responses, whether correct or incorrect, as attempts to learn, and following them with accepting rather than rejecting comments. ("No, you'll have to try again," rather than "How could you make such a stupid error!")
   — Reinforcing or rewarding subject approach responses.
   — Providing instruction in increments that will allow success most of the time.
   — Eliciting learning responses in private rather than in public.
   — Providing enough signposts so that the worker always knows the rate of progress.
   — Providing the worker with statements of your instructional objectives in an understandable mode.
   — Detecting the worker's level of knowledge so as to avoid repetition (and the possibility of a very boring session.)
   — Providing feedback that is immediate and specific to the worker's response.
   — Giving the worker some choice in selecting and sequencing subject matter, thus making positive involvement possible.
   — Providing the worker with some control over the length of the instructional session.
   — Relating new information to old, within the experience of the student.
   — Treating the worker as a person rather than as a number.
   — Using active rather than passive words during presentations.
   — Making use of those variables known to be successful in attracting and holding people's attention, such as motion, color, contrast, variety, and personal reference.
   — Allowing only those instructors who like and are enthusiastic about their subjects (and workers) to teach.
   — Making sure the worker can perform with ease, not just barely, so that confidence can be developed.
   — Expressing genuine delight at seeing the worker.
   — Expressing genuine delight at seeing the worker succeed.
   — Providing instructional tasks that are relevant to your objectives.
   — Using only those test items relevant to your objectives.
   — Allowing workers to move about as freely as their physiology and their curiosity demand.

Modeling _DOESN'T WORK ANYMORE

Another way in which behavior is strongly influenced is through modeling (learning .by imitation). The research on modeling tells us that if we want to maximize approach tendencies in workers, we must exhibit that behavior ourselves. In other words, we must behave the way we want out employees to behave. When we teach one thing and model something else, the teaching is less effective than if we practice what we preach. The message to safety management is obvious.

Attitude Development

Let us summarize what is known about developing attitudes generally and adapting that knowledge to safety:

1. There are three ways in which attitude (and behavior) toward a subject may be influenced: by the conditions associated with the subject matter, by the consequences of the subject matter contact, and by modeling.
2. Exhortation, a procedure used regularly for safety, has seldom been very successful in influencing behavior.
3. Approach and avoidance behavior are influenced by the things we do and by the things we say.

Attitude and Safety

In his book, *Supervisor's Guide to Human Relations,*[50] Dr. Earle Hannaford defines attitude as the potential for action and safety attitude as "a Readiness to Respond Effectively and Safely, Particularly in Tension-Producing Situations." He goes on to state the three components of attitude:

$$\text{Attitude} = \text{learned responses} + \text{habit} + \text{emotional set}$$

and suggests there are four steps in building attitudes:

1. Laying the foundation for the attitude.
2. Personalizing the attitude for the individual.
3. Fixation of the attitude emotional set.
4. Keeping the attitude alive.

Dr. Hannaford's work with attitudes has shown the relationship between attitudes and results in safety (see Exhibit 20.1 A-B). As is readily evident in Exhibit 20.1A, the poorer the safety attitude of the employee, the greater the number of lost-time accidents during the five-year period studied.

The 769 male employees studied came from 47 companies representing a cross-section of various industries—companies with excellent, average, and poor safety records. Exhibit 20.1B shows that as the supervisor's safety attitude test score worsens, the number of lost-time accidents per employee increases. Obviously, the main conclusion to be drawn from this study is that a positive attitude toward safety fosters safe working practices.

CHANGING AN ATTITUDE TOWARD SAFETY

There are a number of current theories about attitude change or development today that might be helpful to the safety professional. Traditionally we in safety have used preaching and teach-

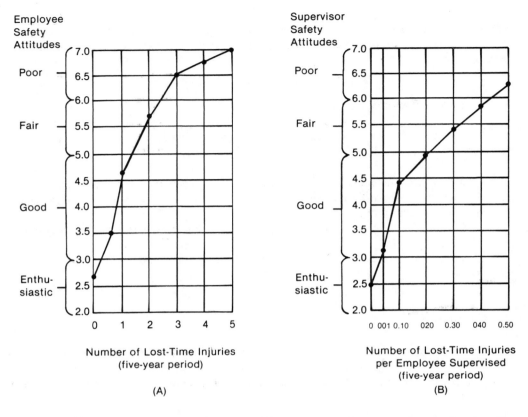

Exhibit 20.1 Relationship of employee's and supervisor's safety attitudes.[50]

ing as our primary methods to develop or change a worker's safety attitude. The current theories suggest to us that the whole process of attitude change is considerably more complex than perhaps we have realized. Following are a few of those theories, and some thoughts about how the safety program might do different things to achieve better results.

First, there is what might be called a "Balance Theory" relating to attitude change. It suggests that a change in attitude toward a thing is influenced not only by the thing itself, but also by other people and their attitude toward that thing, and by your attitude toward them. For instance, if you as a person have a positive feeling and attitude toward another person, your boss (see Exhibit 20.2A), and you both have positive feelings toward a thing (in this case your newly developed safety program), then you are in a balanced situation which requires no attitude adjustment on your part to either the other person or the subject. Attitude change comes only when there is an unbalanced situation.

In Exhibit 20.2B there is such an unbalanced situation. You have positive feelings toward your boss, i.e., respect, general agreement in judgment, etc. You like your newly developed safety program. Your boss thinks it stinks. This is an unbalanced situation that requires some adjustment to bring it back into balance. You will do one or both of two things. You will adjust your attitude toward your boss, beginning to think less positively, or you will think less of your new program and consider changing it in some ways.

Exhibit 20.2C depicts another balanced situation. You have negative feelings toward your boss, who dislikes your new program. Since you are aware of your boss's incompetence, it is balanced; you would expect that jerk would reject your excellent program: "Anyone that stupid wouldn't understand a good program anyway."

Exhibit 20.2D is unbalanced, requiring an attitude change. You dislike your boss, who likes your program. Obviously your program must not be as good as you thought. You'll no doubt want to reconsider much in the program (your attitude toward it has changed); or you'll begin to reconsider your boss (who is perhaps not as stupid as you thought).

The theory suggests then that attitude change will take place only when the unbalanced situation exists for a person. To change a safety attitude, you must then create such an unbalance.

Other attitude change theories give us some further insights. The balance theory is somewhat limited, as it assumes our attitude must be either for or against, not in between; it is limited to the situation between one person and one subject; and it assumes if you like someone you must always agree with that person. Other theories attempt to deal with some of these shortcomings.

The theory of the "Associative Bond" suggests that on any particular subject we choose an opinion somewhere on a scale between total agreement and total disagreement. Attitude change

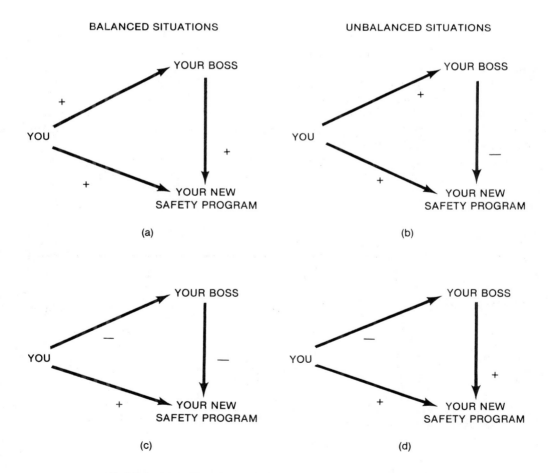

Exhibit 20.2 The Balance Theory of Attitude Change.

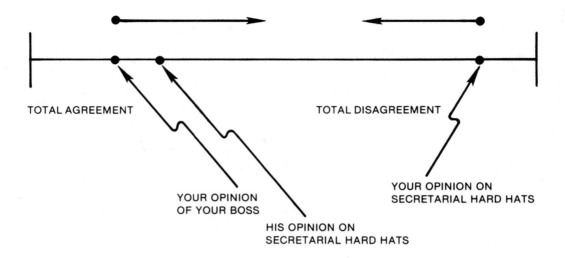

Exhibit 20.3 The Associative Bond

is the process of shifting to the right or left on that continuum. The associative bond works as is shown in Exhibit 20.3.

You have a high opinion of your boss, who decides that there should be a new safety rule in the organization requiring all secretaries to wear hard hats. You disagree totally. At this point, research shows that two things will happen in your thinking: Your opinion of your boss shifts to the left and your position on secretaries wearing hard hats shifts to the right (perhaps they ought wear them at least in the plant).

A third attitude theory that provides us with some insight is the "Ego-Involvement" theory (see Exhibit 20.4). It suggests that as any person selecting a position on any subject not only

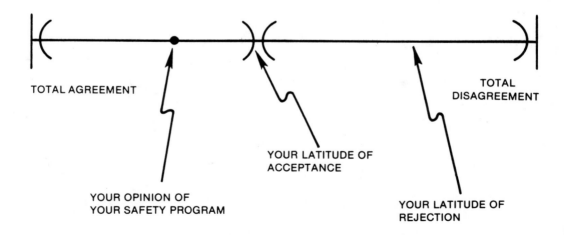

Exhibit 20.4 The Ego-Involvement Theory

selects a crisp position somewhere between total agreement and total disagreement, but also subconsciously selects what is termed "latitude of acceptance" (see Exhibit 20.4). This is that person's comfort range. Each person also has a "latitude of rejection," points beyond which there will not be a shift in attitudes toward that subject.

The theory says to safety professionals that when working on the safety attitudes of others our mission is not to convert them from non-believers into believers; rather, our task is simply to work on widening their latitude of acceptance and narrowing their latitude of rejection. Research also has shown that the more "ego-involved" a person is in an opinion, the narrower the latitude of acceptance and the wider the latitude of rejection (if you made up the safety program you won't shift far from your belief in its greatness).

Probably the most publicized theory relating to attitude change is the theory of "Cognitive Dissonance". The theory, developed by Festinger,[37] suggests that all individuals hold a number of cognitions (things that we "know" to be true; our psychological realities). These cognitions can exist in three types of relationships:

1. Irrelevant—no relationship at all between them.
2. Dissonant—two in opposition to each other.
3. Consonant—in harmony with each other.

Festinger's theory is that individuals ahve a drive to maintain a sense of consonance or harmony among their cognitions. If there is dissonance in the cognitions, there is a need to reduce that dissonance by altering the cognitions (change the attitude).

For example, if you (1) believe totally in ERA, (2) believe for this country's good we must reinstitute the draft, and (3) believe your 18-year-old daughter should not have to serve, you have inconsistent beliefs (dissonance in your cognitions). This dissonance will require you to alter one or more of the these three cognitions. Attitude change comes with dissonant cognitions; it does not happen without this dissonance.

Festinger suggests four situations which usually create dissonance:

1. *Decisional situations.* After a decision, there will often be dissonance, as you consider what you "could have done" or "should have done."
2. *Forced Compliance.* This is when a person is forced into a behavior in opposition to belief or attitude. In a typical safety situation, eye protection is forced on a worker, which requires alterations either to behavior or to attitude so they will be in consonance. Research shows the individual will over time either change attitude (rationalize behavior), or will stiffen resistance against the behavior, forcing discipline or perhaps even dismassal. This will also be rationalized (with the better paying job up to street).
3. *Exposure to New Information.* When a person is exposed to new information so irrefuteable it must be looked at and considered, dissonance is created. This creates either acceptance and change in attitude, rejection and attacking the accuracy of the information, or upset and irritation with you for pushing the information.
4. *Social Disagreements.* When there is disagreement with those you respect the most, dissonance occurs.

These four theories of attitude development and change are well accepted and well documented in research. They say to us fairly clearly that the process of attitude change is rather complex, certainly more complex than traditional safety programs have recognized. Our "awareness campaigns," or "motivational techniques," and our training courses have had little to do with attitude change. It seems highly likely that many of our attempts might well have had negative rather than positive results.

Chapter Twenty-One

Behavior Modification

Behavior modification is a technique which may enable manpower management to modify or eliminate undesirable employee behavior and replace it with behavior that is more compatible with goal attainment. Direct applications to manpower management include training, retraining, compensation, absenteeism, tardiness, and motivation of organizational participants. On a more grandiose scale, behavior modification may provide the missing link to the fusion of individual and organizational goals.

Fred Luthans[92]

In Chapter 20 we found that the attitude toward safety (or anything else for that matter) is developed through: the conditions surrounding it, the ways others react to it, and the consequences of coming into contact with it. And that positive consequences build positive attitudes and behavior while aversive consequences build negative attitudes and avoidance behavior. By stating this, we have already begun to discuss one of the newest management techniques, behavior modification, which may provide managers on all levels with a practical tool for shaping, improving, and motivating behavior of organizational participants.

Behavior modification is not new and is relatively simple to understand. The basic process involves systematically reinforcing positive behavior while at the same time ignoring or exercising negative reinforcements to eliminate unwanted behavior. The end result is the creation of a more acceptable response to the given situation. The technique concentrates on a person's observable behavior and not on its underlying causes.

PRINCIPLES

There are two primary approaches used in behavior modification programs. One is an attempt to eliminate unwanted behavior that detracts from organizational goal attainment and the other is the learning of new responses. In our case we would like to eliminate unsafe acts. This elimination process is called *extinction*. Extinction may be accomplished either through punishment or by withholding the positive reinforcement. The difficulty with punishment, or negative reinforcement, is the possibility of "backlash" reaction and a short duration of extinction. In other words, another undesirable reaction may result, or the punished behavior may reappear in a short time. Under punishment, the behavior is only temporarily suppressed, only to emerge again when the aversive consequence no longer follows. Under punishment the worker learns not to perform the unsafe act in your presence, but will wait until you are gone.

Extinction of behavior may also be accomplished through the withholding of positive reinforcement. If customary rewards for the action in question are no longer granted, then extinc-

227

tion will occur. A simple example would be the "office jokester" who tells one of his typically nonhumorous stories during the coffee break. For this effort, he receives no positive reinforcement from the group, and they immediately turn to another topic. If reward is repeatedly withheld, the bad jokes eventually will be eliminated. Since the use of punishment may only temporarily suppress and not totally extinguish behavior, a logical alternative would be to withhold positive sanctions.

The second major goal of behavior modification is to create acceptable new responses to an environmental stimulus. As in eliminating behavior, reinforcement techniques play the vital role in developing new behavioral responses. Direct reward of responses is the best method to stabilize intended behavior patterns. Eliminating undesriable behavior without providing a new substitute pattern leaves the worker open to learn another undesirable set of responses. The substitution of new desirable responses in place of unwanted behavior is the overall purpose of behavior modification: to leave the individual with new or modified behavioral patterns in place of behavior that was deemed unwanted or unneeded.

Reinforcers

Learning theory carefully explains that reinforcement must be applied according to a detailed, systematic plan. The reinforcement schedule that is used depends on the nature and degree of behavior that the person presently demonstrates.

If the worker is presently not performing as you wish (not exhibiting the desired behavior), obviously no reinforcement is possible. In this case we have two possible choices: (1) We can inform the individual what response would lead to a reward. ("If you perform in the following manner, you will be given a raise.") The worker is now aware of the response that is required to obtain the reinforcement and expectancy of the reinforcement will cause action. (2) The second approach to reinforcement would require new behavior to be shaped from existing behavior. Assume that the desired pattern does not currently exist (and, therefore, cannot be reinforced). To move toward the new behavior, a reward may be applied to responses which closely approximate the desirable behavior. Modification or shaping is accomplished by discriminately employing reinforcement to behavior that is close to what you desire. A closely controlled *reinforcement schedule* is applied when the wanted end behavior appears.

Reinforcement schedule merely means how and when you choose to reinforce. Many types of reinforcement schedules are available. Generally they are classified into a combination of fixed or variable and interval or ratio. A fixed schedule (reinforcement each time the desired behavior occurs), might be employed until the pattern is at an acceptable level. Later, rewards may be administered on a variable (not every time) basis. A decision about whether to use a fixed or variable schedule and whether this schedule should be based on time (interval) or on the number of performances (ratio) must be carefully analyzed. For example, production jobs that can be performed and measured by each piece produced are easily adaptable to ratio schedules. On the other hand, administrative positions are better suited to interval scheduling. Generally, when possible, variable schedules of reinforcement should be used because they tend to provide the greatest resistance to extinction.

Since we want each worker to perform safe acts indefinitely on into the future (with great resistance to extinction), a variable ratio schedule of reinforcement is best. This merely means that we do reward for the safe behavior, but not every time. We reward after six safe acts sometimes, after twenty sometimes, etc. We reward not based on differing amounts of time, but rather on differing amounts of performances (number of safe acts.) A variable ratio schedule of positive reinforcement is the exact schedule that makes gambling so fascinating and that builds the compulsive gambler—knowing there is a reward coming, but not knowing when. Any person who has tried the one-armed bandits knows the feeling produced by a variable ratio schedule of positive reinforcement.

Systems of rewards which are currently being used by industry and other types of organizations can be adapted to a behavior modification program. Money, status, promotion, public recognition, and personal praise are all valid reinforcements of behavior. In addition, much research has been done using some form of tokens as reinforcers. While to date most of this research has been with retarded children, psychotics, slow learners, and juvenile delinquents, there are indications that it might be used more and more with "normal" people. It could perhaps be used for safety. Trading stamps are already used in some safety contests.

A properly administered reward system must determine exactly what behavior is being reinforced. First, you must decide exactly what you want the person to do. Next, you must determine how you will reward the person for achievement. Assuming that the paycheck reinforces the job performance is not necessarily valid. Superficially, you might think of salary as a fixed-interval schedule of reinforcement in the sense that the money is a reinforcing event which is delivered every fourteen or thiry days. Actually, however, money reinforces only the behavior of accepting the paycheck. This performance is reinforced every time it occurs. If salary attainment was more related to job behavior, it would be a better reinforcer. This has been done by many companies, particularly for management compensation, where the bonus has been related to safety achievement. Generally this approach has worked. Its weakness is that the reinforcement comes only once a year rather than more often, because learning theory also holds that the reinforcement should immediately follow the performance. Salary increments granted according to attainment of specific predetermined objectives would incorporate a ratio schedule that would provide more direct reinforcement than salary administered on an interval basis. Here is an example from a production situation. A worker's job consists of installing a taillight on an automobile as it passes along the assembly line. We might attach a small device similar to a taximeter to the machine the worker employs. This would ring a bell after the worker has made a sufficient number of responses and the amount earned would flash on a sign. Under a setup like this, the worker might have to be pulled away from the line for coffee breaks.

APPLICATIONS TO SAFETY

The man behind the theory behind behavior modification is B. F. Skinner[151] (Harvard University), who has been experimenting with control of behavior since before World War II. During the war, Skinner trained pigeons to aim aerial bombs accurately. Later, using his principles of conditioning, he taught pigeons to play ping-pong and to play the piano and he taught many other animals to perform feats that seemed unbelievable. In an article in the *Journal of the American Society of Safety Engineers* (now called *Professional Safety*), Bird and Schlesinger[16] outlined the applications of Skinner's behavior modification technique to safety management:

> Most important from the safety professional's point of view, Skinner's concepts and techniques have direct application to both accident-prevention and safety.
>
> The Center for Programmed Learning in Ann Arbor, Michigan, has recently reported a successful program for reducing back-injuries due to improper lifting among supermarket-employees in a large chain. The chain's supervisors were taught the technique of positive reinforcement as applied to correct lifting-practices. Instead of calling attention mainly to a trainee's errors—as was their usual practice—they concentrated instead on finding a trainee who was lifting correctly and then

gave him attention, interest, praise, and recognition. According to Gary Rummler, the director of the Center, this reinforcement procedure paid off far beyond expectations.

Certainly safe-behavior reinforcement is not entirely new. It has been the basis for prizes, for contests, and for inclusion in programs for training supervisors; it is intuitively practiced in industry as it is in everyday life; yet an explicit and methodical application of the technology can make a big difference.

Safe-behavior reinforcement requires two essential management procedures:

1. Organized and frequent CONTACTS between (a) supervisors and staff personnel and (b) employees—for the dual purpose of first immediately recognizing among employees instances of desired safe behavior, and second promptly directing attention, approval, praise, recognition, and actual words toward any worker who is performing in the desired manner;

2. A REDESIGN of jobs and equipment so that each employee's on-the-job performance can regularly produce both a MAXIMUM of personally satisfying rewards and a MINIMUM of job-related (or personal) punishment.

In sum, safe-behavior reinforcement operates through both effective supervision and careful job-design. The desired safe behavior should produce rewards both from the job itself and from frequent recognition by supervisory personnel.

In order to fully appreciate the dynamic reversal from our existing industrial practices that this program implies, let's review for a moment the motivational direction that many industrial programs have taken through the years.

First,—safety personnel, by their heavy emphasis on unsafe practices and conditions,—injury, doom, death, and other dismal topics—may have illustrated Pavlov's law.

Safety-specialists are like the pediatrician who gives a repeated series of hypodermic shots to a child. The child soon cries at the mere sight of him.

In their contacts with workers, safety-supervisors, by highlighting unsafe on-the-job practices and their effects, may have wrapped the same mantle of doom around themselves.

Secondly, today's safety supervisors may not be giving enough attention to Skinner's principle, since their regular safety-inspections usually focus on unsafe practices and conditions; today's safety specialist is far too likely to deliver "punishment" rather than "reward." Seldom does he emphasize safe behavior, or even mention it in any written report.

Safety personnel in the field are generally trained to concentrate on unsafety in their contacts, communications, and observations. Even their posters and promotional materials reflect the overall negative approach of our safety-motivation effort—by giving major attention to what's wrong, and little attention to what's right in on-the-job behavior.

The card that has been handed to the worker stating (a) that he has been committing an unsafe act that could have injured himself or a fellow-worker and (b) that he ought to think of the consequences in the future makes this point well.

The little card goes further, pointing out that he should look for another fellow-worker who is violating an established safety-practice and pass the card along to him "in the interest of safety."

Blame is a useless concept if we follow the rules of behavior modification, or operant conditioning. People,—safety-specialists and machine-operators alike—are doing "what comes naturally." It does no good to blame them for not doing differently.

The safety-specialist naturally wants to avoid accidents, because they are "punishing" for him. That may sound a bit strange, as we are always advocating that workers should practice safety in their own interests. But the facts are otherwise.

Unsafe acts have entirely different consequences for management than they have for the worker. The worker may be willing to take a chance on an unsafe performance for the personal satisfactions to be gained. However, management (which always deals with the "laws of large numbers") recognizes that chance-taking will lead to predictable injury in the long run.

In contrast with the individual worker who rarely experiences a personal injury, the management team [knows] that the firm will suffer the consequences of a certain number of injuries if unsafe acts and conditions persist.

This analysis suggests that the management team and the worker have quite different points of view about accident-probability. For the firm as a whole, accidents are much more certain than they are for the individual worker. Therefore efforts to get individual workers to accept management's viewpoints on safety are likely to be quite limited in their effectiveness.

As management personnel look at unsafe acts and add up the costs and time-losses, they may well decide that they cannot tolerate unsafe behavior; the cost-benefit ratio is too great. However, the employee's perspective may be just the opposite; as he balances the personal risks involved against the personal gains, he may be willing to take those risks.

Management's Safety-Efforts May Be Misdirected

Unless this fundamental discrepancy in points of view is admitted as a possibility, efforts to increase safe behavior may be misdirected because (1) it is unlikely that employees will completely accept management's view concerning the true seriousness of an accident; (2) management's attempts to change employee-behavior may neglect the very real personal gains which workers actually enjoy from their unsafe acts; (3) these fundamental differences in perspective between management and workers cannot be solved by management's improved communications nor by management's insistence on promulgating its own views while suppressing the personal assertiveness of individual employees.

What is an appropriate strategy? To increase the worker's personal satisfactions that can be gained from safe behavior.

If you can't easily persuade workers to behave safely, can you suppress unsafe acts by discipline and punishment?

Yes, but punishment has limited effectiveness, according to the operant-reinforcer theorists. . . .

Unfortunately, punishment has side-effects which reduce its effectiveness. To summarize:

1. The employee may continue to behave the same way, but try harder to avoid being observed by the supervisor.
2. When the same behavior can lead to both reward and punishment, the employee will be in a state of conflict.
3. Conflict leads to frustration and aggression: the employee is likely to try to take out his frustration by reduced output, substandard quality, damage, waste, or fighting with other workers.

It is true that one way of changing behavior is to make that behavior lead to an unpleasant situation since aversive control—or punishment—involves suppression of behavior by penalizing it. "Punishment" acts the same way as "reward,"—but in the opposite direction.

WHY EMPLOYEES PERFORM UNSAFE ACTS

If both persuasion and punishment are of limited effectiveness, what can be done? The first step is to understand what maintains and supports unsafe behavior.

The worker who regularly performs unsafe acts when he "knows better" is probably following Skinner's law of operant conditioning. The unsafe act has been learned and is maintained because it has been (and continues to be) reinforced by satisfying events.

Behavior that appears "abnormal" (because it courts possible injury) is as much a product of reinforcement as "normal" behavior is; in fact, it is learned right from childhood in exactly the same way: children learn to throw tantrums because their parents provide the right reinforcements at the right time, say the operant conditioners. When parents (a) refuse to give a child what he asks for whenever he cries—only to give in the moment he cries harder, and (b) make a regular practice of refusing to give in unless the child raises the intensity of his crying, the child will soon reach full-blown temper-tantrums.

Unsafe Behavior Is Also Reinforced

There are many reasons for the worker to perform in a way that management would describe as an "unsafe act." Here are some of them:

How Workers Personally Benefit from Unsafety

1. The advantages and satisfaction to be gained by the worker at the particular moment seem greater to him than the disadvantages and dissatisfactions;
2. The unsafe act "makes real sense" to the employee. If he is challenged, he will explain to the foreman exactly why he thinks his way is the most sensible way to do the job. Typically, the older employee will justify himself by saying that he has been doing it that way for years;
3. The unsafe act actually gives the worker personal satisfaction: It may attract the attention of co-workers; gain their approval and admiration; give him either the thrill of taking a chance or the satisfaction of bucking authority—or even paying back an imagined grudge; it may make him feel daring; and it may involve many other personal incentives;
4. To the worker, his "unsafe" act may be perceived as having definite job-related advantages—advantages that include either such monetary incentives as getting his job done sooner, thus increasing his work-output and his take-home pay—especially if he is on piece-work pay—or personal incentives such as avoiding extra effort or fatigue and having more "personal control" over product-quality.

This list of possible employee-satisfaction which can arise from unsafe acts indicates that these acts are probably strongly motivated; they have meaningful consequences for the employee—consequences that he not only enjoys but that actually support and maintain his unsafe behavior.

WHAT IS SAFE-BEHAVIOR REINFORCEMENT?

Any behavior-pattern will sharply increase in probability if it is immediately followed by some event that satisfies a need.

By Rewarding Safety You Get More Safety

Whenever this satisfying-state-of-affairs consistently follows the same behavior-pattern its probability of repetition will increase to a high level. As we have seen, unsafe on-the-job behavior-patterns may often have that characteristic: they are consistently followed by personal need-satisfying events. We repeat that these are both personal and job-related needs.

Management's Rewards Must Fast Follow Safe Acts

Management's plans to control the behavior-patterns of its employees require that satisfying events (rewards) must regularly occur immediately after the desirable behavior. If an employee's safe behavior can consistently be followed by need-satisfaction, management can regularly achieve consistent control in the directions that are desired. As we have seen, however, the worker's safe behavior-patterns seldom result in as wide a variety of personal satisfaction as his unsafe behavior-patterns do.

Technically speaking, unsafe behavior-patterns in industry are actually maintained and controlled by the personally-satisfying events that result from them.

We, as safety-supervisors, instead of attempting to suppress unsafe behavior (by having it followed by punitive discipline), should strive to increase the satisfactions that follow safe behavior.

Reinforcing Safe Acts Has Many Advantages:

1. It removes the unwanted side-effects of discipline, avoidance, conflict, frustration, and aggressive damage;
2. It increases the job-satisfactions experienced by the employee;
3. It changes the nature of the relationship between a foreman and his employees from a "watchdog" and "policeman" to a helpful resource;
4. It creates an atmosphere of mutual reciprocity between supervisor and employee;
5. It increases the probability of safe behavior, rather than reducing the probability of unsafe behavior;
6. It is more direct in its effects.

Skinner and his colleagues have used these principles to teach desirable behavior. They always emphasize desirable behavior—or a close approximation of it.

Several experimenters in dealing with such diverse groups as juvenile delinquents, stutterers, the obese, mental patients, and young children were able to alter behavior by focusing on the desirable behavior and then increasing its advantages to the behavior. These experimenters found that reinforcement had to be given precisely at the time the desired behavior occurred.

A reinforcement, these psychologists say, is a "high-probability behavior"—one that would occur spontaneously if the person were simply left to his own devices and preferences—one that can be used to reinforce (reward) the less-probable habits. . .

MOTIVATING EMPLOYEES TO WORK SAFELY

Every first-line supervisor knows that an unsafe act is "an accident about to happen." The foreman who recognizes these deviations from safe practices does his best to eliminate them.

This recognition is an obvious start towards a solution to the probelm; but it is only a small part of the solution, since there are many ways for an employee to behave unsafely on the job—especially compared with the number of ways for him to work safely. The foreman who spends his time trying to eliminate each unsafe act among each employee will soon run out of time, patience, and energy.

An alternative (but less-obvious) strategy is for the foreman to increase the readiness for safe behavior.

With these two strategies to choose from—(1) the ELIMINATION of unsafe acts, and (2) the BOLSTERING of probabilities for safe acts—the foreman will be better off following the second strategy.

There are many ways he will benefit from this strategy:
1. He will spend more time recognizing and rewarding safe performance, compared to time spent in disciplining employees for unsafe performance.
2. He will strengthen and enhance the importance of management-standards for safe performance.
3. He will focus attention on the importance of safe performance.
4. He will remind employees of the techniques of safe performance.
5. He will be seen by the men who work for him as a colleague interested in their welfare—rather than as a disciplinarian (or "nag").

HOW CAN MANAGEMENT INCREASE READINESS
TO CHOOSE SAFE BEHAVIORS?

Management's final objective is to reduce both the injuries and the property-damage caused by unsafe acts (improper procedures; use of unsafe tools, equipment, materials, etc.).

Safety-Specialists Have Two Main Avenues Open

To increase safe acts and reduce unsafe acts, safety-specialists can follow these two strategies: they can—

1. Increase The Satisfactions
   Associated With Working Safely by—
   a. Increasing the ratio of time spent on recognition of safe behavior compared with the time spent in disciplining employees for unsafe behavior.
   b. Recognizing that safe behavior should be separated from discipline.
   c. Increasing the ratio of (a) recorded commendations to (b) recorded reprimands.
   d. Recording an increasing ratio between safety-observations and inspections that deal with safe behavior rather than with unsafe behavior.
   e. Giving employees tangible recognition and awards for their safe behavior.

    f. Rewarding safe behavior through increased opportunities among employees for job-satisfaction, work-group satisfaction, supervisory-support, and rewards from the company.

    g. Placing increased emphasis on the personal gains that employees reap whenever they work safely.

    h. Placing increased emphasis on the job-gains of working safely.

    i. Using both increased individual participation and workgroup participation in developing recommended safe procedures.

    j. Developing operational procedures not only for identifying but also for rewarding safe practices. For example, each month a specific safe practice (or set of safe practices for doing different jobs) could be publicized.

    k. Making (whenever possible) an actual audit of the occurrences of safe practices—an audit that measures and reports the actual results of the program.

2. Eliminate The Obstacles That Discourage Working Safely:

For each safe practice chosen for program emphasis, the supervisor should:

    a. Determine the obstacles: Time? Effort? Discomfort? Interference with production? Personal or group-dissatisfactions? (Discussions with the workers actually involved should help identify these obstacles).

    b. Reduce the obstacles: Use group-participation techniques; get ideas and suggestions from employees; make these ideas the topic of safety meetings; invite workers' ideas and reactions—especially to any solutions you or your supervisors might come up with.

    c. Evaluate each safety-improvement on a cost-benefit basis.

    d. Discuss each proposed change with the workers involved, and get their agreement and cooperation before making the change.

As we have said previously, the presence of unsafe acts indicates that the employee sees each unsafe act as having more-immediate personal satisfactions than the safe act. The satisfactions and advantages of the safe act need to be increased while the obstacles and dissatisfactions need to be reduced.

Each foreman can develop a list from his own experience. Each month a troublesome safe practice can be selected for study and action. For example: (1) hold safety-meetings and informal discussions with employees; (2) hold meetings among foremen who have the same problems or similar ones.

Employee need-satisfactions can be influenced in two ways (or a combination of the two, since it is a mistake to try to solve all motivational problems using merely one procedure).

These two necessary procedures are: (1) increase the satisfactions and advantages of a worker's safe act by communications, tangible (and intangible) rewards; (2) increase the satisfactions and advantages to workers by changes in job design.

Time spent by safety-professionals on the nourishment of safe behavior will really pay off. Special reinforcement techniques—such as praise, approval, or tangible reward—do not have to be continuous to be effective; but if each job is designed so that safe behavior can be made to yield the most satisfaction—or the least punishment (energy, time, discomfort)—then each job will have its own built-in source of reinforcement. And your firm's overall safety-record will automatically reach new heights year after year—all with lower supervisory costs, but higher morale in every department throughout the entire firm."

This rather lengthy quotation from Bird and Schlesinger summarizes behavior modification applications in safety management.

NON-SAFETY APPLICATIONS OF BEHAVIOR MODIFICATION

There are many notable successful uses that have been made of behavior modification in business. Probably the most successful application of reinforcement theory with dollar payoff has been the work of Ed Feeney, Vice President, Systems Performance at Emery Air Freight. Feeney's process and spectacular results have been documented in a number of business journals. Feeney asks these kinds of questions on preparing to utilize the reinforcement principles.

1. What is the standard of performance?
2. Does the employee know the standard?
3. How well does the employee think he or she is doing?
4. How well does the supervisor think the employee is doing?
5. What aversive consequences of the desired behavior may be suppressing it?
6. What is reinforcing the undesired behavior?
7. What natural or contrived reinforcers are at hand in the immediate work environment to begin reinforcing the desired behavior?
8. What aversive consequences of the undesired behavior are at hand?
9. What learner responses are already available in embarking on a program of progressive approximation to the desired behavior?
10. What schedule of reinforcement is most efficient for developing and maintaining the desired behavior?
11. What reinforcers are available to reward the worker's supervisor for reinforcing the worker's new behavior?

Exhibit 21.1 outlines a number of applications that have been made of the behavior modification concepts and shows the results achieved.

The key concept in behavior modification is that a stimulus is presented after the behavior has occurred. If this stimulus is reinforcing, one can anticipate the occurrence of that behavior again in the future because something was gained from emitting that behavior. If nothing happens (the behavior is extinguished), one can anticipate the frequency of that behavior recurring to be very low or nonexistent. If the behavior is punished (suppressed), we can anticipate that the frequency of the behavior occurring again will also be very low or nonexistent, at least in that situation. Thus the consequence the behavior generates is the key factor as to whether or not it will occur again in the future.

The behavior modification technique begins by an analysis of the stimulus, the behavior, and the consequence. Next, a baseline is established to determine how often, where, when, under what conditions, and to what extent the behavior occurs. This baseline tells specifically when and how often the behavior occurs and, by analyzing the situation, one can ascertain why it occurs. At this point if the behavior is deemed inappropriate as in an unsafe act, an intervention strategy is devised to change the behavior to one which is more appropriate. This is where the importance of the consequence comes into play. If the behavior being emitted is inappropriate but is still being reinforced, the environment can be constructed so that this reinforcement no longer occurs. If the behavior is appropriate but no reward is forthcoming, then the environment must be constructed so that it contains some reinforcement schedule to "pay off" the appropriate behavior.

The traditional approach to safety training and supervisory development is based on the following set of assumptions:

1. The worker or supervisor can be taught new knowledge, skills, abilities, and attitudes.

2. These new inputs will produce a significant alteration in the worker's motivational pattern.
3. These new inputs will be retained even when the worker, or supervisor, leaves the training program.
4. The worker will be able to apply the new inputs to job behavior, performance, and effectiveness for goal attainment.

These traditional assumptions no longer seem valid. A new set of assumptions are needed for effective safety and supervisory training. The new approach suggested is derived from learning theory and, more specifically, from operant conditioning and behavior modification technology. This new approach encompasses the following assumptions:

1. Improved job performance and job effectiveness is needed rather than merely increased inputs.
2. Training should be focused on solving problems and not just on improving the skills of trainees.
3. There should be insistence on empirical evaluation of training instead of accepting the results of faith.
4. Training should emphasize the trainee's job experiences and the organization's goals.
5. Positive reinforcement obtained from actual job behavior should follow all training.
6. Instead of the value assurance that training will be "good for you" there should be a greater emphasis on job performance.

In recent years a great deal has happened in industry in the use of behavior modification principles. In the first edition of this book, after talking about Feeney's work, Bird and Schlesinger's article, and the possible use of behavior modification in safety, we used up pretty much everything known or used.

We have a different picture today. The concepts have now been used and well tested in both safety and non-safety applications. Exhibit 21.1 shows a sampling of non-safety applications. Exhibit 21.2 shows a sampling of safety experiments using positive reinforcement to reduce unsafe behavior. The author had the pleasure of participating in several of these.

POSITIVE REINFORCEMENT EXPERIMENTS

In the spring of 1982, the Association of American Railroads, funded by the Federal Railroad Administration, embarked upon an experiment to learn if the concept of positive reinforcement would be feasible and effective in an industry as different as railroading.

A project study team was formed composed of safety directors of some of the major U.S. railroads, and myself as an outside consultant. In earlier years, this project study team had done major work in administering a survey to several pilot railroads to better understand the perception of workers, supervisors, and three levels of managers of railroad safety programs.

Phase II of this research effort was to determine if some new approach might impact those accidents on the job caused by human error. Traditionally railroad statistics show that a major portion of all accidents are caused by human error.

On each of two pilot railroads a group of between 75 and 150 firstline supervisors were selected to receive experimental training in human behavior, accident causation, human error reduction, positive reinforcement techniques, and employee assessment techniques. The cooperation of their superior was also sought to assure that the techniques taught would be used by the supervisors once back on the job after the training.

Length of Program	Number of Employees Covered/ Total Employers	Type of Employees	Specific Goals	Frequency of Feedback	Reinforcers Used	Results
1969-1976	500/2800	Entire workforce	(a) Increase productivity (b) Improve quality of service	Immediate to monthly, depending on task	Previously only praise and recognition; others now being introduced	Cost savings can be directly attributed to the program
1972-1976	2000/5500	Employees at all levels in operator service	(a) Decrease turnover & absenteeism (b) Increase productivity (d) Increase Improve labor-management relations	(a) Lower level—weekly & daily (b) Higher level—monthly & quarterly	(a) Praise & recognition (b) Opportunity to see oneself become better	(a) Attendance performance has improved by 50% (b) Productivity and efficiency has continued to be above standard in areas where positive reinforcement (PR) is used
1974-76	220/5500	Maintenance workers, mechanics, & first- & second-level supervisors	Improve (a) productivity (b) quality (c) safety (d) customer employee relations	Daily, weekly, and quarterly	(a) Self-feedback (b) Supervisory feedback	(a) Cost efficiency increase (b) Safety improved (c) Service improved (d) No change in absenteeism (e) Satisfaction with superior & co-workers improved (f) Satisfaction with pay decreased
1941-1976	3000/ 13,500	Clerical employees & first-line supervisors	(a) Decrease absenteeism (b) Decrease lateness	Immediate	(a) Self-feedback (b) System-feedback (c) Earned time off	(a) Chronic absenteeism & lateness has been drastically reduced (b) Some divisions refuse to use PR because it is "outdated"
1974-1976	1000	Employees at all levels	(a) Meet EEO objectives (b) Decrease absenteeism & turnover (c) Improve training (d) Increase productivity	Immediate — use modeling & role playing as training tools to teach interpersonal exchanges & behavior requirements	Social reinforcers (praise, rewards, & constructive feedback)	(a) Cost savings can be directly attributed to the program (b) Productivity has increased (c) Worked extremely well in training minority groups and raising their self esteem (d) Direct labor cost decreased
1974	28	Supervisors	Increase supervisor competence	Weekly over 5 weeks (25-hour) training period	Feedback	(a) Improved supervisory ability to give feedback judiciously
1973-1975	1122/1930	Garbage collectors	(a) Reduction in paid man-hour per ton (b) Reduction on overtime (c) 90% of routes completed by standard (d) Effectiveness (quality)	Daily & quarterly based on formula negotiated by city & sanitation union	Bonus (profit sharing) & praise	(a) Citizen complaints declined signficantly (b) City saved $1,654,000 first year after bonus paid (c) Worker bonus was $307,000 first year or $350 annually per man (d) Union somewhat dissatisfied with productivity measure and pushing for more bonus
1972-1976	160/420	Manufacturing employees at all levels	(a) Better meeting of schedules (b) Increase productivity	Weekly	Praise & recognition; freedom to choose one's own activity	Production has increased over 300%

Exhibit 21.1 Results of Behavior Modification Programs in Organizations.

Length of Program	Number of Employees Covered/ Total Employers	Type of Employees	Specific Goals	Frequency of Feedback	Reinforcers Used	Results
1974-1976	350/350	All levels	(a) 96% attendance (b) 90% engineering specifications met (c) Daily production objectives met 95% of time (d) Cost reduced by 10%	Daily & weekly feedback from foreman to company president	Positive feedback	(a) Profit over 25% over forecast (b) $550,00 cost reduction on sales (c) Return on 1900% on investment including consultant fees (d) Turnaround time on repairs went from 30 to 10 days (e) Attendance is now 98% (from 93%)
1974-1976	500/40,000	Clerical, production (tree planters) & middle-level management & scientists	(a) To teach managers to minimize criticism & to maximize praise (b) To teach managers to make rewards contingent on specific performance levels & (c) To use optimal schedule to increase productivity	Immediate— daily & quarterly	(a) Pay (b) Praise & Recognition	(a) Using money, obtained 33% increase in productivity with one group of workers, 18% increase with a second group, an 8% decrease in a third group (b) Currently experimenting with goal setting & praise and/or money at various levels in organization (c) With a lottery-type bonus, the cultural & religious values of workers must be taken into account

Exhibit 21.1 Continued

An experimental design was conceived to establish whether or not the program was effective. On each railroad one division was selected as an experimental division (the division to receive training and implement the reinforcement techniques), and an adjacent, similar division selected as a control division (to receive no training and to do nothing differently).

Measures of Performance

For each of the four divisions as many measures of safety performance were taken as possible both before and after to best judge the results. Measures used were:

1. Knowledge attained by supervisors as a result of the training experience. Before and after tests were used.
2. Reaction to the training and to the technique of positive reinforcement were asked, initially and after three months of using the technique.
3. Performance of supervisors, before and after three months, was measured by asking each to report the number and kinds of activities engaged in the week prior to training, and the week prior to the three months follow-up training.
4. Performance of the work force as measured by safety sampling, a statistically valid sample of worker behavior, sampled before training, after three months, and after six months. Samples were taken in all cases by safety professionals, and the same samplers were used throughout the entire study period to ensure accuracy and uniformity.

Accident results were not used as the measure of whether or not the program was effective. Whether or not accidents happened is simply an invalid measure of effectiveness due to the small size group and the short length of the study. One division's record for six months is not a large enough sample to tell the facts needed to judge effectiveness of any program.

RESEARCHER	YEAR	TYPE REWARD	MEASURES USED	RESULTS OBTAINED	PERCENT IMPROVE-MENT
1. Komaki, Barwick & Scott	1978	Praise & Feed-back 3-4x/wk	% Safe Behaviors	70% up to 95.8% 77.6% up to 99.3%	37% 29%
2. Komaki, Heinzman & Lawson	1980	Feedback Daily	% Safe Behaviors	34.4% up to 68.4% 70.8% to 92.3%	98% 30%
3. Krause	1984	Feedback	Unsafe Behaviors Lost Time Accidents Severity Rates		80% 39.3% 39.2%
4. Sulzer-Azaroff & DeSanta-maria	1980		Accident Freq. Accident Freq. Hazards Hazards	15 to 0 45 to 33	100% 27% 29% 88%
5. Zohar	1980	Tokens for future use	% use of Ear Plugs % use of Ear Plugs % use of Ear Plugs	35% up to 85% 35% up to 90% 50% up to 90%	143% 157% 80%
6. Sulzer-Azaroff	1978	Trading Stamps		Reduction (No more information)	
7. Fox	1976	Trading Stamps		Reduction (No more information)	
8. Calkins	1971	Cash Bonus & Paid Time Off	Vehicle Accidents Personal Injuries Dollars Saved	 $5766 saved — $1899 spent	48.7% 41.0%
9. Hoppe & Terry	1983	Tangible Incen-tives	% Seat Belts Used % Seat Belts Used	36% up to 60% 60% up to 70.3% 65% up to 70%	67% 8% 8%
10. Rummler		Praise & Recogni-tion	Lifting Practices	Improved (No more information)	
11. Smith, Anger & Uslan		Praise	Accident Rate	Improved in 5 crews	
12. Felliner & Sulzer-Azaroff	1984	Praise & Feed-back	% Safe Behaviors	78% up to 86% 79% up to 85%	10% 8%
13. Harrell		Self Monitoring	Number of Accidents	Reduced for 8 mos.; Reversal Phase — Returned to Previous	
14. Komacki	1979	Feedback & Off Time			
15. Sundstrom	1984	Piece Rate to Monthly Rate	AccidentFrequency		29%
16. McKelvey	1973	Financial Incen-tive	Number of Accidents		80%
17. Petersen	1983	Praise & Feed-back	Safety Sampling	Two Railroad Divis-sions for six mos. with control groups	40% 49%
18. Rhoton	1980	Praise & Feed-back	Violation	1-4 per mo. to 0	100%
19. Fox, Hopkins & Anger		Tokens	Accident Frequency		90%
20. Hopkins, Conrad & Smith		Feedback & Money	% Safe Behaviors Housekeeping Rate Housekeeping Rate	60% up to 100% 20 to 90 45 to 100	67% 350% 122%
21. Chokar & Wallin	1984	Feedback	% Safe Behavior	65% up to 81% 81% up to 95%	20% 17%

Exhibit 21.2 Safe Behavior Reinforcement Studies & Results

RESEARCHER	YEAR	TYPE REWARD	MEASURES USED	RESULTS OBTAINED	PERCENT IMPROVE-MENT
22. Uslan & Adelman	1977	Praise	Injury Frequency		50%
23. Earnest	on-going	Teaching Behavior Management	Incidence Rate Dollars Saved	4.0 down to 2.5 (now much lower) $1.5 million (now more)	38%
24. Nasanen and Saari	1987	Visual & Verbal Feedback	Housekeeping Index # Accidents	62% to 75% FB to Formen only 62% to 88% FB to Workers Also	21% 42%
25. Krause, et. al. Mfg.	1980	Visual Feedback	Unsafe Behavior		60 to 80%
Oil	1982	Visual Feedback	Safe Behavior	79% to 94%	19%
Plastic	1984	Visual Feedback	Safe Behavior	79% to 91%	15%
Chem	1984	Visual Feedback	Safe Behavior	39% to 81%	108%
Util	1985	Visual Feedback	Acc. Freq.		26%
Util	1985	Visual Feedback	Back Injuries		67%
Chem	1985	Visual Feedback	Safe Behavior		35%
Util	1985	Visual Feedback	Safe Behavior		25 to 45%
Util	1986	Visual Feedback	Acc. Freq.		42%
Trans	1984	Visual Feedback	Acc. Freq.		66%
35. Petersen	1987	Verbal Recognition	Unsafe Behaviors	Eight Railroad Divisions over 6 mos. with control groups	11% & 28%

Exhibit 21.2 Continued

Results Obtained

Exhibit 21.3 shows the results for these measures as written up by the project committee in the executive summary and as presented to the top executives of each railroad by the consultant. The primary measure looked at to determine program effectiveness was the safety sampling results. As shown in Exhibit 21.3, in the experimental division there was an improvement of performance of 40 percent in one railroad and 49 percent in the other. Changes in both control divisions were insignificant. All measures showed signficant improvement in the experimental divisions compared to the control division counterparts.

As indicated in V in Exhibit 21.3, the accident record, while not looked at as a judge of results, did improve considerably more in experimental than in control divisions. One railroad showed a 51 percent reduction in the accident record for the period of the study compared to the previous period.

It would seem that this training did in fact have an impact in both experimental divisions. The project study committee of the Association of American Railroads stated that the following statements can be made on the basis of the results obtained:

1. It is possible to significantly improve safety performance through educating first-line supervisors to the human behavior problems associated with accidents, training them to apply positive reinforcement techniques in their daily activities and providing support through the organization for their efforts.
2. It is possible to measure safety performance through observations of employee behavior. This may be a much more reliable technique than the use of accident statistics, in the short term, for determining the effectiveness of a company's safety program effort.

		RRI	RRII
I. Knowledge—gain as measured by pre-and post test scores	A. Average pretest score	4.54	3.58
	B. Average post-test score (same items)	23.57	28.50
	C. Gain	519%	769%
II. Reaction to training and use of techniques taught	A. Average of reaction to training session (scale 0-5)	4.38	4.23
	B. Positive reaction to being asked to use technique (scale 0-5)	4.29	4.18
	C. Positive reaction to using technique (3 mos. later 0-5)	4.04	4.05
III. Performance reported by supervisors	A. Reported increase in use of techniques taught—as measured by what supervisors described themselves as doing before and after training.	40%	14%

		% of Unsafe Behavior Observed			
		RRI	RRI Control	RRII	RRII Control
IV. Performance of work force as measured by sampling of worker behavior before and after training supervisors	A. Prior to training	35%	34%	39%	31%
	B. Three months after training	20%	34%	23%	29%
	C. Six months after training	21%	31%	20%	30%
Improvement in performance		40%	8%	49%	3%

V. Accident performance for the two areas was influenced by a number of variables not related to the study. A short-term comparison of reported accidents for the experimental and control groups on both properties shows both "experimental groups" experiencing fewer injuries than did their respective "control groups."

Exhibit 21.3 Results of positive reinforcement program.

The training provided supervisors in both experimental divisions is shown in Exhibit 21.4. Six hours of training was given with a three-hour follow-up session three months later. All training was by the consultant.

Supervisors Required to Use Training

More important than this training, however, is the fact that supervisors were required to use some of the training given when back on the job. In each case the division manager was asked to devise a simple system of accountability where each supervisor had to report weekly as to the utilization of the concepts learned. In each case the supervisors were asked to observe workers as usual and to report to their boss the number of observations made each week.

```
HOUR 1-2      INTRODUCTION
                    Overview of the study
                    Introduction to the Session
                    Pre tests and questions

              WHY TRY A DIFFERENT APPROACH TO SAFETY
                    Updating safety theory
                    —What we've always done vs. what we now know
                    —What an accident is, and what causes them
                         —Unsafe acts/conditions vs. human error
                    —What we can do to prevent accidents
                         —The 3 Es of safety vs. the 3 behavior changers
                    —What other companies and industries have found

HOUR 3        WORKER MOTIVATION AND WHAT WE KNOW ABOUT IT
                    What motivation is
                    —A person doing whatever is necessary to satisfy current needs
                    What needs are
                    —The major theories and concepts
                         —Needs of most workers
                         —Needs change
                         —Motivators and dissatisfiers
                    —What the research shows

HOUR 4        INFLUENCING WORKER BEHAVIOR
                    Your three choices
                    —Motivation—the environment
                    —Attitude change
                    —Changing behavior
                    Motivation—the easiest way
                    —What influences workers usually
                         —The peer group
                         —You and your style of leading
                         —Your credibility (particularly in safety)
                         —Your attitude toward safety (what they see)
                         —How you measure and judge them
                         —What they think your priorities are
                         —The organizational climate

HOUR 5        WHAT YOU CAN DO
                    Understand each worker
                    Understand the group of workers you supervise
                    Use the motivators

HOUR 6        USING POSITIVE RECOGNITION


                    FOLLOW UP SESSION OUTLINE

HOUR 1        REVIEW OF THE SESSION 1 CONCEPTS
                    Motivation concepts
                    The influences on worker behavior
                    Positive recognition

HOUR 2        HOW YOU DID IN THE LAST 3 MONTHS
                    Difficulties
                    Successes

HOUR 3        ENSURING YOU UNDERSTAND EACH WORKER
                    Your task for the next three months
                    —The assessment technique
                    —Continuing positive recognition
```

Exhibit 21.4 Training Session Outline.

Second, supervisors were asked to act following each observation. If the worker observed was working unsafely, the supervisor was asked to deal with the infraction normally. If, however, the worker was working safely, the supervisor was required to act also—to contact the worker and positively reinforce that desired safe behavior, to do something with that worker that might make the worker more likely to work safely in the future.

Each week each supervisor reported on a 3 x 5 card the number of observations made and the number of positive reinforcements given.

After three months, a short three-hour follow-up session was held for all supervisors in the experimental groups. The concepts previously taught were reviewed and discussions held on the reactions to the technique, both good and bad.

Then an outline of some of the concepts taught earlier was provided each supervisor and each was asked, in normal supervision, to look and think daily about one individual and to simply assess that individual using the provided outline (Exhibit 21.5). For the next three months they were to continue reporting observations and positive reinforcements on the 3 x 5 card, and also to report the number of assessments made with the outline. No names were asked for or recorded, but merely the number of times they did in fact stop to consider the needs of an individual.

ASSESSMENT
QUESTIONS TO CONSIDER

In the sessions 3 months ago and today, we talked about a lot of concepts and theories. Use some of these as you look at each individual that works for you:

Which motivator seems to turn him on most?
 Achieving?
 Responsibility?
 Recognition?
 Growth?
 The work itself?
 Advancement?
Which dissatisfier seems to turn him off most?
 Company policy?
 Rules?
 Me?
 My approach?
 Working conditions?
 Money, security?
How important is the peer influence on him?
 Is he a part of the group or a loner?
 Is it a strong cohesive group, or a weak one.?
 Is his group for safety, or do they fight it?

How have I been managing him?
 Authoritatively?
 Participatively?
How should I be managing him if I want to turn him on?
To him, am I credible when I push safety?
 Does he think I know what I'm talking about?
 Does he think I have a right to talk safety based on the past?
What attitude toward safety does he think I have?
What attitude toward safety does he have?
How do I measure him most of the time?
 By the amount of work?
 How safe he works?
 How fast he works?
 Some other way?
How do I reward him? How should I?
What do I reward him for most of the time?
What priority does he think safety is?
Does he think I have?
What is his perception of the climate of our organization?

Exhibit 21.5 Assessment questions to consider.

At the completion of six months the final sample was taken to determine results, as shown in Exhibit 21.3.

The experiment seems to indicate a positive value to the training and to the techniques used.

This original railroad project was done in 1982 and 1983. Due to the results, another similar project was done during 1987. Two different railroads were used, one of the country's largest, which volunteered six divisions (three experimental and three control), and a smaller regional railroad.

The comparison of pre and post test scores showed results similar to the earlier (1983) results. Course reactions also were relatively high, as were supervisory reactions to the concept of positive reinforcement. There were mixed results on supervisory performance. In the large railroad, one division increased its supervisory safety activity by 155 percent during the three month study period, while another division increased activity by only 20 percent. In the smaller railroad there was a substantial increase in both the activities that were required (observing, contacting, and using positive reinforcement) and all other safety related activities. Safety samples showed results similar to the first study.

The bottom line is that positive reinforcement does work. In Exhibit 21.2 you'll notice no instances where positive results were not achieved. The average improvement over the 35 experiments was 59 percent (and usually over short periods of time). It appears this tool is ready to take its place in traditional safety technology.

Chapter Twenty-Two

Building a Safe Environment

Because people, cannot be redesigned to fit the machine they must use in the industrial scene, their tools and their work environment must be redesigned to reduce errors and accidents.

Any discussion of the human approach to safety management would surely be considered incomplete without some mention, however brief, of human factors engineering, even though in many respects this field is somewhat different than those previously discussed. One reason it is different is that more is known and more research has been carried out here than in the other fields discussed, research that has as one of its primary purposes the safety of the operator. Human factors by its very definition is safety related. While human factors engineering is not an exact science, it is a rational approach to the problems of designing and constructing things which people are expected to use so that the user will be less likely to make errors resulting in accidents. It attempts to make machines more convenient and comfortable, less confusing, less exasperating, and less fatiguing.

The discipline has various names. In Great Britain it is called *ergonomics.* Others call it *biomechanics, biotechnology, psychophysics, biophysics, human engineering, human factors,* and *engineering psychology.* Despite the variety of names, there is general agreement that emphasis is placed on anthropology and medicine. To start with, here are some simple definitions of human factors engineering:

1. Engineering for the population that will use it.
2. Designing the system so that machines, human tasks, and the environment are compatible with the capabilities and limitations of people.
3. Designing the system to fit the characteristics of people rather than retrofitting people into the system.

Frank Vilardo[162] has outlined a couple of quick examples showing the possibilities of human factors engineering in design:

> Exhibit 22.1 shows two electric meters. To read the meter on the bottom (the old but still used meter), a man relies heavily on previous knowledge of such meters. But the meter at the top has been redesigned to print-out directly the reading, which requires no special knowledge or interpretation to record.
>
> Reiterating, one of the three functions man serves in a system is that of sensor; Exhibit 22.1 is an example of human factors engineering being used to minimize errors in sensing.

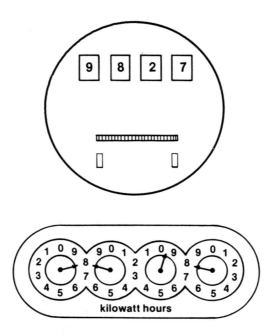

Exhibit 22.1 Human factors use on minimizing errors in sensing.[162]

Two heat-regulation controls are shown in Exhibit 22.2. Control A is not designed with the human operator in mind; Control B, on the other hand, is designed to eliminate errors in the information-processing and machine-controlling functions.

Although these two illustrations are elementary, they present the possibilities that human factors engineering holds.

In any man-machine system, there are tasks that are better performed by man than by machine—and, conversely, tasks that are better handled by machines.

In general, machines can usually perform more efficiently on those tasks that must be performed routinely and rapidly with a high degree of accuracy in a given

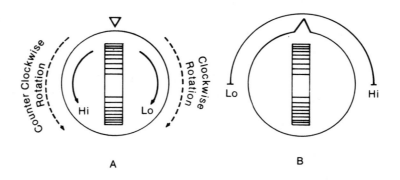

Exhibit 22.2 Human factors use on minimizing dial reading errors.[162]

environment. Men perform better the tasks calling for responsibility and flexibility in addition to tasks that cannot be anticipated. Exhibit 22.3 outlines the uses of man and machine.

Design Considerations. A wealth of data on human factors has been and is being developed by research to determine the physiological limits of the human with respect to his environment: e.g., temperature, noise, humidity, acceleration, light, vibration, pressure, etc.

The literature will usually yield sufficient information to determine whether a given environment will be hostile to the human or degrade his performance. Most of that information is not included here, but is available through the references.

Man excels in	Machines excel in
Detection of certain forms of low energy levels	Monitoring (both men and machines)
Sensitivity to an extremely wide variety of stimuli	Performing routine, repetitive, or very precise operations
Perceiving patterns and making realizations about them	Responding very quickly to control signals
Detecting signals in high noise levels	Exerting great force, smoothly and with precision
Ability to store large amounts of information for long periods and recalling relevant facts at appropriate moments	Storing and recalling large amount of information in short time periods
Ability to exercise judgment where events cannot be completely defined	Performing complex and rapid computation with high accuracy
Improvising and adopting flexible procedures	Sensitivity to stimuli beyond the range of human sensitivity (infrared, radio waves, etc.)
Ability to react to unexpected low-probability events	Doing many different things at one time
Applying originality in solving problems, i.e., alternate solutions	Insensitivity to extraneous factors
Ability to profit from experience and alter course of action	Ability to repeat operations very rapidly, continuously, and precisely the same way over a long period
Ability to perform fine manipulation, especially where misalignment appears unexpectedly	Operating in environments which are hostile to man or beyond human tolerance
Ability to continue to perform even when overloaded	
Ability to reason inductively	

Exhibit 22.3 Man vs. machine.[172]

The comparisons of Exhibit 22.3 should be viewed only as guidelines in designing any particular job or man-machine system, as man is highly flexible and adaptable but man is generally excluded from tasks that are likely to result in a high probability of error. Such tasks are:

1. Perceptual requirements near or beyond the physiological limits or that conflict with established perceptual patterns. These limits are spelled out in Woodson's *Human Engineering Guide.*[172]
2. Response requirements that are physically difficult, conflict with established patterns, or cannot be readily checked or monitored for adequacy. Exhibit 22.4 lists some general response population stereotypes.
3. Decisions that require undue reliance on short-term memory or must be accomplished within too short a time interval in view of other necessary tasks.
4. Tasks that overload the human, resulting in imbalanced workload/time distribution, or do not permit adequate or timely monitoring of the system.
5. Communication requirements that conflict with other activities.

If an operator misreads a poorly designed display and operates the wrong control, or the right control in the wrong direction, safety may be jeopardized and the system effectiveness degraded, if not lost entirely. Many accident reports would classify this as an "unsafe act" or "human error" when it is really a design error. Retraining of the operator would not prevent recurrence of the series of events that led to the accident. Much of the effort in occupational safety has

---

- Handles used for controlling liquids are expected to turn clockwise for off and counterclockwise for on.

- Knobs on electrical equipment are expected to turn clockwise for on or to increase current and counterclockwise for off or decrease in current. *(Note:* This is opposite to the stereotype for liquids.)

- Certain colors are associated with traffic, operation of vehicles, and safety.

- For control of vehicles in which the operator is riding, the operator expects a control motion to the right or clockwise to result in a similar motion of his vehicle, and vice versa.

- Sky-earth impressions carry over into colors and shadings: light shades and bluish colors are related to the sky or up, whereas dark shades and greenish or brownish colors are related to the ground or down.

- Things which are further away are expected to look smaller.

- Coolness is associated with blue and blue-green colors, warmness with yellows and reds.

- Very loud sounds or sounds repeated in rapid succession, and visual displays which move rapidly or are very bright, imply urgency and excitement.

- Very large objects or dark objects imply "heaviness." Small objects or light-colored ones appear light in weight. Large, heavy objects are expected to be "at the bottom." Small, light objects are expected to be "at the top."

- People expect normal speech sounds to be in front of them and at approximately head height.

- Seat heights are expected to be at a certain level when a person sits down.

Exhibit 22.4 General population stereotype reactions.[172]

been directed toward training. The approach to the machine has been to guard hazards—many of which are simply products of deficient engineering design.

HOW PEOPLE SHOULD BE USED

In this brief overview of human factors, we might consider how people are used and what design considerations we should have for them. Vilardo[162] summarizes what is known:

> One of the functions man serves in a man-machine system is that of a sensor, an information seeker. Contrary ot the popular notion, man has something like 12 to 13 senses—not just five. [Some of these are given in Exhibit 22.5.]
>
> An information display is a device used to gather needed information and to translate such information into inputs that the human brain can perceive.
>
> Two general classes of information displays—pictorial and symbolic—are utilized.

Symbolic displays present the information in a form that has no resemblance to what it is measuring. Some examples are a speedometer, a thermometer, a pressure gauge, and an altimeter. In pictorial displays, the geometrical and spatial relationships are shown as they exist. Maps, pictures, and television are examples of pictorial displays.

The two most common types of symbolic displays are the visual and auditory. Much study has been given to the design characteristics of these types of displays and some general principles have emerged:

Principle of Simplicity

The purpose for which the display is to be read will dicate its design, but as a general principle, the simplest design is the best.

Principle of Compatibility

The principle of compatibility holds that the motion of the display should be compatible with (or in the same direction as) the motion of the machine and its control mechanism. For instance, a pointer that moves to the right to show an increase should have its corresponding display output value.

Principle of Arrangement

As the design of the display is important, so too is its location or arrangement with other displays. A poor arrangement of displays can be the source of error.

Sometimes dials must be arranged in groups on a large control panel. If all the dials must be read at the same time, they should be pointing in the same direction when in the desired range. This will reduce check-reading time and increase accuracy. In Exhibit 22.6, for instance, configuration *A* is better than *B*.

Principle of Coding

All displays should be coded or labeled so that the operator can tell immediately to what mechanism the display refers, what units are measured, and what the critical range is.

Sensation	Sense organ	Stimulated by	Originating
Sight	Eye	Some electromagnetic waves	Externally
		Mechnical pressure	Externally or internally
Hearing	Ear	Some amplitude and frequency variations of the pressure of sur-surrounding media	Externally
Rotation	Semicircular canals	Change of fluid pressures in inner ear	Internally
Falling and rectilinear movement	Muscle receptors	Muscle stretching	Internally
	Semicircular canals	Position changes of small, bony bodies in the inner ear	Internally
Taste	Specialized cells in tongue and mouth	Chemical substances dissolvable in saliva	Externally on contact
Smell	Specialized cells in mucous membrane at top of nasal cavity	Vaporized chemical substances	Externally
Touch	Skin mainly	Surface deformation	On contact
Vibration	None specific	Amplitude and frequency variations of mechnical pressure	On contact
Pressure	Skin and underlying tissue	Deformation	On contact
Temperature	Skin and underlying	Temperature changes of surrounding media or of objects contracted	Externally and on contact
		Mechanical movement	
		Some chemicals	
Cutaneous pain	Unknown but thought to be free nerve endings	Intense pressure, heat, cold, shock chemicals	Externally on contact
Subcutaneous pain	Thought to be free nerve endings	Extreme pressure and heat	Externally and on contact

Exhibit 22.5 Survey of man's senses and the physical energies that stimulate them.[150]

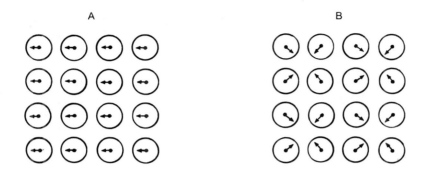

Exhibit 22.6 Principle of arrangement.[162]

REDUCING RESPONSE ERRORS

Any display-response that calls for a movement contrary to the established stereotype is bound to produce errors. The designer is calling for errors by asking the operator to change, in this or that unique situation, a behavior pattern that can be described as habit.

Research bears out these principles:

Principle of Compatibility

Control movement should be designed to be compatible with the display and machine movement.

Principle of Arrangement

This principle provides for the grouping of elements or components according to their function (those having related function are grouped together) and for grouping in terms of how critical they are in carrying out a set of operations. It also suggests that each item be in its "optimum" location in terms of some criterion of usage (convenience, accuracy, speed, strength to be applied.)

Also, items used in sequence should be in close physical relationship with each other, and items most frequently used should be placed closer than items less frequently used.

Principle of Coding

Whenever possible, all controls should be coded in some way. A good coding system uses shape, texture, location, color, and operation. Also all controls and displays should be labeled. Labeling is crucial where the operators change often or the equipment is shared.

OUR PHYSICAL CAPABILITIES

One basic approach to industrial safety relates the physical characteristics and capabilities of the worker to the design of the equipment and to the layout of the workplace. In order to carry out this approach several different types of information are needed, namely, a description of the job, an understanding of the kinds of equipment that will be used, a description of the kinds of people who will use the equipment, and finally, the biological characteristics of these people. In general, the first three items (job, equipment, and users) can be defined both accurately and easily. The biological characteristics, however, are often overlooked or ignored in our own design.

The anthropometirc, or human body size, data needed consist of various heights, lengths, and breadths which are used to establish the minimum clearances and spatial accommodations and the functional arm, leg, and body movements which are made by the worker during the performance of the task. These measurements are defined in detail in Woodson's *Human Engineering Guide.*[172]

We also need data relating to the range, strength, speed, and accuracy of human body movement. Knowledge of human strength in single maximum exertions and over prolonged periods of time is important for the design of equipment in order to ensure maximum safety in operation. The specific kinds of strength data needed vary from task to task, but might include the maximum weights that can be lifted or carried, or the maximum forces that can be exerted in operating a wheel, lever, or knob during single applications and over extended periods.

Dr. Ross McFarland,[95] a pioneer in this field, offers this example of human factors work on safety:

> At this point it might be advantageous to present one concrete example of the application of biomechanical data to the solution of a specific problem which is of great concern to the safety engineer, namely, the large numbers of accidental injuries which result from the lifting, carrying and handling of various objects. Such accidents are the principal cause of compensable work injuries, and within this category they represent some 22.6% of all disabling injuries.
>
> More specifically, handling accidents were the source of 13.9% of all fatal or permanent total disabilities, 9.6% of all permanent total disabilities, and 28.5% of all temporary total disabilities. Many of the accidents are directly attributable to incorrect practices on the part of the worker. To help alleviate this problem, a brief survey was made of existing sources dealing with this topic, and the following general information has been obtained.
>
> In lifting, the most important determinant of the force exertable is the distance of the feet from the grasping axis. Lifting force is greatest when the weight lifted is in the same vertical plane as the body, and decreases sharply as the weight moves away from the body plane. The best height for lifting, as in vertical pull, is at or slightly above the level of the middle fingertip, measured with the subject wearing shoes and standing erect with arms hanging by his sides. Middle fingertip height varies from about 25 inches in men 65 inches tall to 31 inches in men 75 inches tall, averaging about 28 inches. Above this height, lifting power decreases very rapidly; below it, more slowly. With a load near the floor, lifting force may be only 3/4 to 4/5 that at the best height.

Height lifted from floor feet	Maximum weight lifted, pounds
1	142
2	139
3	77
4	55
5	36

Exhibit 22.7 The maximum weight that can be lifted to various heights by men of the 5th percentile in lifting ability.[95]

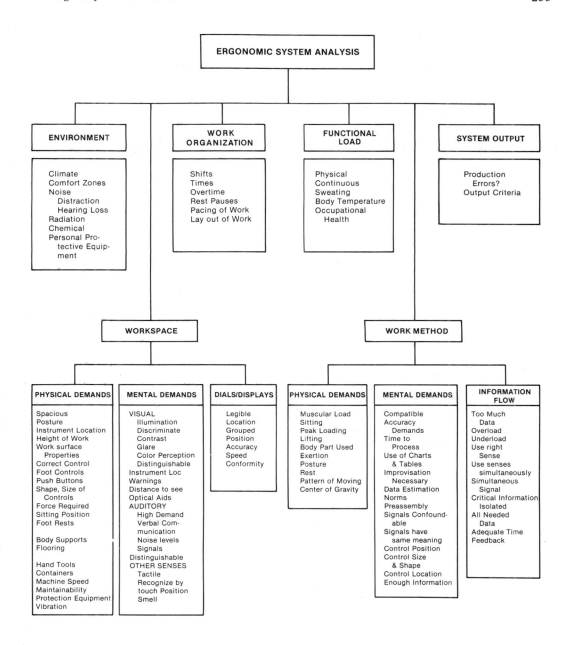

Exhibit 22.8 Ergonomic System Analysis.

In general "leg lift," with back vertical and legs bent, affords a stronger vertical pull than "back lift," with legs straight and back bent, although at lifting heights from 5 to 20 inches. Whitney found no appreciable difference between the two methods. Leg lift entails so much less risk of back injury that it is certainly preferable.

The heaviest weights that can be lifted to various heights by men of the 5th percentile in lifting strength has been calculated by Emanuel and Chaffee. All weights were lifted in a metal case roughly 25 inches long, 11 inches high, and 6 inches wide with handles at either end. The form of lifting was essentially a "leg lift." In using these data, two points should be kept firmly in mind

(1) the subjects were healthy, young males, and
(2) the tests were not repeated to the point of fatigue. [See Exhibit 22.7.]

Women, when measured by their ability to lift a weight, demonstrate about 55 to 65 per cent as much strength as do men.

Relatively little research has been done on carrying though Muller, et al. have established that a "normal" man can move about 200 tons a day through a horizontal distance of 1 meter, or 50 tons through a vertical distance of 1 meter. The method of carrying should permit the maintenance of erect posture and a normal gait insofar as possible, since any departure from these conditions will become progressively inefficient and physiologically costly.

For loads of 20 to 60 pounds a yoke has been found to be most efficient (though this may sometimes be impractical) with the hand carry, one load in each hand, the next best. For heavy weights, the physiological cost is less to carry one load a given distance than to carry two loads, each weighing half as much, the same distance in two trips.

In general, the more compact the container the more easily it can be carried. For bulky equipment the center of gravity should not be more than 20 inches from the carrier's body, and when such articles are as big as 30 inches to a side their weight should not exceed 20 pounds.

While the above statements on lifting and carrying are by no means as definitive and comprehensive as could be wished, they represent a synthesis of the currently available data, and they illustrate the kind of potential contribution that human engineering, specifically biomechanics in this case, can make to industrial safety.

McFarland's comments indicate the kind of information that is available in the human factors research, that for the most part, we in safety management are not using. Obviously, we can barely penetrate the human factors knowledge in this book. We cannot even mention all the high points here. However, we should be aware that such a field exists and that it has rich resources for us.

Another approach to assist is shown in Exhibit 22.8. This Ergonomic System Analysis from *Human Error Reduction*[125] by the author attempts to categorize six areas the safety professional might routinely look into in building the safe work environment.

Note: This subject is covered in greater detail in Techniques of Safety Management, Third Edition.

Chapter Twenty-Three

Building Working Relationships

We are interested in the reactions that take place between people, the relationships that are set up between people as a result of our communicative process. We might call these reactions of these relationships inter-personal relations, which merely means the situation that is set up between people when they say something.

It isn't the words that pass between the two that are of such importance as it is the feeling, the reaction, the interaction that is set up between two people.

Don't you see that the communications that take place between people do set up the bridge for understanding or tear down the bridge for understanding? That is what we mean by this business of inter-personal relations.

C. E. Smith is a speech before the annual meeting of the
American Mutual Insurance Engineers, 1956

In Part VI we are looking at how we can obtain and maintain safe behavior in our employees. In doing this we have looked briefly at some of the research and literature from the behavioral scientists and from safety management, at whether or not we can predict safe behavior and thus screen out any people who will exhibit unsafe behavior, and at how we can change behavior through our traditional approaches to training and through a learning theory or behavior modification approach. This chapter will examine how we can maintain safe behavior. We will examine only one small aspect of it, however: interpersonal relations, that aspect of safe behavior maintenance that comes from the relationship established and maintained between the worker and the supervisor (or for that matter also, the relationship between a supervisor and the safety manager).

We know these relationships are crucial to success. We know from the research that the effectiveness of our safety programming (such as safety talks, training, inspections, competitions) is, in the eyes of the worker at least, highly dependent on this interpersonal relationship. While there is no literature on it, I rather strongly suspect a great deal of what is accomplished in safety by the supervisor depends on the interpersonal relationship between the supervisor and his or her boss (or the safety manager). If these things are true, we would be remiss not to examine, however briefly, the techniques there are today to understand these relationships and to systematically improve them. We will concentrate on one technique rather than try to summarize the infinite number of approaches that might be possible to improve interpersonal skills. The technique is *Transactional Analysis* and its theory is deceptively simple.

TRANSACTIONAL ANALYSIS

Transactional analysis (TA) was introduced into the language of psychotherapy in the 1950s. As early as 1969, at least one company had incorporated it into a training program. Yet today—

while that original company considers its program *so* successful that it has progressively expanded it to encompass *thousands of employee trainees*—only a few companies have actually tried using Transactional analysis for improving organizational effectiveness.

Transactional analysis grew out of psychoanalysis as a method of group therapy, a means of treating individuals who are unable to form relationships with others. Transactional analysis is a means of analyzing the stimulus/response patterns that take place between individuals. These patterns are viewed as a series of interpersonal "transactions." The asusmption is that understanding these transactions can enable individuals to improve their interactions to make them more emotionally mature and productive in terms of contributions to a group effort.

The analysis is based on a breakdown of the "person" into three elements:

The Child represents the responses and training of the very early years. The Child uses a passive-aggressive techinque to obtain gratification.

The Parent represents the conditioned response learned or copied from authority figures (parents). *The Parent likes to be right.*

The Adult state is the objective, rational thought process, a *computer* basing its output on the lessons of experience. *Feeling and emotion are not part of the Adult ego state.* Only the Adult thinks; the other ego states merely feel or react.

These three ego states together form a person and all three are present in everyone. An individual can exist in any of these three basic states at any time. The ego state in command at a particular moment governs the individual's reaction to a particular situation.

Analyzing the three ego states of a person is called *structural analysis*. Examining the interaction between two individuals is called *transactional analysis*. We are dealing with two human beings, each with a Child, Parent, and Adult. In other words, when two people are alone in an interaction there are actually six "people." Here are some examples of different kinds of transactions:

1. Between two Adult ego states:
    Safety Director: "That investigation report was due this morning."
    Middle Manager: "Let me check on it."
2. Between a Child and an Adult:
    Safety Director: "What are we going to do? That investigation report was due this morning."
    Middle Manager: "I just couldn't get to it. I'll have it by noon."
3. Between an Adult and a Child:
    Safety Director: "That report was due this morning."
    Middle Manager: "You always blame me when something goes wrong."
4. Between a Parent and a Child:
    Safety Director: "Get that investigation report in within an hour."
    Middle Manager: "Don't blame me; its not my fault."

You can perform transactional analysis on yourself in your relationships with anyone. For example, suppose your boss tells you to work all day Saturday to check out some new equipment for safety before a test run Sunday, and you have been planning for several weeks to meet some old friends for the football game (game of the year, of course). Do you (1) turn to one of your colleagues and say, "Why do these things always happen to me, anyway? What's the use in making plans around here? Somebody else could just as easily have been picked." (2) Jump up from your chair and say to your boss, "That's not fair! You have no right to expect me to change my personal plans for Saturday when you give me such short notice!" (3) Reply to your boss, "That's going to cause a problem for me. You see, I've been planning for several weeks to meet some old friends Saturday whom I haven't seen in three years. Is it possible to get someone else to work Saturday? Or perhaps, I can come in early Sunday morning. What do you think?" Perhaps you can remember reacting to similar situations in each of these three different modes of behavior. According to TA, everyone has these three typical modes of behavior: the Parent ego state, as typified in answer (2), the Adult ego state, as in (3), and the Child ego state as in (1). Recognizing our own Parent, Adult, and Child states is the first step in applying TA to our daily living. Let us look at each in some more detail.

The Child State

The Child in us is that body of data that was recorded and stored in our brains when we were little. It reflects from how we responded internally to what we saw and heard at that time. These recordings are mostly feelings (and conclusions about ourselves based on those feelings) such as frustration, inadequacy, and helplessness. They contain the early recordings of joy, curiosity, imagination, spontaneity, and the excitement from new discoveries which were also part of our childhood. In the example above, it was the Child state than answered sulking away from the boss. We can spot our Child state when we find ourselves whinning, sulking, throwing a tantrum, or abandoning ourselves to the joy of a pleasurable new experience.

The Parent State

This is that body of data, also recorded and stored in our brain, that comes from our observations about the way our mother and father (and other important "Big People") behaved. It is based on external events that occurred essentially in the first five or six years of our life. It is what we learned about "the way the world really is!" In those tender years our overriding assumption was that "THEY" were right. In the Parent state we act as our parents would. Your Parent state would respond to your boss by lecturing on what is or is not a fair and on the proper way to treat you. Our Parent state lectures, moralizes, points its finger righteously or accusingly, teaches, and "lays down the law." We can recognize our Parent state when we are scolding, finger pointing, lecturing others about what is wrong or correcting somebody.

The Adult State

The Adult state is that part of you that figures things out by collecting and looking at the facts. The Adult state is your computer which you use to estimate probabilities and to make decisions based on facts. Everyone has an Adult which is capable of making assessments about outside reality. When we told the boss why working overtime would create a problem and suggested some alternatives to consider, the Adult state had taken charge.

The first step is to recognize each of the three states within yourself and to be able to identify who is in charge at certain times. For instance, our Parent state is the one that tells us internally such things as: "You must ...," "You ought to ...," "You shouldn't ...," "Don't ever ...." Our Parent tape plays back such old familiar recordings as: "If you want something done right, do it yourself." "Big boys never cry ..." "Idleness is the devil's playmate." "A penny saved is a penny earned." Any time you find yourself talking to yourself (either out loud or under your breath) and using the word *YOU,* your Parent is very likely addressing your Child. For example, when you say to yourself, "That was a dumb thing for YOU to do," your Parent is scolding your Child. Our Child state is the voice that says: "I want what I want when I want it." "Try and make me!" "Wow!" "Great!" "Drop dead!" Any time we are experiencing feelings or emotions (happiness, sadness, fear) our Child is participating in the experience in some way. Our Adult operates on facts based on what is true today. It is the voice within us which says things like: "What's going on here?" "*Now* I see why this happened the way it did." "What part of me came on just a few seconds ago—my Parent? Adult? or Child? Why did I react just the way I did?"

We can become acquainted with ourselves by listening to these three different voices inside ourselves or by paying attention to the feelings bubbling up inside. Naturally, we cannot hear the voices or directly experience the feelings occurring within other people, but we can become skilled in spotting the states in others by watching for the kinds of cues shown in Exhibit 23.1. As we become more skilled in spotting these cues in others, we will become more aware of some of them in ourselves. We can also get some cues from others about how we come across to them by identifying how they react to us. For example, suppose a supervisor reacts to something you have just said with, "Well, I was just trying to find out what the hell you wanted on that dumb investigation report." You might learn something about yourself by replaying what it was you said (or how you said it) immediately prior to the hurting-complaining response. Possibly the supervisor thought you were "putting down" or scolding.

Everyone has the three ego states operating at all times. The goal of TA is to strengthen the Adult in each of us so that we can not only ask, but also answer, questions like: "What part of me is coming on? Are these data true, appropriate, and reasonable for this situation I'm in?" To put it another way, TA provides us with a means of making our Adult state more often ready to respond to the problems of daily living and the frustrations of the job. This does not mean that we are to do away with our Parent and Child states. It would be dull without them. It does mean, however, that we want to be able to examine them. To the extent that our Parent and

	Parent ego state	Adult ego state	Child ego state
Voice tones	Condescending, putting down, criticizing, or accusing	Matter-of-fact	Full of feeling
Words used	Everyone knows that . . . You should never . . . You should always . . . I can't understand why in the world you would ever	How, What, When, Where, Why, Who, Probable	I'm mad at you! . . . Hey great! (. . . or any words that have a high feeling level connected with them)
Postures	Puffed-up, supercorrect, very proper	Attentive, eye-to-eye contact, listening and looking for maximum data	Slouching, playful, beat-down or burdened, self-conscious
Facial expressions	Frowns, worried or disapproving looks, chin jutted out	Alert eyes, paying close attention	Excitement, surprise, downcast eyes, quivering lip or chin, moist eyes
Body gestures	Hands on hips, pointing finger in accusation, arms folded across chest	Leaning forward in chair toward other person, moving closer to hear and see better	Spontaneous activity, wringing hands, pacing, withdrawing into corner or moving away from laughter, raising hand for permission

Exhibit 23.1 Cues to ego states.[51]

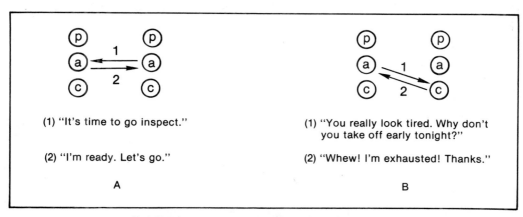

(1) "It's time to go inspect."

(2) "I'm ready. Let's go."

A

(1) "You really look tired. Why don't you take off early tonight?"

(2) "Whew! I'm exhausted! Thanks."

B

Exhibit 23.2 Complementary transactions.

Child tapes are out-of-date and unknown, we will be ruled by them. To the extent that we can learn about them, we will be more effective with the people who work with us.

Transactions

As we indicated earlier, a *transaction* is an exchange of words and related behavior between two people. When we see each person involved in the exchange as having a Parent, Adult, and Child state, we are able to draw an accurate diagram of what happens in the transaction. The Parent, Adult, or Child in one person will always be answering the Parent, Adult, or Child in the other person. Any conversation is a series of transactions, one exchange after another. Transactions can be Adult to Adult, Adult to Parent, Adult to Child, Parent to Parent, Parent to Adult, Parent to Child, Child to Parent, etc.

Examples A and B in Exhibit 23.2 are simple transactions. The arrows indicate who is saying what to whom. In each of these examples, the lines are parallel or uncrossed. As long as the lines in a transaction remain uncrossed, the conversation can go on indefinitely with no breakdown in communication. For this reason this type of exchange is called a *complementary transaction.*

In Exhibit 23.3, the safety director Adult is asking the Adult of the supervisor, "May I see your inspection results?" Instead of the Adult responding with something like, "Yes, here it is," the supervisor's child responds angrily, "I've already shown it to my general foreman." The supervisor has reacted as if the safety director has made an unreasonable demand. The communication about the supervisor giving the safety director inspection results has broken down. The transaction has become crossed. A second rule of communication is: Whenever the transaction is abruptly diverted rather than simply completed, look for a crossed transaction. In this situation the safety director has a choice of response to this crossed transaction, either with the Parent scolding the supervisor for being so owly, or with the Child showing anger, or with the Adult giving more information to the Adult of the supervisor: "I know you did, Bill, but his office is in the other building and it would be easier for me if I could just see your copy." Probably the supervisor's adult will respond to this last comment. Thus the transaction is uncrossed and the business at hand can be completed.

Following is an example of a *duplex transaction.* When we diagram it, the transaction looks like Exhibit 23.4.

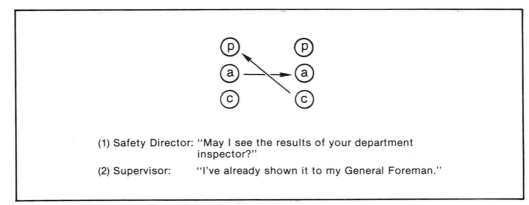

(1) Safety Director: "May I see the results of your department inspector?"

(2) Supervisor: "I've already shown it to my General Foreman."

Exhibit 23.3 A crossed transaction.

On the surface, this series of transactions appears to involve a supervisor's adult asking for suggestions from the safety director and the safety director's Adult, in turn, appears to be replying. There is, however, a second, or hidden, level of communication occurring. At this hidden level the supervisor is saying, "I'm helpless to solve my own problems so I need a wise person (Parent) like you to solve them for me." The safety director, in turn, responds on the level of, "Yes, I recognize my wisdom and will be happy to give you advice." But then in the third and fifth parts of the transaction, the supervisor rejects the advice apparently asked for. If we look only at the surface level of this transaction, it does not make sense. However, if we look at the hidden level, we can begin to see that the basic purpose of the transaction was to reject advice rather than to receive it. By rejecting the ideas of the safety director, the supervisor is able "to prove" superiority in an underhanded way. This type of transaction is called a

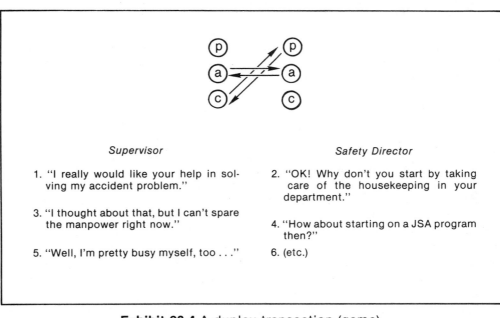

Supervisor   *Safety Director*

1. "I really would like your help in solving my accident problem."

2. "OK! Why don't you start by taking care of the housekeeping in your department."

3. "I thought about that, but I can't spare the manpower right now."

4. "How about starting on a JSA program then?"

5. "Well, I'm pretty busy myself, too . . ."

6. (etc.)

Exhibit 23.4 A duplex transaction (game).

duplex transaction because it involves two levels of communication, an apparent surface level and a hidden level.

Duplex transactions are commonly called *games*. A game is composed of three elements: (1) a series of transactions, (2) a gimmick (a hidden level of communication), and (3) a payoff. Many games are played over and over with the same theme. Some of these have been given names for easy recognition. The above game is the "Why Don't You—Yes But" game. This game ends with nothing more to be said. The supervisor may later admit having thought of all the suggestions, but the Child was used to gain attention from the safety director's Parent. This is the gimmick. The payoff is having proved that the parent cannot teach anything new. Games like this one have become so deeply rooted in our behavior patterns that they have become institutionalized. Everyone knows the game and the rules.

Other standardized transactions include "rituals," (such as "How are you?" "Fine, thanks. And you?") and "intimacy," a straightforward, game-free relationship between two people, which is extremely rare.

Games are utilized to avoid the intimacy of authentic relationships, so if nonproductive games can be recognized, people can make progress toward more authentic and productive interactions. While we cannot delve into games here, anyone interested ought to read *Games People Play*[15] by Eric Berne, who popularized TA with his witty volume. Anyone interested in the field should read it or one of his other books on the subject.

Another key book on TA is the best seller *I'm OK–You're OK*[51] by Thomas A. Harris, which is a step-by-step explanation of TA, and a discussion of its application in numerous social relationships. It is a good starting point for the layman. Dr. Harris explains that your behavior toward and reactions to other people are based largely on specific assumptions you make about yourself in relation to others. For most people, these assumptions were made very early in life by their early Adult ego states. They were the result of the young person needing to make some sense out of the many hours of confusing and often contradictory data recordings in all three ego states, but particularly in Parent and Child tapes. These assumptions about yourself in relation to others are called *life positions*. They can be classified into one of the following four categories:

1. *I'm Not OK and You're OK.* This is the life position held by most people. It is a conclusion based mostly on early negative or *Not OK feelings* about yourself rather than on a more comprehensive later review of data available in all three ego states. It is, therefore, a "felt decision" rather than a thought-out decision. It is a decision made at a time in your life when you saw and felt a clear distinction between yourself and the Big People around you who could do so many things that you could not do. Many people continue to make this distinction between themselves and others in their grownup years. People in this position often feel low or depressed. Much of their energy is spent in getting away from other people.
2. *I'm OK and You're Not OK.* This is a distrustful life position. It usually results because a person was beaten or brutalized by grownups when small. The people who base their lives on it spend much of their energy getting rid of other people.
3. *I'm Not OK and You're Not OK.* This is a despairing life position. From it, life appears hopeless. People in this position use up much of their energy getting nowhere with other people or with their own lives.
4. *I'm OK and You're OK.* This is an Adult decision. It is made after the individual has had a large number of OK experiences with others and concludes: "*All* persons are important (OK). You and I are important (OK)." Much empirical testing is essential before an individual fully accepts this conclusion. This is the only healthy life position even though

an individual who is in it continues to experience some Not OK feelings. However, know-ing enough to figure out what is happening with the Adult, it is possible to get on with others and with living life. (I'm OK and You're OK) rather than surrendering to depres-sion, despair, or the impulse to make the world less Not OK by getting rid of others.

These four life positions can be visualized as follows. How you see yourself can be plotted on an axis with "I'm OK" at one end and "I'm Not OK" at the other end. For example:

I'm OK	I'm Not OK
(Axis on how I see myself)	

Similarly, the way in which you see others can be plotted on a second axis with "You're OK" at one end "You're Not OK" at the other end, for example:

You're OK	You're Not OK
(Axis on how I see others)	

Or, as shown in Exhibit 23.5.

Updating

At any time we can making a conscious decision to throw out the old Not OK position and replace it with the "I'm OK and You're OK" position, the only *healthy* life position. This gets us off the hook of the past and gives us freedom of choice in the present. It is done by strengthening the Adult ego state.

In his book, Dr. Harris[51] describes the experiments of Dr. Wilder Penfield, a Canadian neuro-surgeon. Dr. Penfield, using an electrical stimulus, was able to trigger recorded speech and feel-ings that were stored like tape recordings in the patient's brain. Not only was the memory of an event recalled, but the whole experience was relived by the individual. For example, one in-dividual not only recalled an early experience that involved her walking past a bakery, but she also began whistling a tune that she associated with the experience and reported actually smell-ing the aroma of freshly baked bread. The development of the concept of three ego states was, in part, based on these experiments. The Parent and the Child are permanent recordings in the

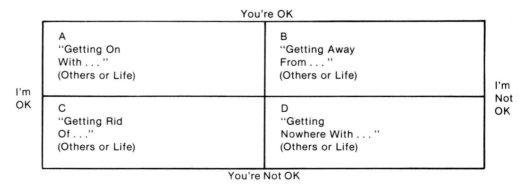

Exhibit 23.5 Life positions.[51]

brain. These tapes are never erased. They can, however, be updated by the Adult. Strengthening your Adult sufficiently to maintain an I'm OK—You're OK life position requires regular systematic exercise. Your Adult gets this exercise through updating your old Parent and Child tapes. This updating involves using your Adult to sort out which of the old Parent and Child recordings are still appropriate and valid for you in *today's* living and which are no longer appropriate or true. By sorting out these old tapes, you can produce new data about yourself which you can now store in your Adult and use for living more fully in the present. This can be done by learning the Parent cues from Exhibit 23.1 and by practicing spotting these cues in yourself whenever they occur.

We can use certain Adult questions in this process of updating our Parent and Child recordings. These questions include: "Is it now true?" "Does it apply today?" "Is still appropriate?" "Where did I get that idea?" "What is the evidence for its validity today?" Through this process of monitoring your old Parent and Child recordings, you can begin sorting out portions which you want to retain for your daily living today. This becomes your updated or OK Parent tape which you will need to sustain an "I'm OK—You're OK" life position. You will also have identified old recordings you will want to turn off whenever they begin to play because they are no longer appropriate or sensible today.

Here is a six-step procedure to help assess your transactions:

1. *First ask yourself, "How do I hurt?"* It is first essential that you recognize when you are hurting. This hurt can be any Not OK feeling, however small, that you feel. Next you state exactly the word or phrase that describes how you hurt. For example, "I feel stupid." or "I feel put down." or "I feel rejected."
2. *Next ask yourself, "Which of my ego states hurts?"* It is not likely that your Adult will hurt. It is more likely that your Child is the part of you that is hurting. Or it is possible that some old Parent tape has been trampled on, thus causing a hurt to enter there before your Parent sent the "ouch" on down to your Child.
3. *Next ask yourself, "Who did it?"* Who triggered your hurt? Did you hurt yourself? Or was it someone else who hurt you?
4. *Next ask yourself, "Which ego state did it?"* Compute which ego state of the other person initiated the hurt. If you did it yourself, was it your supercritical Parent beating up on your Not OK Kid again? Or if it was another person, was it Parent, Adult, or Child that gave you the hurt?
5. *Next ask yourself, "Why did it happen?"* Why did you give yourself or why did the other person give you this hurt? Was it accidental? On purpose? Was there a particular reason the statement was made? What made it hurt.
6. *Finally, ask yourself, "What do I do differently?"* Ask yourself, "Instead of just sitting here smarting, what else can I do right now? What can I do differently next time I meet this person or someone like him? What can I do the next time I meet someone who says or does the same thing?"

Updating your Child and Parent tapes enables your Adult to take charge of figuring out what happened when your old Parent and Child data create problems for you.

HOW CAN WE USE ALL THIS?

What does all this mean to the safety manager and what does it have to do with the effectiveness of the safety program? Only this: the safety program depends on other people doing things. It depends on top management doing some things, middle management doing other things, and supervisors doing still other things. How do we get individuals to do their jobs on safety? By getting them to want to do what we want them to do. How do we do this? First, by understanding as best we can each person with whom we work; what that person is thinking and on what assumptions decisions are being based. And by understanding ourselves, by understanding better our assumptions, by updating our Parent and Child tapes, and by spending more of our side of relationships in an Adult ego state.

Our Assumptions

Underlying most of our attempts at solving safety (people) problems are collections of unstated assumptions. All assumptions that relate to rightness, wrongness, value, and "properness" are Parent assumptions. Assumptions about what will or will not work or what will or will not happen under certain conditions are Adult assumptions. These can be tested in reality. Assumptions about how others are feeling or about their "OKness" or "Not OKness" in relationship to ourselves are Child assumptions. When we find ourselves making assumptions about another person, a situation, or ourselves, we should figure out which ego state we are using. Our assumptions and the ego state we are using to form them can both very much influence the outcome of situations and relationships.

"Oughtmanship" is sometimes used by both superiors and subordinates in their reaction to situations. For instance,

> "If management around here wants to solve the safety problem, they *ought* to show more interest and concern for the employees!"
> "Employees around here *ought* to be interested in working safety. There's no excuse for accidents!"
> "You employees *ought* to be more safety conscious in your work! Don't you know our record stinks?"
> "Supervisors *ought* to make it easier for the employees to talk to them. We have lots of good safety ideas, but they'll never listen to us!"

Each of these statements is a Parent statement. They each make a judgment or assumption about other people being responsible for the problem or its solution. This conveniently takes the speaker off the hook by placing the total responsibility for taking action on someone else's back. "Oughtmanship" does not lead to problem solving since it involves the Parent rather than the Adult. Only the Adult can make decisions and solve problems. The Parent and Child only react to problems. This is not to say that there is no room for the Parent ego state in work situations and relationships. An organization without a stated safety policy (Parent) often seems chaotic and without direction. But we must deal primarily with our Adult ego state if we are to have meaningful relationships with other people.

BUSINESS APPLICATIONS OF TRANSACTIONAL ANALYSIS

Transactional analysis is now being used primarily in training programs for employees who deal with the public—in banks, airlines, and retail establishments. Where TA is being applied to management personnel, it is usually included as one behavioral technique (along with role playing, job enrichment, simulation, and the like) within an elaborate training program.

Several *banks* have experimented with TA in the training of tellers, the employee group most frequently in contact with customers. Banks have also found that TA increases the number of options available to a trust officer for response to a situation. At one stage in the trust officer's dealings with the bereaved, acting as the Parent is the most effective; at another, the Adult is more effective. The officer must be sensitive to the changing requirements of the client's ego states.

Transactional analysis provides the trainee with insight into the underlying causes of customer behavior. It can provide the ability for an appropriate response that will improve the chances of Adult-Adult interaction. A large retail chain has worked up a highly sophisticated program including slides and cassettes, integrating TA into a formal training package which is distributed to stores. A major airline has put almost 7,000 public-contact employees and their supervisors through a TA training program, teaching them the basic concept of TA and gives them a general understanding of several games. While some first-line supervisors feel uncomfortable with psychological concepts, experience with TA seems to have evoked positive response with such groups in a number of cases, and many first level supervisory personnel have found the vocabulary of TA comfortable and indicate that they enjoy trying it out in their own private relationships. A food-processing organization is using TA in its counseling of employees for conflict resolution, the counselor applying it in a variety of ways, depending on the particular problem at hand.

At present, there is a scarcity of hard data to demonstrate to management the success of TA. Results can only be observed but, with continued applications, the future may bring more definitive results. Those companies presently using TA feel it is validated by the enthusiasm with which it is received by trainees and by frequent indications that it is a behavioral technique which is retained by relatively unsophisticated personnel and incorporated into their own vocabulary and daily relationships. Where management accepts this as sufficient validation, experience has been that the concept can then be extended from the public-contact employees to their supervisors and to many other internal employee groups.

So far, TA has had only limited exposure in business groups. There are signs that it may become a valuable technique for improving organizational effectiveness. It has not been used in a strictly safety situation as yet. It is included here as one technique to assist in the improvement of interpersonal relations.

FEELINGS

We might, in this brief discussion of some aspects of interpersonal relations, look at another author's thoughts and see how they apply to safety management. Harry Levinson, in his book *The Great Jackass Fallacy*,[35] discusses the role of feelings in management and in motivation in industry. He states there are four major feelings with which every human being must deal: love, hate, feelings about dependency, and feelings about one's self-image. He states whenever any of the four feelings are exacerbated by the organization, a person becomes ill. The symptoms may be physical, psychological, or both. He further suggests that these feelings result in accidents. Let us look at each.

Love

Whenever a person is deprived of sources of affection in the work situation or of the opportunity to give affection, the symptoms will begin to appear. For instance, if there is a change in a person's work process or if a person is removed from a compatible work group, physically transferred, fired, or even promoted, there will be changes in the ability to attach to other people and to familiar and preferred objects.

Hate

Whenever anger is unnecessarily increased by unfair criticism, exploitation, poor planning, attack, or simply when a person is made to feel like a fool, the result will be some kind of symptom by which people act against their own interests or those of the organization, or both, such as performing unsafe acts.

Dependency

Whenever people are made more dependent than they can allow themselves to be, or when they are made to feel helpless, frustrations arise which increase their anger. Paternalism only increases the sense of dependency for the very word implies that the employer is a father and the employees are children. Such a relationship makes it impossible for people to see themselves as adults in their own eyes. Most safety programs today are highly paternalistic.

Self-Esteem

Whenever self-esteem is lowered by contempt or by arbitrary changes in a person's work patterns, work relationships, or the ability to be in charge of what is happening without having to consult with superiors, the result is a feeling of being treated like a jackass (Levinson's words). Lowered self-esteem stirs up anger within an individual, who if conscientious and unable to sabotage the product or the organization, will commit personal sabotage which results in the increasing incidence of illness. Sickness, absenteeism, and accidents must be viewed as withdrawal phenomena requiring investigation.

Levinson's thoughts infer that not only are interpersonal skills important in achieving results in safety, they also (if bad) can be causes of accidents.

INTERPERSONAL RELATIONS AND OSHA

Before leaving this subject, which up to this point has been directed only to the interpersonal skills in a supervisor-subordinate relationship within a company, we might briefly look at the interpersonal relations situation created by OSHA.

Much has been written about OSHA since its passage. Periodicals are filled with articles about what the act is, how to comply, what is happening, what progress there is, etc. The only area not discussed is the area of what happens when the compliance officer comes to a business, that is, what is *really* happening vs. what appears to be happening, and how does that relate to the safety director's reason for existence. What interpersonal relations are taking place and how can they be bettered?

Theoretically, the act is for the sole purpose of making the worker safe. The compliance officer, then, has the sole purpose of making the worker safer. To do this, there is an elaborate set of standards by which every company visited is inspected for violations. In the final analysis, it is the inspector's job to find "things wrong" as defined by the standards being "right."

Consider the interpersonal dynamics of this situation. The compliance officer's job is to find things wrong in the company. The safety director's job is to make sure there are few things wrong. At two extremes the compliance officer can find:

1. Few things wrong and have few fines.
2. Many things wrong and have many fines.

If 1. occurs, the safety director is perceived by management as having done a good job. But how is the compliance officer perceived? If an inspector's job is to find things wrong, the level of rule infractions discovered determines how well the job is perceived as being performed, (and a low level may be used as a negative comparison with the compliance officer at the next desk). Under 1. above, the compliance officer involved is not perceived as having done a good job. If 2. occurs, the opposite is true. The safety director, obviously, has not done a good job in the eyes of management. The compliance officer has.

Both of these people theoretically have the same objective in mind, the safety of the work. And yet, with the same objective, they end up at counter purposes. Ludicrous, of course, but nonetheless true. It can be no other way unless:

1. Management does not make any judgment of the safety director's value to the organization based on what the compliance officer finds.
2. The area director (and on up the hierarchy) does not make any judgment of the compliance officer's value to the Bureau of Labor based on what he does or does not find wrong in his contacts.
3. Both the safety director and the compliance officer ignore how they are measured and what their bosses think of them.

The first two possibilities above seem to be highly unlikely. Hierarchical organizations seem to need concrete measures of employees' worth. We all live with some numbers game, which either we learn to manipulate or we are manipulated by it. The third possibility seems to be contrary to the way people are. So, the safety director and the compliance officer, striving for the same goal (theoretically), are, in fact, almost natural enemies because they are a threat to each other. If either does well the other is hurt.

There is one other alternative, for either or both to do less than best—mediocrity. And this is no doubt happening. The compliance officer finds much more than is reported. The safety director chooses to leave things unattended. Both save face and neither is too badly hurt. Of course, the worker may be hurt.

The interpersonal dynamics of OSHA are contrary to its objectives. They have to be. In the jargon of TA, industry (management) is the Child, the Bureau of Labor is the Parent and our feelings get in the way of progress. It is only through an Adult-Adult relationship that progress can be achieved. Penalties serve to reinforce the Parent role of government and the Child role of industry. For example, suppose you are a corporate safety director and in the course of an OSHA inspection the compliance officer says to you, "Your lift truck training program must not have been very effective. I've seen several driving violations." What would your reaction and response be? Would you (a) turn to one of your colleagues and say, "This SOB is really unfair to me. I had a good course. It's not my fault they aren't following what I said"; (b) snap back, "If you had any experince at all you'd know that an occasional infraction is human nature"; or (c) state, "Yes, I saw them too. I'll have to emphasize those areas harder in next week's class."

Recalling our discussion of TA, the Parent ego state is typified in answer (b) above, the Adult ego state in (c) above, and the Child ego state in (a) above. Exhibit 23.6 diagrams this

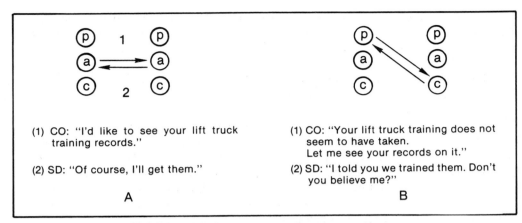

Exhibit 23.6 Complementary transactions.

exchange. Obviously, example **A** in Exhibit 23.6 is a productive complementary transaction, while example **B** is not too productive. In example **B**, the compliance officer has already alienated the safety director, and it will be difficult to achieve a good relationship again between them.

In Exhibit 23.7, the compliance officer's Adult is asking the Adult of the safety director, "May I see your results?" Instead of responding with Adult by saying something like, "Yes, here it is," the safety director's Child responds angrily with "Don't you believe me?" The safety director has reacted as if the compliance officer's Parent has made an unreasonable demand when in reality that is not true. The communication about the safety director giving the compliance officer the inspection results has broken down. The transaction has become crossed. Now the compliance officer has a choice of response to this crossed transaction: Using the Parent to scold the safety director or using the Child to react angrily, or use the Adult again and give more information to the Adult of the safety director: "I believe you, Bill, but I only wanted to see the program as an example for other companies." Probably the safety director's Adult will respond to this last comment. Thus the transaction is uncrossed and the business at hand can be completed.

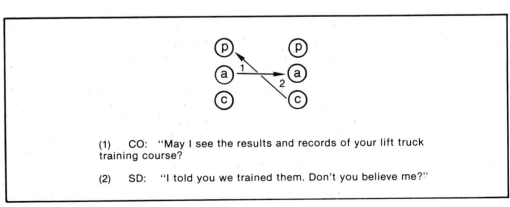

Exhibit 23.7 A crossed transaction.

The difficulty in an OSHA-Management transaction is that it almost invariably starts out from a Parent-Child relationship. The compliance officer is there for the purpose of finding things wrong—of checking on the safety director or on management. It takes tremendously strong individuals on both sides to overcome this. It is notably hard on the person who must begin in the Child state, the management representative, because when in the Child state we are guided by our emotions, not by logical thought.

Now, in this context consider the four feeling areas Levinson spoke of. Consider, for instance that

The boss' affection toward the safety director does not usually increase if the compliance officer has really done a competent job.

The pride in other safety accomplishments is often severely battered by the compliance check.

There is often unfair criticism of the safety director following the compliance check.

There is often a severe feeling of helplessness by the safety director before, during, and after the compliance check.

Any company is, in most cases, completely dependent on the whims of the compliance officer.

The self-esteem of the safety director is often bruised after the check.

All or any one of the above can give rise to feelings that are very strong. And feelings influence our attitudes, and our attitudes influence our actions.

In the situation as it exists now, it is difficult to view **OSHA** favorably and difficult to honestly predict great things from its advent. We can, perhaps, think out our own feelings and determine their cause. We can attempt to keep our relationships on an Adult-Adult plane and particularly to remember this when the OSHA compliance officer comes.

RECOGNIZING OUR OWN STATE

A first step might be to recognize each of the three ego states within ourselves and to be able to identify which one is in charge at certain times. We can become acquainted with ourselves by listening to these three different voices or by paying attention to the feelings bubbling up inside. As we become more skilled in spotting these cues in others, we will become more aware of some of them in ourselves. The goal of **TA** is to strengthen the Adult in each of us. TA provides us with a means of making our Adult state more often ready to respond to the problems of daily living and the frustrations of the job.

Chapter Twenty-Four

SBO - Safety By Objective

The popular management system, Management by Objectives (MBO), provides an excellent vehicle for a program. MBO, first described by Peter Drucker and later popularized by George Odiorne, provides an opportunity for all personnel to contribute to job goals and encourages the setting of checkpoints to measure progress. MBO is ends rather than means oriented. Such a system facilitates achievement motivation, particularly for jobs that can be designed and controlled by the subordinate. Positive reinforcement is inherent in the achievement of objectives. Material rewards, in terms of salary increase, bonus, and promotion, also come from the appraisal by results.

Fred Luthans[91]

A number of safety professionals believe that the total effectiveness of safety programming today is seriously reduced by our inability to integrate safety into the regular management systems, or to effectively relate safety performance to corporate goals. Too often the safety professional still only asks for (or hopes for) "management support." Too often this ineffective position is because management is not convinced that safety is a managment controllable cost—controllable by planning, by organization, and by management direction and control.

There are some reasons for our position in the eyes of our managers. Consider these as perhaps the main ones:

1. We are not using (or even developing) the needed tools to quantify our objectives or our progress. Our literature has discussed our weaknesses in this area at length. While some new techniques have been devised and tested, their use is still so limited in comparison to the use of Z16.1, that we cannot really report much progress here. Most safety professionals agree that Z16.1 is ineffective as a measure, as a diagnostic tool, or as a communication to management. Yet for many reasons (trade secrets, not taking time to understand new systems), we as a profession are not improving our measures of line performance, our communications to management, or our diagnostic tools.

2. We are not tapping the human resources in our organizations. Modern management is incorporating the new knowledge from the behavioral sciences into their newer methods and techniques:

    A. Job enrichment is an exciting new concept based on Herzberg's research which bears out the fact that responsibility, achievement and recognition are true motivators and should be built back into jobs if we really hope to "turn on" the employee to the work.

    B. Many of our supervisors, according to McClelland's research,[93] are high achievers, and we must use totally different techniques to motivate them. They will not be directed; they are self-directing people. While we have not discussed high achievers here, the reader might wish to pursue this area further.

C. The most effective supervisors are those who are "employee-centered," who create an atmosphere in which employees can work and who remove the obstacles in their way, according to research from Likert.[89]

D. Interpersonal relations training is common today and effectively (according to Argyris' data[7]) bridges the gap between mature people's needs of independence, deep interests, equality rather than subordinancy, and management's historical principles which treat them like children (close supervision, specialization).

E. Participative management has been shown to be effective by many studies.

These and many others indicate the marked changes that are and will be taking place in American industry. There is little in the literature about safety techniques that is built on new research, on new safety principles, or on handling the "new" worker of today. In general, management is now beginning to recognize that employees can contribute markedly to organizational effectiveness. Systematic techniques are being used to tap these human resources, but so far, safety has not developed comparable systematic techniques.

3. We are not coping with group resources in safety. It was proven over forty years ago in Mayo's Hawthorne Studies that an employee's decision as to whether or not to produce is controlled by group norms more than by any other single determinant of behavior. Management today is coping with the phenomenon of the group through participation-sharing of decisions, improved communications, and group methods improvement. Safety has not begun to cope with this phenomenon even though we can see every day that the employee's decision about whether or not to work safety is also determined by the group attitude.

4. We are not relating safety goals to management objectives. Some companies are succeeding here today, but they are the rare ones. Few companies set achievable safety goals as they plan corporate goals. Few companies have adapted management's planning system to safety planning. Few safety techniques lend themselves to this kind of projection and planning.

MANAGEMENT BY OBJECTIVES

Management by Objectives (MBO) has achieved great recent popularity in management circles, not only in industry but also in hospitals, schools, etc. One survey recently showed 23 percent of polled companies using an MBO approach. It is used widely in this country and others.

MBO is an umbrella concept covering a multitude of objectives. According to some business historians, Douglas McGregor[98] (Theory X-Theory Y) was responsible for the MBO concept, although he did not use the term. McGregor was looking for a method of performance appraisal which was better than what was available at the time (1957). Perhaps equally responsible was Peter Drucker,[34] who contended that objectives are necessary in every area where results and performance directly affect the business. (This also means safety). One of the leading advocates of MBO today is George Odiorne.[122] Odiorne states that MBO is a process whereby superior and subordinate managers jointly identify common goals, define each individual's major area of responsibility in terms of expected results, and use these measures as guides for operating the unit and assessing the contributions of each of its members.

John Humble[64] is the leading advocate of **MBO** in Great Britain. He describes **MBO** as a five-part process:

1. Agree with me what is expected of me.
2. Give me an opportunity to perform.
3. Let me know how I am getting on.
4. Help and guide and train me.
5. Reward me according to my contribution.

According to George Strauss[152] of the University of California, **MBO** has a number of attractive features. Its emphasis on specific goals makes performance appraisal more objective, and even limited subordinate involvement in goal setting tends to make the goals more realistic and palatable to the individual. It is a step toward a systems view of management, linking individual goals to those of the organization as a whole and linking strategies to objectives. General Mills, Minnesota Mining & Manufacturing. Honeywell, PPG, and Kimberly-Clark are a few of the companies which have experimented with **MBO**, according to Strauss.

It is enjoying success in so many companies for these reasons:

1. It brings about goal-directed behavior.
2. It is an organization-wide method of allocating responsibility for the corporate goals.
3. It fosters participation in goal setting and decision making.
4. It provides current, quick, regular feedback and reinforcement.
5. It sponsors planning at all levels.
6. It measures results while allowing freedom of decision and of action at lower levels.
7. It fosters imagination and creativity in even the largest organizations.

All of these are essential elements in today's organizations. All of these are essential in today's safety programming.

SAFETY BY OBJECTIVES

Historically our safety programs have failed in some of these essential elements. Many programs are far from producing behavior which could be considered goal-directed. Responsibilities, even with written policy, are often unclear. Participation in goal setting and decision making is almost nonexistent. Feedback and reinforcement is slow and not often connected to the amount of effort expended in safety (especially when the number or severity of accidents is the measuring stick). Planning is minimal and while results are often measured, freedom of decision or control is seldom left to the lower levels. And finally, imagination and creativity are rare commodities in most safety programs. The principles of **MBO**, adapted to safety programming (**SBO**), can overcome some of these failings.

The SBO System

Freely adapting Humble's **MBO** approach to safety programming, these are the steps of **SBO**:

1. *Obtain management-supervision agreement* (*with safety staff consultation*) *on objectives.*
   In the installation stages of an **SBO** program the agreement will emphasize not only re-

sults objectives but also activities objectives (what means, tools, and resources are to be used as well as results). Once under way, only objectives are agreed to.

2. *Give each supervisor an opportunity to perform.* Once agreements are completed, supervisors should be left alone to proceed with their action plan. Require only progress reports.

3. *Give the supervisors feedback.* Based on the quantified objectives, (either result or activity objectives must be quantified) you must provide regular, current, and pertinent feedback so the supervisor can adjust the plan as needed.

4. *Help, guide, and train.* Both management and safety staff fulfill this role. Safety staff provide the technical and safety technical expertise while management provides the managerial help—training at the outset, helping when asked, and guiding when indicated.

5. *Reward according to progress.* This requires a reward system that is geared to the progress made toward agreed objectives. The various managerial rewards should be used: pay, status, advancement, recognition.

These are the elements of the SBO system. Within this broad framework any company can develop an SBO system that is right for it. This system, just like its parent MBO, works. Here is how SBO was installed at one company:

1. *Background.* Safety programming at the company in past years had been traditional, utilizing periodic inspections, occasional spurts of directed activities including film showings, meetings, etc., with results of frequency rates considerably above national averages, severity rates slightly below average, and insurance costs slightly above the average.

2. *Physical Improvements.* As with most companies, the passage of OSHA created some stir at the company including the need for considerable expenditures in order to come up to standard. These improvements came through a process of developing corporate safety standards, and the formation of a special safety maintenance task force to make the physical improvements. After these improvements the new SBO program was introduced (Exhibit 24.1) following the completion of a 10-hour supervisory training program.

Year prior to SBO	Little activity early in the year. Programs of physical improvement inaugurated about the sixth month of the year. Physical improvement targeted for completion by the first quarter of year one

Year one	First quarter—Completion of physical improvement
	Second quarter—Program installation including:
	   • Survey of all employees on effectiveness of strategies in a safety program
	   • Ten-hour supervisory training program to teach the various strategies (activity goals)
	   • Selection of strategies by supervisors for use for the remainder of year one
	   • Agreement on strategies and on their quantification between supervisor and his general foreman
	   • Feedback on rating of investigation reports
	Third and fourth quarters:
	   • Program in effect
	   • Objective setting for year two

Year two	Full SBO program in effect (results and activity goals)

Exhibit 24.1 SBO schedule.

3. *Training.* The training course provided training in safety techniques which gave supervisors the background for the agreements required in step 1 of the SBO program. Specifically, the course included:
   — An introduction session to set the stage followed by the introduction of an employee safety survey. This survey fulfilled two basic purposes: it asked employees to express each effective approaches to ideas on safety and to rate the interest of management, supervisors, peers, and themselves.
   — A session to teach the concept of muiltiple causation and to practice investigating accidents for cause. Accident investigation was the only required strategy for supervisors.
   — Several sessions to teach the techniques of inspecting, sampling, holding meetings, job safety analyses, STOP series, special emphasis campaign, and other proven techniques in safety for possible strategies.
   — A final session to introduce SBO and a message of endorsement of the SBO concept by top management.
4. *SBO Installation.* The SBO program began at the final supervisory safety session. Each supervisor was provided with a profile of the results of the employee safety survey for the department and for the company as a whole. This data was provided as an aid in selecting comfortable strategy that would be acceptable to the group. For instance, from this data the supervisors know that their employees either liked or disliked (in relation to the companywide norms) 5-minute safety talks, posters and campaigns, etc. They also had a feeling for their credibility level compared to all the other supervisors. Interpretation of this data was also given to the supervisors.
5. *Objective Setting.* Armed with this data, and having practiced each strategy both in the sessions and between the sessions, each supervisor then set as an objective the required strategy of investigating accidents for multiple causes and also two of the optional strategies included in the training, for the remainder of the calendar year. Additionally supervisor and coach agreed to a schedule for carrying out the strategies. Objectives for performance in both quantity and quality (all quantified) were agreed to at this point. In accident investigations, for instance, objectives were set on timing and quality of report (quantifiable through a rating system).

   The objective setting in the second year would be quite different from the first year of the SBO program. In the installation year the overriding corporate objective was for each supervisor to "do something"—to get started in activities that they had not done before. *Activity* goals were set in succeeding years when the competence was built and the familiarity was present, the objectives agreed to would take the form of *results* goals.
6. *Leave Them Alone.* The next step in SBO was to give each supervisor the opportunity to perform. The supervisors were left alone to carry out their agreed on strategies. The only requirement was a monthly report.
7. *Let Them Know.* The next step was to let the supervisors know how they were doing. Coaches received monthly reports and provided feedback. The coach also surveyed and sampled regularly and provided feedback from this. Safety staff provided a summary report of all supervisors showing progress toward agreed goals.
8. *Help Them.* Help, guidance, and training during the installation year was available to each supervisor on request and the results were shown in monthly reports. This was highly individualized the first year depending on needs.
9. *Progress Report.* Finally, the key step of rewarding the supervisor according to progress toward agreed goals was handled in the first year by tying progress toward objectives to management bonus (one of the influencing factors). In later years the progress towards

goals would be more results-oriented and the quantification of safety results would be done by a costing system. Immediate costs would be assigned to each accident, this information would be fed back monthly to supervisors, and management bonus would be influenced by the relation of actual costs to planned costs of injuries.

Very briefly, this is an SBO program in action. Supervisory reception was excellent—supervisors were carrying out their strategies and providing input to safety staff where there was no interest before. Given the tools and training, one general foreman, formerly apathetic, showed more creativity in his coaching plan for his supervisors than ever seen in an activity. Several other coaches also exhibited rather remarkable turnarounds in attitude toward their safety responsibilities. Exhibit 24.2 shows the monthly comparisons for the three months prior to the installations of the program through the supervisory sessions.

Column 1 in Exhibit 24.2 is merely a count of cases per million man-hours to a physician. The figures here perhaps are not comparable to any other industry due to a management decision years ago to send all cases for outside care, since the care facility was almost adjacent and financially attractive. Column 2 is roughly comparable to frequency rate and column 3 to severity rate although in both cases the record keeping is more liberal than the Z16.1 (ANSI) standard resulting in larger numbers. The important point in these figures is the trend. In the first month following program introduction we found an immediate reduction in all three indicated. These reductions continued and improved in the next two months. These figures are early but seem to indicate the progress hoped for. However, SBO provides long-term improvement, not short-term, so we did not expect results until somewhere in the second year of the program.

OTHER CASE STUDIES

In the first edition of this book, published in 1975, the SBO concept was somewhat new and tested in only a few places (as in the above example). In the years since a number of companies have accepted the concept and are currently utilizing SBO. The SBO[126] approach has been in-

Month	Doctor cases per million man-hours	Lost time cases per million man-hours	Days lost per million man-hours
Three months prior to installation of program	169.8	53.05	159.15
Two months prior	270.3	50.05	100.10
One month prior	138.3	53.76	391.70
Program installation month ...	246.2	54.70	218.80
First month of program	163.80	22.33	44.67
Second month of program	196.55	24.56	24.56
Third month of program	131.16	24.60	147.60

Exhibit 24.2 Early accident trends.

Type Organization	Size	Type	Results
Food processor	15,000 employees	Delimited, with both mandatory & optional	36% and 40% red. yrs. 1 & 2
Food processor	10,000 employees	Delimited, with both mandatory & optional	30% and 25% red. yrs. 1 & 2
Contractor	3,000 employees		67% red. in 2 years
Brewery	15,000 employees	½ limited ½ no limits	64% red. 3 years
Oil Production	3,000 employees		50% red. 1st year
Railroad	4,000 employees	Limited	50% red. 1st year
Metal Manufacturer	3,000 employees	Select from a list of 12	50% red. 1st year

Exhibit 24.3 SBO Results.

stalled in a variety of different industries: brewing (Adolph Coors Company), Chemical (DuPont Corporation); railroading (Frisco, Union Pacific, Santa Fe, Chicago Northwestern); paper industry (Hoerner-Waldorf); and others. The results, of course, are not uniform. There are some successes and some less successful, as much depends on the installation, the commitment of management, and the meaningfulness of the objectives that are set.

For the most part however, SBO works, and continues to work once installed. One of the above companies reports a 75 percent reduction in the frequency rate (as they figure it) in the first three months of the program. One reported a six month claim savings of $2.3 million. One reported a 67 percent reduction in frequency continuing after three years.

Safety by Objectives was introduced in 1978, at Salt River Project,[1] an Arizona electric and water utility company, with amazing results. Previous to SBO at Salt River Project, the lost-time accident rate averaged over 2.25 cases per 200,000 hours worked, with some years exceeding 3.0. In 1978, the year in which SBO was actually introduced, the lost-time accident rate was an abysmal 3.16. With SBO fully implemented in 1980, the rate dropped to 1.03, which was good enough to win the American Public Power Association (APPA) award as the best in the country for APPA member companies. This was not a fluke because Salt River Project Electric Utility Company lowered this further to 1.01 in 1981. Salt River Project has maintained this outstanding success in 1982 and 1983 with rates of 1.01 and 0.95 respectively.

Exhibit 24.3 shows some other results.

SBO is a workable tool which harnesses some of our new behavioral insights and systematizes safety techniques, building them into normal management procedures. We see these advantages:

1. It utilizes our most modern proven techniques of safety management.
2. It taps human resources, particularly the brains and energies of our supervisors.
3. It fosters participation in the safety program.
4. It brings goal-directed behavior.
5. It provides current, immediate, relevant reinforcement.
6. It facilitates planning in safety.
7. It fosters creativity.
8. It incorporates responsive tools and decision-type information.
9. It also works.

Part VII

Current Trends

Chapter Twenty-Five

The Subjective Injury

At one time our primary problem in safety was the traumatic injury—today it is the subjective injury.

D. C. Petersen

A phenomenon that has been particularly interesting to observe over the years is the increasing number of subjective injuries that have come under Workers' Compensation. To illustrate, our national figures in early years showed a preponderance of cuts and bruises from machine accidents, slips and falls. In the late 1950s and early 1960s, the mix of injuries changed to include a large percentage (almost 20 percent) of overexertion claims. By the late 1960s the mix was even more skewed to manual material handling (a large part again overexertion) injuries. By the 1970s, some studies began to show 30–40 percent back strains, and other things began to enter in to the mix. Soft tissue injuries, primarily of the wrist and shoulder, became big in the early 1980s. Psychological stress related illnesses emerged in the mid-1980s. In 1980, stress was an almost non-existent problem, but by 1982 it had grown to 11 percent of the Worker's Compensation health payout. In 1985, it was the single largest occupational disease payout under the Workers' Compensation system.

One might wonder why the above changes are occurring. At one time our primary problem in safety was the traumatic, the objective injury. Slowly but very obviously and systematically this has changed. Between strains of various types of stress related illnesses, the subjective injury has become a major problem.

Why have strains become so prevalent? It is not because we lift more today. We have more labor saving devices in industry today than ever before. Between automation, ergonomic improvements and the shift from basic to service industries, we should have fewer strains than today than ever before. The opposite is true.

First some definitions: By the objective injury we mean those that are obviously real and easy to diagnose by a physician. By subjective injury we mean those impossible to accurately diagnose, where a physician must depend to a large degree upon what the patient says to determine diagnosis. Back pains, soft tissue pains of the wrist and shoulder, and psychological symptoms seem to be the primary subjective injuries of today. Why do they continue to increase year after year when the work is easier and easier?

One might suggest we should look elsewhere for the answer. Perhaps the worker of today is less able physically than in earlier years. While perhaps true in the 50s and early 60s, it probably is not true in recent years in the light of our national emphasis on health and personal exercise. If we are more fit and work less hard, then why more pains and strains?

Perhaps the best place to look for answers is in the psychological environment that has been created in this country. In earlier years our Worker's Compensation system paid benefits to injured employees that did not totally replace their lost income. In some states there were longer waiting periods before payments started than today. It was psychologically and financially painful as well as physically painful to the injured. In the last twenty years this has changed markedly. Benefits have increased to the point where in some states the injured employee makes more money injured at home than working on the job. In addition, in some states there is also the right to sue. Is this a factor? Let's look at the record and at one study done in this area.

Exhibit 25.1 describes the phenomenon best. The chart shows the accident record from 1954 to 1982 for a large federal installation, a U.S. Naval shipyard. The trend is remarkable. From 1954 to 1974 the lost workday injuries per 200,000 workhours remained low—almost always below the 1.0 mark. From 1974 to 1982 (and to the present) it has consistently not only been above that mark, but constantly climbing and is five times as great as it used to be. What is more remarkable is that this chart almost mirrors the record of all other U.S. Naval shipyards and is very similar to the record of most other federal agencies.

The real question is why the change after 1974? Could it be related to the policy change in 1974—when an injured employee would receive no reduction in pay when injured instead of a reduction? Further analysis of the data shows most of the increase is in the subjective injuries. What this costs the taxpayers is shown in Exhibit 25.2. This chart offers no proof of a connection between the new policy and the results, but it certainly suggests one.

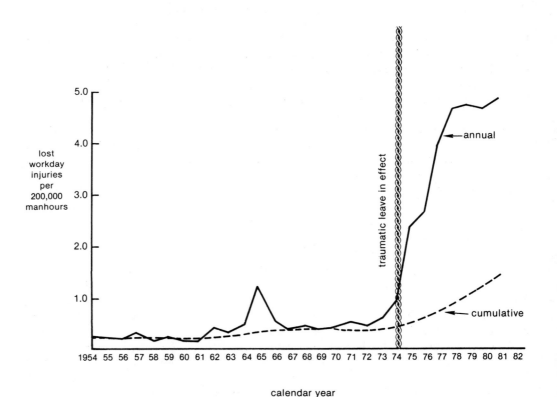

Exhibit 25.1 Accident frequency rate of U.S. Naval shipyard.

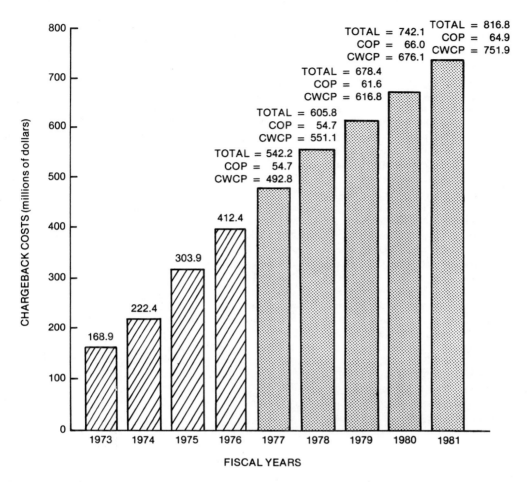

Exhibit 25.2 Office of Workers' Compensation Programs Chargeback to Federal agencies, total Federal Government Costs, fiscal years 1973-1981 with Continuation of Pay (COP) Costs and Total Costs 1977-1981.

THE METALS STUDY

One study has been made in this area. Dr. Leon Robertson[139] (Yale University) made an epidemiological study for a large multiplant metals company in 1982. He looked at three plants in three different states for eight years to see what things correlated with the accident record. He studied age, experience, sex, military experience, marital status, previous number of jobs, weight-height ratios, and several external factors. His most clear findings were that in each case following an increase in Worker's Compensation benefits over inflation increases, the following year showed a significant increase in injuries (and primarily subjective injuries). To quote from the study:

> It is clear that the increased costs accruing from injuries toward the end of the decade occurred in large part from an increase in workers presenting sprains, strains, and pain as symptoms when such symptoms were considered severe enough to be relieved from work. In an attempt to learn whether subjective injuries were con-

entrated among particular jobs or workers, a regression analysis of the worker attributes and job histories in relation to percent of a worker's total subjective injuries was undertaken.

Of the worker attributes, only age was significantly correlated to percent subjective injuries. The younger workers more often reported such injuries. The percent of subjective injuries increased as a function of the length of time the workers were in a given department. This result could be an indication of strain as a chronically developing problem resulting from time worked with heavy materials or it could be the artifact of some other factor that increased the reporting of these symptoms toward the end of the period studied.

Two external factors that potentially could have influenced the trends were considered—changes in Workmen's Compensation and citations by the Occupational Safety and Health Administration (OSHA). During part of the period of the study, Workmen's Compensation maximum payment per day for injury increased more rapidly than the rate of inflation. The amount of compensation was different among the states and increases occurred at different times and at different rates in each state. Connecticut changed Workmen's Compensation at about the inflation rate until 1979 but increased it 63 percent in 1980. New York's compensation changed at a slower rate than inflation until 1978 but it increased 72 percent in 1979. Wisconsin's Workmen's Compensation was 26 percent above Connecticut's and 36 percent above New York's at the outset and increased each year, jumping 63 percent in 1976 alone.

OSHA inspections resulted in citations for violations on seven occasions—two in the New York plant and five in the Wisconsin plant. Although fines were reduced after negotiation with OSHA, management took the incidents seriously and made an effort to increase compliance with OSHA standards.

To measure the potential impact of these factors, the data were fitted to the following regression equation:

$$R = a + b_1 W + b_2 O + b_3 L + b_4 M$$

where R = ratio of actual injuries to those expected from exposure to particular jobs by workers with particular attributes

W = ratio of maximum Workmen's Compensation payment to the GNP price deflator for personal consumption expenditures

O = 1 if OSHA citation during the year, otherwise 0

L = 1 if OSHA citation the year before, otherwise 0

M = 1 if OSHA citation two years before, otherwise 0

b_i = increment in R per increment in the i-th variable

a = constant

This equation allows for possible longer term as well as immediate effects of OSHA citations but assumes that responses to changes in Workmen's Compensation would occur in the year that the compensation increased. Exhibit 25.3 presents the results of the analysis. Injuries are higher than expected as Workmen's Compensation is higher relative to inflation and lower than expected in the year following an OSHA citation. The OSHA effect is temporary, however, and does not show up in the second and third years after the citation. The probability that either effect is due to random fluctuation in sampling is very low. More than 70 percent of the variation is explained by these factors indicating a good fit to the data.

The middle and bottom parts of the table present the results for subjective and objective injuries separately. In the case of subjective injuries, the effect of Work-

All Injuries	Coefficient	t	P
Workmen's Comp./GNP Deflator	0.42 ± 0.12	3.46	0.003
1 Year After OSHA Citation	-0.39 ± 0.10	-3.83	0.002
2 Years After OSHA Citation	0.02 ± 0.12	0.20	0.50
3 Years After OSHA Citation	0.06 ± 0.13	0.47	0.50

Constant = 0.79, R = 0.84, Standard Deviation of Residuals = 0.18

Subjective Injuries	Coefficient	t	P
Workmen's Comp./GNP Deflator	0.37 ± 0.09	4.05	0.001
1 Year After OSHA Citation	-0.05 ± 0.08	-0.66	0.50
2 Years After OSHA Citation	-0.05 ± 0.09	-0.57	0.50
3 Years After OSHA Citation	-0.03 ± 0.10	-0.27	0.50

Constant = -0.16, R = 0.76, Standard Deviation of Residuals = 0.13

Objective Injuries	Coefficient	t	P
Workmen's Comp./GNP Deflator	0.05 ± 0.14	0.38	0.50
1 Year After OSHA Citation	-0.34 ± 0.11	-2.96	0.009
2 Years After OSHA Citation	0.08 ± 0.14	0.56	0.50
3 Years After OSHA Citation	0.09 ± 0.14	0.60	0.50

Constant = 0.95, R = 0.63, Standard Deviation of Residuals = 0.20

Exhibit 25.3 Regression Analysis of actual injuries per expected number in a year in relation to Workmen's Compensation and OSHA citations.

men's Compensation is large and the effect of OSHA citations is not significant. In the case of objective injuries, Workmen's Compensation has no significant effect but there is a temporary reduction in the year following an OSHA citation.

These results involve substantial numbers of injuries. The coefficient of 0.42 in relation to Workmen's Compensation means that if the ratio of Workmen's Compensation to the inflation index increases from 1 to 2, there will be a 42 percent increase in injury claims above the number expected from job and worker characteristics. During the 1973–1980 period, the ratio of Workmen's Compensation to the gross national product (GNP) price deflator increased from 0.96 to 1.48 in Connecticut, from 0.90 to 1.21 in New York, and from 1.22 to 1.84 in Wisconsin. An OSHA citation resulted in an average 39 percent fewer injuries than expected but only during the year after the citation.

The results of Workmen's Compensation poses a dilemma. We do not know to what extent the increase in subjective injury claims associated with increased Workmen's Compensation represent false claims, claims for back problems that have origins from other than the work place, or claims that are legitimately work-related but were not reported when workers could not afford to be away from work before Workmen's Compensation was increased. Although there are some indicators of malingering that physicians may use in the diagnostic process, no one, including a physician, can say with absolute certainty that a person is or is not suffering pain or strain. Back problems are not uncommon among white collar as well as blue collar workers and their origins are often obscure.

Perhaps more challenges to suspect cases would lower the incidence of claims. Such challenges put the physician in an ethical dilemma of not being a full advocate for the patient and the company in the position of questioning the veracity of its employees. The costs in morale and litigation must be carefully weighed in considering a policy of challenging claims.

Another possibility is to increase use of the limited work program. This involves having the worker who is ambulatory, but may not be able to do heavy work, report for work that is within the worker's capability while injured. During the period studied, only 1299 days of limited work were performed—8.7 percent of the total days lost. Most of the limited work days were in the New York plant (1033) for objective injuries.

The effect on objective injuries of OSHA citations, masked in the overall trend by the countervailing effect of increased subjective injuries, suggests possible actions that companies can take to reduce injuries. Apparently plant management is able to reduce injuries substantially when there is externally generated incentive to do so. A limited literature suggests that greater management attention to hazardous conditions and unsafe practices can reduce injuries. In this case, apparently the disincentive of threatened fines by OSHA increased management attention to the problem. Incentives to reduce hazards could be motivated by the company more frequently than by OSHA.

Conclusion

The major effect on injury trends has been the increase in Workmen's Compensation relative to inflation that made absence from work more affordable. Greater challenge to the more suspect claims of pain or strain injuries might reduce their frequency but the subjective nature of pain, along with the accompanying ethical and morale implications and costs of litigation, raise substantial barriers to the success of such challenges. Greater use of the limited work program might discourage such claims.

SUBJECTIVE INJURY CONTROL

Dr. Robertson began to look at some of the control elements for subjective injuries when he talked about challenging claims and light duty programs. Probably our controls to the problem could be listed in two categories: those things we can do proactively and the things we can do reactively.

Proactive Controls

Perhaps the first and most effective proactive control for the subjective injury (at least those that are questionable) has already been discussed in this book in some detail in Chapters 12 and 13. In Chapter 12 Chris Argyris' conflict theory was presented and in Chapter 13 the various conflicts and reactions to conflict were discussed. Some of those reactions were the fake injury and the "Monday morning" injury, exactly what we are talking about in this chapter. The answer to the problem, a proactive answer, was suggested: to build a psychological climate in which the workers buy into the system, and are sufficiently satisfied and motivated so as to be less likely to take advantage of the compensation system, and less likely to try to screw the company. In Chapter 10 we also looked at those things that build such a climate. Probably of most value are the ten areas identified by Rensis Likert:[89]

> Confidence and trust
> Interest in subordinate
> Understanding of problems
> Training and helping
> Teaching how to solve problems
> Available required resources
> Information they wish
> Ideas and opinions
> Approachability
> Recognition

Other proactive controls for the subjective injury (particularly the real ones) fall into the area of employee selection, proper training in the proper way to work or to lift, preconditioning exercises, and proper location design (ergonomics). Exhibit 25.4 shows some general guidelines for back strain control that fit in these areas, while Exhibit 25.5 suggests some for the soft tissue injury problem. Each of these are extremely complicated subjects, worthy of entire texts; in fact, such texts are available to the reader. Similarly, the proper design of work area and work stations is worthy of considerable additional information beyond the scope of this section. Exhibit 25.6 suggests some of the approaches in proper task design.

Reactive Controls

After the alleged injury a number of things can help to control the cost outgo. Perhaps the starting point is to maintain contact with the injured from the beginning on. Many organizations require the first line supervisor to accompany the injured to the hospital or clinic on the first

---

1. Preconditioning exercises
   A. Anterior hip flexibility
   B. Posterior lumbar spine flexibility
   C. Abdominal muscle endurance
   D. Pelvic girdle position control
2. Avoid imposing stress on a curved lumbar spine
   A. When standing — maintain pelvic girdle balance
   B. When bending and lifting, keep lumbar spine straight while bending knees and hips
3. Minimize disc stress while performing work tasks
   A. Use proper techniques
      1. Standing posture — pelvic balance, head balance, shoulders square
      2. Sitting posture — pelvic balance, use back rest
      3. Carrying — keep weight close to body
      4. Avoid twisting the body while lifting
      5. Maintain good body balance to avoid slipping
      6. Avoid "jerky" movements
   B. Select strong people when appropriate
   C. Design the work station and work task

Exhibit 25.4 General guidelines for prevention of low back problems.[120]

A. Carpal Tunnel Syndrome
   1. Use preconditioning exercises
   2. Use progression
   3. Avoid bent wrists when possible
   4. Avoid pinch grip and extensive use of thumbs
   5. Avoid repetitive forceful use of fingers and thumb
B. Epicondylitis
   1. Use preconditioning exercises
   2. Use progression
   3. Avoid extensive pinch grips, especially with forearm promated or supinated movements
   4. Avoid forceful, repetitive elbow pronation and supination movements
   5. Avoid forceful, repetitive elbow flexion movements with forearm pronated
C. Shoulder Joint Impingement Syndrome
   1. Use preconditioning
   2. Use progression
   3. Avoid sudden shoulder joint movements
   4. Keep elbows to the side as much as possible
   5. Avoid repetitive, extensive shoulder joint movements

Exhibit 25.5 General guidelines for prevention of upper extremity syndromes.[120]

TASK DESIGNS

Eliminate MMH (mechanize)
       *Hoists, lifts, balancers, chutes, etc.

Modify container
       *Tare weight (steel vs plastic)
       *Handles, grips
       *Smaller containers
       *Avoid wide, bulky objects

Modify task/workstation (better lever arms)
       *Change MH type
        **2 people/container
        **slide not lift
        **lower not lift
        **push not pull
        **push or pull, not carry
       *Horizontal location
        **Initial
        **Final (carrying, clothing)
       *Vertical location
        **Initial (Don't put on floor or over shoulder)
        **Final
       *Avoid twisting
       *Don't be a jerk
       *Support back

Increase recovery time
       *Working rest
        **Vary posture, muscles
        **Decrease movement frequency
       *Idle rest

Exhibit 25.6 NIOSH Guidelines for manual material handling. [79]

visit and to maintain contact. A modified (or light) duty program often helps in cost containment; getting the person back on the job (even light duty) usually speeds the return to the regular job. In addition, some form of claims management is usually essential.

One of the most important aspects of containing costs in Workers' Compensation is the successful management of claims. Usually claims are managed by the insurers with little investigation or effort from the insured.

Once claims have been made, a team approach consisting of plant safety and health staff, an occupational medical expert, an insurance adjuster and a defense lawyer might be considered. Not all cases will require the team and as eventually the nurse and claims people will learn the defense strategy and many cases can be handled totally by them. Initially, an overall strategy needs to be developed as follows:

1. Case selection (based upon predetermined criteria).
2. Case analysis.
3. Case strategy.
4. Defense including use of experts.

These items are fairly straight forward. In analysis of cases, many times important items are overlooked or not used appropriately. The analysis should be an ordered evaluation of all medical aspects of case including:

1. Presence and extent of injury/illness claimed.
2. Mechanism of injury/exposure to hazardous substance claimed.
3. Correlation of exposure.
4. Recommendation for further data gathering, etc.
5. Alternative etiologies (causes).

Deciding whether to fight a claim and *how* to fight the claim are critical.

In sophisticated cases such as cumulative trauma, the use of experts in all phases of the program can be essential. Selecting an expert means finding someone who is trained and experienced in occupational medicine.

Chapter Twenty-Six

Stress

By 1985 the problem had grown to the point where it has become our number one occupational health payout.

Center for Disease Control

As indicated in the last chapter, the subjective injury has become a problem of major proportions in recent years.

The newest, and potentially most costly, of the subjective injury problems in the Worker's Compensation payout as a result of illness due to psychological stress on the job. While this problem is large enough to deserve much greater treatment than we can give it in one chapter, at least we can overview the problem and suggest some control areas.

THE PROBLEM

The problem is large. Facts and figures abound and are written about constantly, so perhaps a quick look at what is happening might be a starting point. Exhibit 26.1 shows some revealing trends in the fatality rates in the United States. In 1900, our top killers were not stress related illnesses, but by 1970, four of the top five killers were stress related. These shifts are beginning to cost industry in Worker's Compensation also. In 1980, sterss claims were minimal in this country. By 1982, the problem had grown to where 11 percent of our Worker's Compensation health payout had become stress related problems; and we began to learn something about the problem. The charts in Exhibit 26.2 suggest that this problem is different from our typical occupational health problem. Where most occupational diseases happen to older males (more exposure), stress claims happen to younger females.

By 1985, the problem was our number one occupational health payout under Worker's Compensation. The Center for Disease Control stated:

> There is increasing evidence that an unsatisfactory work environment may contribute to psychological disorders.
>
> Studies show the factors contributing to unsatisfactory jobs include lack of control over working conditions, non-supportive supervisors or co-workers, limited job opportunities, role ambiguity or conflict, rotating-shift work, and machine-paced work.
>
> Mental stress can produce such illnesses as neuroses, anxiety, irritability, amnesia, headaches and gastrointestinal symptoms.
>
> Average medical costs and indemnity payments in 1981–1982 for these forms of mental stress actually surpassed the average amounts for other occupational diseases.

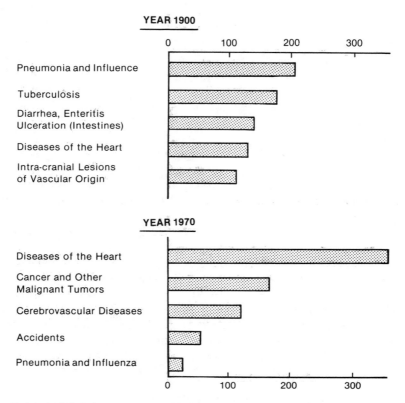

Exhibit 26.1 Top causes of death (rates per 100,000 people).[41]

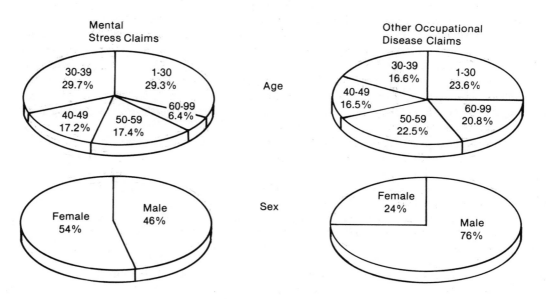

Exhibit 26.2 Differences in age and sex between stress compensation claims and other occupational disease claims.

The trends discussed by the Center for Disease Control indicate clearly the direction for the future. Without some legislative intervention, the problem could be massive for industry. One of the reasons for this is the variety of diseases caused by stress that could end up compensable. For instance, here's a partial list:

— coronary heart disease	— diabetes
— hypertension	— gout
— ulcers	— migraines
— colitis	— glaucoma
— anxiety	— epilepsy
— depression	— hemmorhoids
— allergies	— asthma
— arthritis	— acne
— cancer	— back pain

The reasons for the connection between stress and these illnesses can be better understood when we examine the body's responses to stress.

HUMAN RESPONSES TO STRESS

First a definition:

> *Stress.* A term used mostly in physics to mean strain, pressure, or force on a system. When used in relation to the body cells, it describes the effects of the body reacting, that is, the buildup of pressure, the strain of muscles tensing; stress is taken to mean a fairly predictable arousal of psychophysiological (mind-body) systems which, if prolonged, can fatigue or damage the system to the point of malfunction and disease.

Our natural physiological responses to stress cause us physical problems. Those responses (see Exhibit 26.3) were completely appropriate in another age. For the caveman faced with a stressful situation, the physiological responses prepared him for action to deal with the stress either to fight or for flight as discussed in an earlier chapter.

This fight or flight reaction has helped ensure our survival and continues to do so; no amount of relaxation training can ever diminish the intensity of this innate reflex. Stress is physical, intended to enable a physical response to a physical threat; however, *any* threat—physical or symbolic—can bring about this response. Once the stimulation of the event penetrates the psychological defenses, the body prepares for action. Increased hormonal secretion, cardiovascular activity, and energy supply signify a state of stress, a state of extreme readiness to act as soon as the voluntary control centers decide the form of the action, which in our social situation is often no "action" at all. Usually the threat is not real, but holds only symbolic significance. Our lives are not in danger, only our egos. Physical action is not warranted and must be subdued, but for the body organs it is too late: what took only minutes to start will take hours to undo. The stress products are flowing through the system and will activate various organs until they are reabsorbed back into storage or gradually used by the body. And while this gradual process is taking place, the body organs suffer.

NATURAL RESPONSE	ORIGINAL BENEFIT	TODAY'S DRAWBACK
Release of cortisone from the adrenal glands	Protection from an alergic reaction to dust from a fight	Cortisone destroys resistance to cancer, other illnesses
Thyroid hormone increases in the bloodstream	Speeds up body's metabolism providing extra energy	Shaky nerves, exhaustion, jumpiness, insomnia
Release of endorphin from the hypothalamus	Provides a potent pain killer so can't feel wounds incurred	Aggravates migraines, backaches, even pain of arthritis
Reduction in sex hormones testoseronec progesterone	Decreased fertility, helps in overcrowding; and loneliness	Anxieties and failures in sex; fustration, irritation
Shutdown of the digestive tract	Blood diverted to muscles, more power for fight	Dry mouth, bloating, diarrhea, discomfort, cramps
Release of sugar into the blood, with insulin to metabolize it	Quick energy supply to escape danger	Diabetes, hypoglycemia
Increase of cholesterol in blood from liver	Long distance energy to the muscles for fight or flight	Coronary, heart problems
Racing heartbeat	More blood to muscles and lungs for the fight	High blood pressure, strokes
Increased air supply	Extra oxygen to the lungs for the fight or flight	Dangerous for smokers
The blood thickens	More capacity to carry oxygen fight infection from wounds	Strokes, heart attacks

Exhibit 26.3 Natural responses to stress.

We can categorize the damaging effects of stress as either (1) changes in the physiological processes that alter resistance to disease, or (2) pathological changes, that is, organ system fatigue or malfunction, that result directly from prolonged overactivity of specific stress organs. There is a gastrointestinal system response, a brain response, a cardiovascular response, and a skin response; thus the connection to so many diseases.

By definition, stress is an arousal reaction to some stimulus—be it an event, object, or person—characterized by heightened arousal of physiological and psychological processes. The stimulus that causes this arousal reaction is the *Stressor* (see Exhibit 26.4). Some of the stressors are work related, but many are not. Therein lies the problem for industry. When stress induced illness does occur, was the cause at work or elsewhere? This is the question regarding most subjective injuries.

Psychosocial	
Adaptation	Overload
Frustration	Deprivation
Bioecological	
Biorhythms	Noise
Nutrition	
Personality	
Self-perception	Anxiety
Behavioral patterns	

Exhibit 26.4 Stressors

Stressors may be divided into three general classes:

1. *Psychosocial causes.* These stressors are a function of the complex interaction between social behavior and the way our senses and our minds interpret those behaviors. In other words, much of our societal stress is determined by the meanings that we assign to the events in our lives. Different individuals are likely to interpret differently, or to assign different meanings to, the same situation. This explains why each person's pattern of societal stress is unique.
2. *Bioecological causes.* These stressors basically are biologically related and may arise out of our relationship with our environment. They are only minimally subject to differing interpretations.
3. *Personality causes.* These reflect the dynamics of an individual's self-perception and characteristic attitudes and behaviors which may somehow contribute to excessive stress.

SIGNS OF STRESS

Stress has a cumulative effect. It can affect one's personality, job performance, health, and home life. Usually before health problems appear, certain personality characteristics and behavior patterns emerge. For instance:

- Abrupt change in typical behavior.
- Rapid mood changes.
- Overly suspicious.
- Excessive use of alcohol.
- Overhostility.
- Extreme defensiveness.
- Frequent illness.
- Excessive nervous habits.
- Accident prone.
- Taking of unnecessary chances.
- Obsessive about working.
- Sleep disturbances.
- Decrease in work performance.
- Depression.
- Use of excessive violence.

PSYCHOSOCIAL CAUSES

As indicated earlier, causes can be psychosocial, biological or personality induced.

There are four psychosocial processes that appaer to be most connected to stress: adaptation; frustration; overload; and deprivation.

1. *Adaptation.* This is the tendency of the body to fight to restore equilibrium in the face of forces which upset it.

   In the early 1960s Thomas H. Holmes and Richard H. Rahe[137] attempted to discover if change did have major effects upon human health. *Generic change*—that is, change resulting in either positive or negative consequence—was the focus of their research efforts.

   From these efforts emerged the Social Readjustment Rating Scale (SRRS), first published by Holmes and Rahe in 1967. This scale originally listed 43 specific life events and each item carried with it a weighting indicative of the amount of stress to be attributed to that item. The weightings are determined by the sample populations being tested. The weighting units were called Life Change Units (LCUs). The most highly weighted life event was the death of a spouse (100 LCU) and the lowest weighted event was minor violations of the law (11 LCU).

The SRRS is administered by asking the individual to indicate how many of the 43 items he/she has experienced over the past 12 months. A total LCU score is then obtained by adding up the LCU's for all of the items that have been checked. This scale has proven to be a remarkable predictor of physical and mental illness for a two-year period after the accumulation of the stressors.

The SRRS is shown in Exhibit 26.5. If you end up with a score of 100 units there is a 37 percent likelihood of a serious illness or accident in the next two years; 200 units a 51 percent chance; 300 units a 79 percent likelihood. Thus there does seem to be a statistical relationship between the events experienced and illness or accident.

2. *Frustration.* Frustration occurs when we're blocked from doing what we want to do, whether it is a certain behavior we wish to perform or a goal we wish to attain. Emotionally, we respond to frustration with feelings of anger and aggression and with the nervous and hormonal responses that accompany these emotions. Frustration, then, causes the stress response, and in a highly technological, urban society this source of stress should be recognized so that it may be dealt with.

3. *Overload.* Overload, which means the same as overstimulation, refers to the state in which the demands around you exceed your capacity to meet these demands.

The four major factors which contribute to the excessive demands of overload are: time pressures, excessive responsibility or accountability, lack of support, and/or excessive expectations from yourself and those around you. Any one or a combination of these factors can result in stress from overload.

Within the work environment such things as deadlines (time pressure), excessive responsibility and accountability, lack of managerial or subordinate support, and excessive role expectations from self, supervisor, or subordinates can all create overload. Task overload occurs when the work environment places demands upon the individual beyond that person's available resources.

4. *Deprivation.* This is the stress response caused by states of boredom and/or loneliness.

Deprivational stress has been defined as "the internal bodily reaction to cognitive understimulation," that is, our body's response to boredom (the reaction to monotonous, unchallenging tasks) and to loneliness (a state of emotional deprivation).

In affluent societies advanced technology relieves humans from many tasks, but the human time and interest in those tasks are often replaced with the boredom of watching a machine do the work. Highly repetitive or insufficiently challenging tasks can result in a state of distress.

BIOECOLOGICAL CAUSES

Three classes of stimuli which are biologically relevant and which may play a role in distress are: biological rhythms, nutritional habits, and noise pollution.

1. *Biological rhythms.* This concept focuses on three cycles:

— A 23-day *physical* cycle. This is thought to be related to muscle protein levels, muscle tone, and the metabolism of various neurohormones. It governs physical strength, endurance, energy, and resistance.
— A 28-day *emotional* cycle, based on hormonal levels, this is thought to exist in both men and women. It governs nervousness, sensibilities, feelings, moodiness, cheerfulness, general temperament, and creative abilities.

Below are listed events which occur in the process of living. Place a check in the left-hand column for each of those events that have happened to you during the *last 12 months*.

Life Event	Point Values
_____ Death of Spouse	100
_____ Divorce	73
_____ Marital separation	65
_____ Jail term	63
_____ Death of close family member	63
_____ Personal injury or illness	53
_____ Marriage	50
_____ Fired from work	47
_____ Marital reconciliation	45
_____ Retirement	45
_____ Change in family member's health	44
_____ Pregnancy	40
_____ Sex difficulties	39
_____ Addition to family	39
_____ Business readjustment	39
_____ Change in financial status	38
_____ Death of close friend	37
_____ Change to different line of work	36
_____ Change in number of marital arguments	35
_____ Mortgage or loan over	31
_____ Foreclosure of mortgage or loan	30
_____ Change in work responsibilities	29
_____ Son or daughter leaving home	29
_____ Trouble with in-laws	29
_____ Outstanding personal achievement	28
_____ Spouse begins or stops work	26
_____ Starting or finishing school	26
_____ Change in living conditions	25
_____ Revision of personal habits	24
_____ Trouble with boss	23
_____ Change in work hours, conditions	20
_____ Change in residence	20
_____ Change in schools	20
_____ Change in recreational habits	19
_____ Change in church activities	19
_____ Change in social activities	18
_____ Mortgage or loan under	17
_____ Change in sleeping habits	16
_____ Change in number of family gatherings	15
_____ Change in eating habits	15
_____ Vacation	13
_____ Christmas season	12
_____ Minor violations of the law	11

Score: _____

After checking the items above, add up the point values for all of the items checked.

Exhibit 26.5 Social readjustment rating scale.[137]

— A 33-day *intellectual* cycle. This cycle varies with hormonal changes and is thought to be related to intelligence, memory, mental alertness, logic, quickness, reasoning power, and ambition.

All three cycles are theorized to start at birth and to continue rhythmically throughout.

While the theoretical basis for biorhythms is sound, the mathematical analysis is speculative and based on the assumptions that all three cycles start at birth and that if there is any illness or other event that alters the cycle, it must somehow reregulate itself. As yet there is little conclusive scientific evidence to either support or refute those assumptions.

2. *Nutrition.* Certain nutritional habits actually contribute to stress. Certain foods can add to the stress of everyday life, for instance:

— Caffeine.
— Vitamin depletion.
— Low blood sugar.
— Too much salt intake.
— Smoking.

3. *Noise.* Noise can produce a stress response in one or more of these ways:
— By causing physiological reaction, that is, by stimulating the sympathetic nervous system.
— By being annoying and subjectively displeasing.
— By disrupting ongoing activities.

PERSONALITY CAUSES

The term *social engineering* refers to the willful altering of lifestyle and/or general environment in order to modify exposure to stressors.

Social engineering is the first alleviation technique because it attacks stress at the most logical and often most effective point—at the source—which is the stressor. It is *not* the primary goal of social engineering to change the nature of the stressor. For example, if your job is stressful because your boss is overly demanding, it is probably useless to attempt to change your boss. Therefore, social engineering entails modifying *your* position in relation to the stressor (the boss), not the stressor itself. In this example you could change the amount of time you spend with the boss, change bosses via a transfer, or even consider changing jobs. These are only three social engineering maneuvers that you could make. Only you can ascertain which technique is most desirable.

Many highly stressed people are under the misconception that they *must* live a stressful life. The reality of the situation is usually that they choose to live a stressful life because it is more externally rewarding (that is, offers more money or prestige) than a less stressful lifestyle; don't know of (or haven't searched for) any alternatives to the present lifestyle; or fear an unknown alternative more than they fear the effects of the present lifestyle. Social engineering strategies are designed to reduce stress by taking the path of least resistance. The specific techniques of social engineering discussed in this chapter parallel the psychosocial and bioecological causes of stress (see Exhibit 26.6).

Researchers and clinicians have known for years that if in a given situation a person's perception is of helplessness and of certain failure, this perception will virtually ensure failure in that

---

Adaptive Stress

    (a) Establish routines when possible.
    (b) Use time-blocking techniques.
    (c) Establish a "mental health" day.
    (d) Remember that a vacation doesn't always mean relaxation.
    (e) If possible, avoid or minimize other changes during periods of massive change.

Stress from Frustration

    (a) Use the Goal Alternative System model to find new alternatives to your frustrated goal.

Overload

    (a) Practice time management and set priorities.
    (b) Avoid overloading situations—avoid overcommitments by learning to say no.
    (c) Delegate responsibility.
    (d) Reduce the task into manageable parts.
    (e) Enlist the aid/support of others.
    (f) Accept fallibility.
    (g) Determine optimal stress level.
    (h) Avoid exposure to stress.

Deprivational Stress

    (a) Plan ahead to avoid potentially stressful situations.
    (b) Realize your vulnerability to deprivational stress.
    (c) Find relaxing activities which are not overly complex or ego-involved.
    (d) Remember that boredom does not equal relaxation.

Bioecological Stress

    (a) Monitor your biorhythms.
    (b) Use nutritional engineering.
    (c) Avoid exposure to noise.

Exhibit 26.6 Summary of social engineering strategies.

---

situation. This concept has been referred to as the "self-fulfilling prophecy": the likelihood of your failure at some task will be greatly increased if you imagine yourself as failing even before the task in question has begun. The converse of this relationship is true as well: that is, if you imagine yourself succeeding at your task, your probability of success will be greatly enhanced.

Just as self-perception affects task behavior, it can greatly affect the stress response and the eventual course of disease.

The greater degree to which persons perceive themselves in control of a situation, the less severe their stress reaction. This suggests that feeling helpless and feeling a lack of sufficient power to change one's environment may be a fundamental cause of distress. Thus, anything that adds to the feeling of self-control is likely to reduce the severity of the stress reaction.

In 1977, NIOSH[119] examined the health records of 22,000 workers in 130 occupations in Tennessee. Four occupations had a higher than expected incidence of stress-related disorders, and 12 of those were especially high; 13 occupations had fewer stress-related disorders than expected. The top twelve, in order are:

— Laborer
— Secretary
— Inspector

- Clinical Lab Technician
- Office Manager
- Foreman
- Manager/Administrator
- Waitress/Waiter
- Machine Operator
- Farm Owner
- Mine Operator
- Painter (not Artist)

The bottom thirteen (not in order) are:

- Sewer Worker
- Checker, Examiner
- Stockhandler
- Craftsman
- Maid
- Farm Laborer
- Heavy Equipment Operator
- Freight Handler
- Child Care Worker
- Packer, Wrapper
- College or University Professor
- Personnel/Labor Relations
- Auctioneer, Huckster

A quick analysis of these 25 job categories suggests some proof of the control issue. The most stressful occupations (laborer, secretary) were those where there was little self control on the job. For the most part the individual is under the direct control of someone else all day in these occupations. It is the feeling of lack of control that leads to the stressful situation.

OTHER PERSONALITY VARIABLES

We have known for years that specific patterns of behavior can adversely affect your health. The prime example of such behavior patterns is the now famous list of cardiovascular risk factors which include smoking, lack of exercise, obesity, and high fat diets. It has been clearly demonstrated that the consistent practice of one or more of these behaviors will increase your susceptibility to premature heart disease.

It is possible that far more general behavioral traits could affect your health. The way in which you *generally* interact with your environment may predispose you to stress and related disease. Evidence strongly suggests that the manner in which you choose to interact with your surroundings can play a major role in determining whether you develop premature heart disease.

Two cardiologists, Myer Friedman and Ray Rosenman, in the normal course of treating their patients, noticed some recurring behaviors among patients, especially in relation to how they dealt with time. They noticed an extreme anxiousness of the patients in the waiting room, and the fact that their conversations constantly centered around time, work, and achievement.

From their contact with coronary patients, Friedman and Rosenman formulated a construct of action-emotion behavior patterns that seemed to embody the coronary-prone individual. They referred to this construct as the Type A Personality, and it included the following characteristics:

1. An intense sense of time urgency; a tendency to race against the clock; the need to do more and obtain more in the shortest possible time.
2. An aggressive personality that at time evolves into hostility; this person is highly motivated, yet may lose his/her temper very easily; a high sense of competitiveness, often with the desire to make a contest out of everything; the inability to "play for fun."
3. An intense achievement motive, yet too often this "go for it" attitude lacks properly defined goals.
4. Polyphasic behavior, that is, the involvement in multiple and diverse tasks at the same time.

During a series of impressive research studies known as the Western Collaborative Studies, the Type A behavior pattern was shown to precede the development of coronary heart disease in 72 to 85 percent of the 3,411 men tested. These results strongly suggest that a Type A personality may be predictive of the eventual onset of premature heart disease. It should be noted here that recent research has questioned this conclusion.

PERSONALITY ENGINEERING

Exhibit 26.7 suggests some strategies for personality change.

"Positive verbalization" refers to the process whereby you reinforce your self-image by pointing out some positive aspect about yourself.

Learning to accept compliments is another strategy designed to improve your self-esteem. When someone gives you a compliment, simply accept it *without* the traditional statement of humility.

The entire area of assertiveness training can do wonders for improving self-esteem. Assertiveness training has been found to be an effective strategy for substituting positive, self-assertive behaviors and perceptions in the place of passive, withdrawing, or generally inhibited behaviors and perceptions.

Poor Self-Esteem and Depression
    (a) Verbalize your positive qualities.
    (b) Accept compliments.
    (c) Practice the Assertiveness Ladder.

Type A Behavior
    (a) Utilize time management.
    (b) Reduce ego involvement.
    (c) Use the Goal Path Model for planning.
    (d) Practice concentration.
    (e) Engage in thought stopping.

Anxious Reactivity
    (a) Engage in thought stopping.

Exhibit 26.7 Summary of personality engineering strategies.

Assertiveness is a positive and productive expression of yourself. On a behavioral continuum between passivity and aggression, assertiveness falls between the poles.

The urgency of time is a major problem of concern for the Type A personality. Much of this time urgency stress is self-created; Type A's rush around, often needlessly. This aspect has been called the "hurry-up" sickness.

Time management is an effective strategy to reduce the stress of time urgency.

JOB STRESS

While stress is clearly holistic, job and off-job stress cannot be separated or diagnosed separately, thus our control problems become extremely difficult. We can hope that there will be some legislative relief, which has happened in some states. Even where stress related illnesses are not covered under Worker's Compensation, perhaps as safety professionals we ought to institute some control programs, for illnesses from job related stress are real; there is a clearly demonstrated connection.

Perhaps our best shot is to concentrate on what we can change, the reduction of on-the-job stress. The NIOSH study mentioned earlier suggests the need to assess the worker's perception of feeling lack of control, and programs are in order to give workers a greater feeling of control. (See the earlier chapters on participation and involvement for suggestions in this area.) Special concentration should be placed on those job categories identified as particularly vulnerable to stress problems.

What are some of the primary stressors on the job? Dr. Michael J. Smith, Chief of the Motivation and Stress Reaction Section of NIOSH, outlines the main ones:

- Job Satisfaction.
- Physical Conditions.
- Organizational Factors.
- Workload.
- Work Hours.
- Work Role.
- Work Task.
- Career Development.

RELATIONSHIP TO ACCIDENTS

These stressors have been linked to heart attack, ulcers, and other illnesses in studies of air traffic controllers, railway dispatchers, and other occupations. As a result of those findings, an increasing number of workers are claiming Worker's Compensation benefits for disabilities caused or aggravated by psychological stress.

Studies have also linked stress to accidents. One scientist found that persons going through a divorce had twice as many traffic accidents as expected during the six months before and after the divorce itself. NIOSH[119] surveyed a group of 1,500 workers and found that 4.8 percent of those in low stress jobs reported a work injury within the past year, compared to 9.7 percent of those in high stress jobs. Off the job 13.7 percent of the low stress workers had injuries, compared to 22.4 percent of the high stress workers.

STRESS CONTROL

Stress is a multidimensional phenomenon, and if its detrimental effects are to be reduced, the individual's entire style of living must change to some degree. There must be a reduction in the stressfulness of the environment and this must be accompanied by an attempt to change some stress-producing personality characteristics. A technique, or better yet, several techniques of relaxation must be mastered, and nutritional and exercise patterns must likewise be altered.

Success in the control of stress and tension demands a *holistic* approach. Some of the complexities are diagrammed in Exhibit 26.8 from the book, *Controlling Work Stress,* by Matteson and Ivancevich.[110]

The 1979 Surgeon General's report, *Healthy People* (Califano), identifies fifteen priority prevention areas for improving the health and well-being of Americans. A companion publication, *Promoting Health/Preventing Disease,* identifies specific objectives to be accomplished by 1990 (U.S. Department of Health and Human Services, 1980). Stress is recognized throughout the list of objectives. For example, the report states that:

— By 1985, a methodology should have been developed to rate the major categories of occupation in terms of their environmental stress loads.

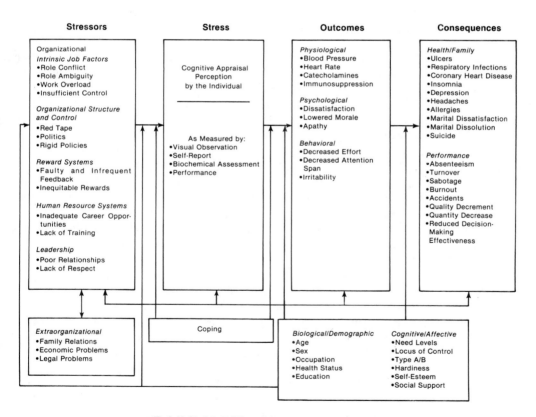

Exhibit 26.8 The stress reactions.[110]

ORGANIZATIONAL STRATEGIES

Goal Setting   Performance Feedback Systems
Participative Decision Making   Role Specification
Job Enrichment   Employee Surveys
Work Scheduling   Training Programs
Culture Change   Wellness Programs

INDIVIDUAL STRATEGIES

Cognitive Appraisal   Breathing
Restructuring   Progressive Relaxation
Transcendental Meditation (TM)   Biofeedback
Relaxation Response   Exercise
Social Support Systems   Diet
Employee Assistance Programs   Cranial Electrotherapy Stimulation
    Rehearsal

Exhibit 26.9 Stress control program elements.

— By 1990, the existing knowledge base through scientific inquiry about stress effects and stress management should be greatly enlarged.
— By 1990, of the 500 largest U.S. firms, the proportion offering work-based stress reduction programs should be greater than 30 percent.

It is now generally accepted that stress is a significant part of performing in a job for an employer or even for oneself (as an entrepreneur). Furthermore, empirical evidence points toward the notion that excessive job stress is associated with negative health consequences. Despite the general acceptance of the pervasiveness of stress and the growing empirical evidence about stress's effects on health, only a limited amount of research rigorously evaluates the effectiveness of stress management programs within organizations.

There is not only confusion surrounding the topic of stress (and even what is meant by the word), but there is also uncertainty about what should be classified as a stress management intervention in an organization. Stress management intervention programs include any effort initiated by the management of an organization that focuses on job-specific stressors (for example, work schedules, decision making, and job design) and is intended to reduce any assumed negative outcomes and consequences associated with stress. This definition includes both programs that attempt to remove stressors from the work environment and those that attempt to match an individual's needs, goals, aspirations, skills, and abilities with job tasks and environment.

As shown in Exhibit 26.9, the elements of a stress control program in an organization fall into two categories—those things the organization can do and those things the organization can sponsor that the individual can do. It is interesting to note that many of the organizational strategies are the exact same strategies that we have discussed earlier in this book for injury control, subjective injury control, productivity and quality improvement. They have to do with improvement of the quality of work life: goal setting, participation, job enrichment, improving the culture and climate, performance feedback, specification of roles, responsibilities and accountabilities (defuzzying things), surveying employee perceptions, training programs, wellness programs, and employee assistance programs.

The reason for the similarity is obvious. Since stress arises from mature adult human beings, who feel as if they are able to control their own behavior, finding themselves in a work environment where they are totally under the control of someone or something else, then the answer to stress problems is to exert less control over those individuals, or rather to shift that control from organization to self. An outline of one company's training program in stress management is shown in Exhibit 26.10.

Some of the individual strategies are a little different.

Cognitive appraisal—or perspective taking, as it is sometimes called—essentially involves asking the question, Am I taking stressors too seriously? Cognitive restructuring involves understanding the relationship between thoughts (beliefs), feelings, and behavior.

Cognitive appraisal and restructuring may both be helpful in reducing stress after an event has taken place. Rehearsal, on the other hand, is a technique that can improve your capability for dealing with stressful situations *before* they happen. In a relaxed state, you visualize an impending situation that you may find stressful. You then run through the situation in your

Time in Program		Topic Areas Covered
30 minutes	I.	Traditional Beliefs About Stress
		A. Facts
		B. Fiction and myths
30 minutes	II.	Diagnosing Your Own Stress
		A. What is your stress IQ?
		B. A personal diary approach
		C. Why you are sometimes inaccurate?
20 minutes	III.	Positive Stress
		A. What is positive stress?
		B. Stress and the inverted-U concept
30 minutes	IV.	How Stress Influences You
		A. Mind
		B. Body
		C. Behavior
		D. The issue of cause-effect
30 minutes	V.	Stress and Health
		A. What we know
		B. What we think we know
		C. What we want to know
40 minutes	VI.	Sources of Stress
		A. Dramatic life events
		B. Hassles and uplifts
		C. Family factors
		D. Job stressors
40 minutes	VII.	Individual Differences as Moderators
		A. Personality
		B. Demographics
		C. Congruence/noncongruence
80 minutes	VIII.	Individual Coping Methods
		(Demonstrations)
		A. Relaxation
		B. Goal setting
		C. Exercise/diet
		D. Cognitive approaches
60 minutes	IX.	Designing Your Own Program
		A. Preparing a contract
		B. Review the contract
30 minutes	X.	Reviewing the Program

Exhibit 26.10 A stress management training course outline.[110]

mind, picturing yourself dealing with any problems or unpleasantness that may arise and successfully overcoming them.

Transcendental Meditation (TM) is one of the less complicated forms of meditation and, contrary to many people's beliefs, requires no particular philosophical commitment, no intricate postures or body positions, no strange or exotic diets, and no external trappings of any kind. TM employs a *mantra*—a Sanskrit word meaning "sacred counsel"—which is a single word or sound that the meditator concentrates upon to shut out other distractions. The meditative process itself is quite simple. Selecting a setting with a minimum of distractions, the meditator assumes a comfortable position, concentrates on the mantra, and enters a state wherein both physical and mental relaxation are at a peak.

Dr. Herbert Benson[14] of the Harvard University Medical School believes that, just as there is in everyone an inherent stress response, there is also its opposites, which he calls the "relaxation response." Developing the ability to summon forth this relaxation response is what the Benson Technique is designed to do. It is a simple form of meditation very similar to TM social support has four categories: emotional support (esteem, affect, trust, concern, listening), appraisal support (affirmation, feedback, social comparison), informational support (advice, suggestions, directives, information), and other support (aid in-kind, money, labor, time, and environmental modification). People who can serve as sources of social support at work include supervisors, co-workers, subordinates, and customers or other nonorganizational members with whom the employee might have contact. Nonwork support sources include family (immediate and extended), friends, neighbors, care-givers (ministers, for example), health professionals (physicians, psychologists, counselors), and self-help groups (Alcoholics Anonymous, Weight Watchers).

The knowledge that breathing can directly affect a person's emotional state has been with us for a long time. Proper breathing involves taking deeper, longer breaths known as "abdominal" breathing. Abdominal breathing can be practiced virtually any time and place, although it is helpful to set aside three or four specific times during the day to practice. This will serve to help reinforce proper breathing behavior as well as lessen any tension that has accumulated. One of the most popular relaxation techniques, progressive relaxation, was originally developed by Edmund Jacobson. The technique, sometimes called the "Jacobson Method," is based on the theory that if you relax those muscles that have accumulated emotional tension, your mind will relax as well, through a reduction of autonomic nervous system activity. As the name implies, it is a technique wherein you progressively relax each of your muscle groups, starting with the feet and moving upward. To practice progressive relaxation, you assume a comfortable position lying down in a quiet environment. You start by relaxing your toes, then your feet, your legs, and so on until all the muscle groups of your body are relaxed.

The potential role of biofeedback as an individual stress management technique can be seen from the kinds of bodily functions that can, to some degree, be brought under voluntary control. These include brain waves, heart rate, muscle tension, body temperature, stomach acidity, and blood pressure. Most, if not all, of these processes are affected by stress. The potential of biofeedback, then, is its ability to help induce a state of relaxation and restore bodily functions to a nonstressed state.

Recent preliminary research suggests that an approach known as "cranial electrotherapy stimulation"—or CES, for short—may hold a great deal of promise. CES is a procedure that has been used successfully in the treatment of severe stress-related disorders, such as extreme anxiety and depression, particularly when those disorders are associated with alcoholism and/or various other chemical dependencies. CES involves the induction of a relaxed state by the application of a low-intensity electrical current to the cranial area, provided by a specially designed electronic apparatus, which can affect brain wave activity.

One organizational strategy is the Wellness Program becoming so prevalent in American industry. This is discussed in the next chapter.

Chapter Twenty-Seven

Wellness Programs

Some estimates suggest that as much as 70 percent of all our illnesses and over 50 percent of our medical costs relate to our lifestyle.

Dr. Brock Weisenberger

In the last chapter on stress, we talked about a holistic approach to control the problem, an approach that looks at and deals with an entire lifestyle, which recognizes that the lifestyle contributes to the problem and conversely the solution must incorporate much more than what happens on the job.

The holistic concept is, of course, valuable way beyond stress control. The concept is a part of medicine, of occupational health and of occupational safety. The health (physical and psychological) of a person is one factor in staying well, being injured, being stressed out, etc.; therefore, the growth in wellness programs in industry in the 1980s. There is a strong trend in industry toward adopting wellness concepts and programs, of dealing with the whole worker, whatever that might mean. What wellness is and means might vary somewhat depending upon to whom we are talking. Dr. Brock Weisenberger,[168] in a recent National Safety Council teleconference, defined it like this:

> The concept of wellness varies from person to person. We all think we know what wellness is but it's difficult to verbalize. Certainly from a business' standpoint we would perceive that if an employee is "well" they are productive and on the job. We know that there are other manifestations of health: A wholesome family life, pursuit of relaxing recreational activities, enjoyment of friends, pursuing educational ambitions, our spiritual life, etc.
>
> One thing is certain—more and more corporations are considering that the health or wellness of their employees is a very important factor in the success of their company. They also are realizing that for most health problems the corporation and the employee have the same vested interest in improvement.

Mark Wright,[173] Director of Health and Safety for Adolph Coors Company, a leader in this area, defines it like this:

> Two different and opposite concepts of disease have evolved through the years. The first, the Platonic view, treats disease as a separate entity that sneaks up and attacks people, more or less at random. Each disease has a name and specific symptoms—basically, an identity all its own.
>
> The second concept, the Hippocratic view, looks at disease not as an external force, but as a deviation from the norm. Most of the time, most of the people are

not ill. But, under certain conditions a person is unable to cope with his or her environment or environmental insult. The body cannot adjust and the result is illness.

How does that boil down to our day-to-day lives? Traditionally, medicine "cures" us. As individuals, we take little or no responsibility for our lifestyle or habits. We go along with our overweight, overdrinking, smoking habits and shun any medical advice until we're in the throes of a crisis. Then we expect the medical community to bail us out—pills, surgery, whatever it takes to "fix" us and make us well again. Unfortunately, what we attempt to "fix" are only the symptoms, not the root or cause of the problem.

Need proof? Take a look at the costs of health care—now 10 percent of our gross national product and escalating at the rate of 15 percent per year.

"Well" medicine, on the other hand, takes the opposite approach. The individual is responsible for his/her own lifestyle and habits. Each individual manages his/her physical fitness, nutrition, stress, spiritual anchoring, etc. When a person is ill, that person looks for the cause of the problem, instead of at the runny nose or hacking cough. The medical profession works as a partner in resolving the illness, rather than as a "fix-it" handyman.

One of the better illustrations of the wellness continuum is by John Travis, founder of the Wellness Resource Center in Mill County, Calif. Basically, what John shows in this continuum (Exhibit 27.1) is the scope from sickness to nonsickness to a well state of being.

Moving from the center to the left shows a progressively worsening state of health. Moving to the right shows increasing levels of health and well being. Traditional medicine, as we said, typically "cures" symptoms of disease, and stops at the midpoints—where there is no discernible illness or injury. Wellness starts at that point and moves the individual as far to the right as he/she is willing to go. Wellness, then, becomes the state wherein an individual is physically and mentally capable of maintaining his/her highest quality of life through self-responsibility and self-management.

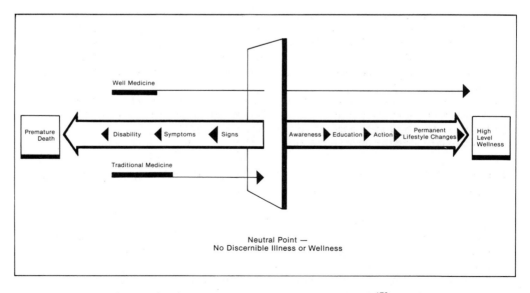

Exhibit 27.1 The wellness continuum.[173]

Assumptions of the Old Paradigm of Medicine	Assumptions of the New Paradigm of Health
• Treatment of symptoms	• Search for patterns and causes, plus treatment of symptoms
• Disease or disability seen as thing, entity	• Disease or disability seen as process
• Professional is authority	• Professional is therapeutic partner
• Placebo effect shows the power of suggestion	• Placebo effect shows the mind's role in disease and healing

Exhibit 27.2 Paradigms.[173]

More and more evidence supports the fact that our American society is moving from the nonsickness point to a state of well-being. Take a look at the sales figures from Nike, Adidas, New Balance or any other running shoe. Look at the popularlity of magazines like Runner's World—unheard of 10 years ago. Marilyn Ferguson states in *The Aquarian Conspiracy* that one of the paradigms we're currently experiencing is the shift from old medicine to health or wellness; from symptoms to causes; from disease as an entity to disease as a process (Exhibit 27.2).

Administrative Functions	Basic Functions
Program Objectives	Medical Treatment
	Allergies
Policies & Procedures	Communicable Diseases
	Dermatitis
Resources & Guidelines	Medical Surveillance/Monitoring
	Cardiovascular Diseases
Staffing	Hearing Conservation
	Respiratory Disease
Job Responsibilities	Environmental Evaluation
Facilities	Job Placement
Equipment	Recordkeeping
Supplies Costs	Health Counseling
	Alcohol/Drug Abuse
	Mental Health
	Nutrition
	Prenatal Health Care
	Illness/Absence Control
	Immunization
	Physical Fitness
	Rehabilitation
	Smoking Cessation
	Health Promotion
	Accident Prevention
	Back Injury Prevention
	Vision Conservation
	First Aid
	Medical Emergency Preparedness
	Cumulative Trauma Disorders
	Workers' Compensation
	Case Management
	Data Processing & Statistical Analysis

Exhibit 27.3 Possible functions of occupational health programs.[9]

What then is included in the so-called Wellness Programs? Dr. Bradford L. Barick,[9] writing in *Occupational Health and Safety* magazine, suggests the functions shown in Exhibit 27.3.

THE NEED FOR WELLNESS PROGRAMS

Some estimates suggest that as much as 70 percent of all our illnesses and over 50 percent of all our medical costs are related to our lifestyle. For instance, look at the factors related to coronary heart disease, the leading cause of death in America:

1. Cigarette smoking
2. Blood fats
3. Family history
4. High blood pressure
5. Obesity
6. Diabetes
7. Sedentary habits
8. Personality

Each above increases our likelihood of heart attack. Most are lifestyle related.

ELEMENTS OF A WELLNESS PROGRAM

Here again we'll run into some disagreement, but the following elements are a part of a number of current programs:

1. An exercise/fitness program
2. A health appraisal program
3. A substance abuse program
4. Smoking cessation programs
5. Nutrition programs
6. Stress control
7. Asymptomatic approaches
8. Employee Assistance Programs

We'll look at each briefly in this chapter. Dr. Barick in his article describes some of the established programs (see Exhibit 27.4).

Exercise

In many companies wellness programs grew out of the exercise component. Dr. Linda Marks[108] describes the component writing also in *Occupational Health and Safety* magazine:

> Physical fitness programs focus on activities which are related to fitness, that is, the promotion of cardiovascular function, stress reduction and prevention of obesity. Many large corporations are providing employees with in-house physical activities centers. Health officials in smaller industries are encouraging their employees to exercise in a safe and effective manner outside the workplace.

Tenneco:	Developed an extensive health and fitness center including a large variety of supervised exercise and cardiovascular fitness programs.
Burlington, Ind.:	Focused upon a "healthy back" program designed to relieve pain and prevent further injury and time away from work.
Johnson & Johnson:	Provided health promotion facilities. Their "Live for Life" program includes resources to help individual employees to determine how to improve their lifestyle.
Ford:	Initiated cardiovascular risk intervention program through which extensive research has been conducted. Thousands of employees have participated in various model programs and cafeterias have changed to offer a "heart healthful" menu.
Xerox Corp.:	Provides in addition to its exercise labs, a health maintenance program aimed at employee fitness.
Kimberly-Clark:	Invested $2.5 million on facilities—an olympic-size swimming pool, a running track, various exercise areas and equipment rooms. The three stage prevention program includes a physical exam and health risk profile, a conference to establish personal health goals and an orientation to the facilities.
Sentry Insurance:	Has established a comprehensive approach to wellness for employees and family members. "Flex-time" allows ample time to use fitness facilities which include the fitness center, pool, gym, racquetball courts, indoor driving range, cross country jogging paths and ski trails. Supervised programs and classes are provided on related health and fitness functions.
CIBA-Geigy:	Developed a wellness program based on seven comprehensive communication objectives to provide employees clear directives and incentives on improving health.
Continental:	Offers special programs each month aimed at a particular subject in addition to sponsored counseling for health functions.
Revco:	Implemented an employee exercise program designed to reduce workers' compensation costs. Warehouse workers take time out to stretch, bend, twist and turn twice each day.
Wausau Insurance Co.	Developed a series of 20 stretch exercises for its corporate clients called "Just Minutes a Day." It was created by Jennifer Locke, a former member of the U.S. Ski Team and registered nurse. The philosophy is that workers need stretching and muscles conditioning just as athletes do to perform properly.
Johnson Wax:	Employed a specialist with a master's degree in exercise physiology to develop the "Johnson Mutual Benefit" Association (JMBA) Fitness Program. It includes scheduled classes on aerobic activities and related health functions. The medical department administers stress tests. Records are maintained on results — weight loss, pulse rate, diet, blood pressure, mileage, etc.
Baxter Travenol Laboratories, Inc.	Provides a convenient exercise facility with equipment and professional services to promote employee physical activity. Employees receive a fitness assessment, individual counseling, a program manual and health education.

Exhibit 27.4 Established wellness programs.[9]

An assessment of physical activity patterns should be included in any physical examination. These screening activities are a poor predictor of fitness but an accurate prediction of who is *not* physically fit. It is reported between 50 and 70 percent of the adult population in the United States is not physicaly fit. Physical fitness offers benefits for the employer and the employee alike. Physical exercise can benefit one through improved well-being, improved appearance, reduced cardiac risk factors and other physiological and psychological benefits.

The benefits of physical fitness and regular exercise are widely accepted by industry and the general population. The physically fit employee has been found to have less absenteeism, fewer accidents resulting in back injury, improved attitudes and higher productivity and morale.

It should be kept in mind that these fitness programs are not without some hazard. For instance, a company sponsored fitness program might open up some liability to the organization:

— Sports injuries covered under Worker's Compensation
— Joggers injured by vehicles on public roads could be covered under Worker's Compensation.

Some organizations have found these types of problems to be their most serious losses in terms of payout. Dr. Marks describes why and shows some of the possible problems (see Exhibits 27.5 and 27.6):

When discussing physical fitness, the occupational health nurse recognizes that employees fall into four distinct categories. The first group is composed of persons who participate in a regular, safe exercise program. Persons in this group would be considered physically fit.

The second group could be described as "physical fitness fanatics" who engage in a wide variety of physical activities which are likely to be unsafe due to lack of preparation and consideration of the needs of the body. Included in this category are the healthy young men, in basically good physical shape, who enjoy working out with their friends. They frequently push beyond their work capacity and do not adhere to needed warm up and cool down activities. This group is composed of individuals who impulsively decide to take up bull riding or hang gliding on a sudden whim. Without adequate training and preparation they suffer a multitude of injuries which definitely affect their work performance.

The third category is composed of the employees who feels good because of his physical activities and insists on continuing his daily jog or run despite advanced cases of tendonitis, stress fractures or other muscular skeletal injuries. Neglect of injuries acquired through overzealous activities not only increases absenteeism and results in lasting physical damage and disability but also increases cost of hospitalization insurance. This person could be called the "physical fitness zealot."

The fourth group of employees maintains a sedentary lifestyle in which the most strenuous activity is walking to and from the employee parking lot or getting up to change the channel and open the refrigerator door during the commercial. This employee is quite likely overweight, mildly hypertensive, under high levels of stress, hardworking and describes his job as being rapid-paced with much internal pressure. It is likely that this employee smokes, drinks a great deal of coffee and feels unable to take a few minutes time for himself. He seems puzzled following his annual physical when the physician advises him to get more exercise. The typical scenario would show the physician telling him the same thing annually for the past five years.

Sport/Fitness Activity	Examples of Potential Injuries or Risk	Prevention: Examples of Equipment Safety	Environmental Risks/Hazards
Baseball	Head, Jaw, Eye, Arm (Epiphyses Upper Humerus), Brachial Plexus	Protection: Eyes, Head, Face (Catcher); Proper Clothing and Footgear	Wind, Climate Changes: Heat, Humidity; Cold, Air Pollution
Basketball	Eye, Back (Strain), Ankle Sprains; Dental Injury	Protection: Proper Clothing; Footgear; Mouthguard	Heat; Hard Surfaces
Bicycling	Head, Knee, Back Strain; Torsion of the spermatic cord; Hyper/Hypothermia	Protection: Head, Eye (Dust, etc.); Proper Clothing, Wind, Sun and Cold Protection	Wind, Weather Changes; Heat/Cold, Rough Terrain, Slippery Surfaces, Traffic, Animals
Bowling	Back Strain, Joint Stress; Blisters	Proper Clothing and Footgear	Rough, Slippery Surfaces
Boxing	Head, Eye, Ear, Nose, Face, Chest, Dental Injuries, Dehydration	Protective Headgear; Properly Fitting Gloves; Footgear	Hard Surfaces
Dancing: Ballet	Ankle Sprains; Fx./Dislocated Toes; Fx. Phlanges	Properly Fitting Footgear	Hard, Uneven Surfaces
Modern	Upper, Lower, Back Strain; Hamstring Strain; Shin Splints		
Fencing	Eye, Face Injuries	Eye, Face, Body Protection	Heat, Slippery Surfaces
Gymnastics	Spinal Arch Fx., Back Strain; Nerve Injuries: Ulnar, Radial, Sciatic, Lateral Femoral; Pubic Fxs.	Adequate Padding, Properly Fitting Clothing and Footgear	Heat, Hard, Rough, Slippery Surfaces
Golf	Injury to Head, Eyes, Back; Elbow Flexor Strain, Hyper/Hypothermia	Sun, Eye Protection; Proper Footgear and Clothing	Wind, Climate Changes, Other Golfers
Handball	Nerve Injuries to Hand; Blunt Injuries, Radial/Ulnar Artery; Occulsive Arterial Disease; Eye Injuries	Proper Footgear and Clothing	Hard, Rough, Slippery Surfaces
Jogging; Running	Low Back/Gluteal Strain; Stress Fx. Tibia/Fibula; Shin Splints, "Runner's Knee," Hamstring Strain; Flexor/Extensor Tenosynovitis; Foot, Arch, Toe Injuries; dehydration, Hyper/Hypothermia	Proper Clothing, Footgear; Weather Protection	Wind, Climate Changes; Increases in humidity, Cold/Heat; Hard, Rough, Uneven Surface, Traffic; Animals
Hockey: Field	Head, Eye Injuries; Soft Tissue Abrasions; Contusions, Dehydration, Hypo/Hyperthermia	Head, Face, Eye, Leg Protection, Proper Clothing, Footgear	Wind, Climate Changes; (Heat/Cold), Rough Surfaces
Ice	Head, Eye Injuries; Upper/Lower Extremity Fxs.; Soft Tissue Abrasions; Dental, Larynx/Tracheal Injuries; Hypothermia	As Above, Also Internae Mouth/Hand Protection	Wind, Climate changes; Cold; Uneven Rough Surfaces
Motorcycling	Head Injuries, Back Strain; Multiple Trauma, Hyper/Hypothermia	Protective Clothing, Headgear, Boots, Gloves	Weather, Wind, Climate Changes; Rough, Slippery Surfaces; Motor Vehicles
Racket Ball	Eye Injuries, Back Strain, Forearm Tenosynovitis	Eye Protection, Proper Clothing and Footgear	Hard, Slippery Surfaces
Riding (Horseback)	Head Injury; Spinal, Clavicle, Pelvic Fxs.	Proper Clothing, Boots, Crash Helmets	Wind, Climate Changes, Unpredictable Terrain/Horses

Exhibit 27.5 Examples of sports/fitness activities, body/environmental risks and prevention measures.[108]

Sport/Fitness Activity	Examples of Potential Injuries or Risk	Prevention: Examples of Equipment Safety	Environmental Risks/Hazards
Scuba Diving	Inner Ear Rupture; Hypothermia; Increased CO_2, Disorientation	Protective Headgear, Wet Suits, Check Equipment, "Buddy" Systems for Safety (Ex: O_2 and Breathing Devices)	Air/Water Temperature Changes, Pressure/Perceptual Changes
Skating, Ice	Back, Pelvis, Wrist, Ankle Sprains, Frostbite/Hypothermia	Eye Protection (Outdoor); Proper Clothing and Proper Skate Size	Wind, Climate Changes (Outdoor); Rough, Weak, Ice Conditions
Skiing: Cross Country	Back; Knee Strain, Eyes (Sun Glare), Frostbite, Dehydration	Wind, Weather, Sun Protection; Proper Clothing; Hands, Foot, Head Heat Conservation; Safety Bindings	Wind, Climate Changes; Rough, Rocky Terrain;
Downhill	Arm, Leg Fx.; Knee Injuries; Frostbite, Hypothermia	Proper Size Skis, Pole, Boots	Sudden Change Snow Conditions, Unfamiliar Terrain
Skiing, Water	Back, Arm, Shoulder Strain; Wind and Sun Burn, Hyper/Hypothermia, Eye (Sun Glare), Dehydration	Personal Flotation Device; Eye Protection, Safety Plan for Pick-Up; Experienced Boat Driver	Wind, Climate Changes, Other Skiiers, Boats
Snowmobiling	Head and Facial Injuries, Eye (Sun Glare); Upper and Lower Extremity Fractures, Dehydration; Hypothermia	Protective Clothing; Headgear, Boots; Hands, Foot, Head Heat; Conservation	Climate, Wind Changes; Unfamiliar Terrain; Other Sportsmen
Soccer	Head, Knee (Medical Cartilage) Injury; Back Strain; Nerve Damage (Peroneal); Dental Injuries	Proper Clothing and Foogear; Mouthguard	Wind, Climate Changes; Rough, Slippery Terrain
Swimming	Otitis Media; Fungal Infections; Radial Nerve Injuries, Dehydration, Hyper/Hypothermia, Water Aspiration	Proper Pool Hygiene and Maintenance Schedules; Ear Plugs	Wind, Climate Changes; Unsanitary Pools; Underwater Conditions (Outdoor); Currents, etc.
Tennis	Eye Injuries; Torn Knee Cartilage; Tennis Elbow (Overstretched Extensors); Ulnar Nerve Injury; Tennis Toes (Sub-Nail Hemorrhages); Dehydration; Hyperthermia	Protect Eyes, Head from Sun; Proper Clothing and Footgear	Wind, Climate Changes; Rough Playing Surfaces
Walking	Stress fractures, Blisters, Sprains; Hypo/Hyperthermia	Protective Clothing and Footgear, Head Protection from Cold/Heat	Wind, Climate Changes; Uneven, Slippery Surfaces, Traffic Hazards; Animals
Water Sports: Boating (Motor; Sail)	Eye Injury (Sun Glare) Dehydration; Hyper/Hypothermia, Overexposure, Substance Abuse and Related Accidents; Drowning	Personal Flotation Devices; Head Protection from Cold/Heat, Eye Protection (Sun); Safety and Fire Equipment; Rescue Plan	Climate and Wind Changes; Other Boaters; Underwater Hazards
Wrestling	Below Neck Paralysis; Axillary Nerve Damage; Herpes Simplex and Contact Viral Infections	Hard Surfaces, Unpadded Posts	Hard, Slippery Surfaces

Exhibit 27.5 Continued.

Exercise Devise and Equipment	Pros	Cons
Trampolines	Indoor Advantage	Increase stress on leg muscles/tendons; Dangerous: If have calf muscle strain or tendonitis of achilles region; Cause serious neuro injury (acute cervical spine flexion); Injury due to falls on frame; Does not give good aerobic exercise; Does not help lose weight.
Inversion Boots for "Upside Down" Hanging	Some recommend for relief from sitting/standing	Increases head, eye vessel pressure
Jump Ropes	Excellent cardio-vascular exercise, Good for coordination, agility; Useful adjunct to weight reduction program	More demanding on middle-aged (see Doctor first) Some find boring.
Treadmills	Builds cardio-vascular fitness; Indoor advantage, Need to do 20-30 minutes, three times weekly for bnefit	Need to see doctor before starting program
Stationery Bicycles	Excellent cardio-vascular condition; Need 15 minutes at a time, three times a week; Uses large muscles, trunk and legs; Good for people with back problems (no impact).	Can be tedious — give up too easily; Early soreness in thighs
Rowing Machines	Builds, strengthens upper extremities; Excellent for back muscles; Good for abdominal muscles; Helps muscle tone; increases endurance	Difficult for persons with poor endurance; Can overexert
Chest Pull-Ups	Develops shoulder muscles, breasts (firms up)	No cardio-vascular benefits
Small Wheels	Good for stomach, arms and shoulder muscles	Can increase blood pressure and heart rate; Avoid: coronary disease, bad back
Stomach Trimmers	None — are "gimmicks"	No data to support localized trimming; Can restrict/press on abdomen
Twisters	Works abdomen/hip muscles; Gives ankle mobility	More of a skill — less for muscle development or cardio-vascular endurance
Progressive-resistance; Machines with weights, pulleys and elastic-type devices	May improve strength; Stress skeletal muscles	Do little for heart-lung capacity
Miscellaneous: Leg lifts and pull-ups; Hand grips	See above	Not for cardio-vascular fitness; May increase systolic blood pressure
Hot Tubs	Therapeutic effect warm H_2O; Meets psychological need for social interaction	Dangerous: Effects of overheated H_2O (dehydration, hyperthermia); bad effects of alcohol/drugs indulgence; Drownings, Genital Herpes, Pseudomonas folliculitis, Trichomonas, Otitis Media

Exhibit 27.6 Pros and cons of exercise/fitness devices and equipment.[108]

The ocupational health nurse should recognize that all four groups need nursing intervention. The first group of ideal employees who exercise regularly in a safe manner certainly merit praise and recognition for their continued fitness. The physical fitness fanactics in the second group need assistance in recognizing and implementing safe exercise patterns. The third group needs assistance in recognizing when it is not safe to continue exercising. All who exercise regularly should recognize the common danger signals to ensure their personal safety. The nurse can use Exhibits 27.5 and 27.6 to assist the employees in selecting a safe activity which meets their lifestyles and needs. In reviewing and planning physical fitness activities, all employees should examine the appropriateness of each activity and its safety.

Health Appraisal

In a Health Appraisal Program the various risk factors of an employee are identified. The employee completes a questionnaire which covers current medical status, past medical history, family history, social history, etc. Lifestyle questions are included, plus numerical data such as height, weight, blood pressure and cholesterol levels. The information is then computer analyzed and related to the twelve leading causes of death in this country. The result gives the individual's current health status, how the status relates to the health status of the average healthy person of the same sex and age, and, if needed, points out ways of changing health risks. The individual can then be directed to whatever changes might be indicated in the appraisal: smoking withdrawal, nutritional counselling, safety, alcohol usage, exercise program, stress management, etc.

Smoking cessation programs, nutritional programs, substance abuse programs, and stress management programs are all different components of many wellness programs. Substance abuse and stress management are covered in different chapters in this book due to the complexity of these subjects. A brief look at Employee Assistance Programs will be given in the chapter on substance abuse.

Asymptomatic Approaches

In the previously mentioned national teleconference, Dr. Weisenberger[168] discussed asymptomatic approaches:

> Part of a wellness program addresses finding diseases in the asymptomatic person. Obviously it would be expensive and wasteful to look for all diseases during the asymptomatic period. The decision whether to include a disease in any surveillance program is based upon the following criteria to determine whether such a program would be worthwhile:
>
> 1. The disease must occur frequently enough to be looked for.
> 2. There should be some inexpensive and relatively easily performed test available that would detect the disease during the asymptomatic phase.
> 3. There must be an effective *treatment* available.
> 4. This treatment introduced during the asymptomatic phase must yield better results than treatment introduced during the symptomatic phase.
>
> For example, cancer of the colon would qualify as a disease that meets the above criteria. It is the second most common cancer in men and the third most common

cancer in women. There is an inexpensive test available—the Stool Occult Blood Test. There is very effective treatment, and lastly, if treatment is introduced during the asymptomatic phase the cure rate is better than 90 percent. This cure rate falls to approximately 50 percent if introduced during the symptomatic phase. Disease of the gall bladder should *not* qualify. Although it occurs frequently, the test is not inexpensive. There is an effective treatment available but the treatment introduced during the asymptomatic phase yields no better results than treatment introduced during the symptomatic phase. In other words, you can wait until the individual becomes sick and then treat them and achieve the same results."

EMPLOYEE ASSISTANCE PROGRAMS (EAPs)

Many of the problems which impact health and safety on the job originate outside the work environment. The changing family structure, single parents and alcohol/drug problems are some of the major problems in our society which affect employers and employees resulting in increased absenteeism, accidents, increased costs of health care or declining productivity. EAPs are emerging as one of the ways industry can reduce the cost of these problems.

The problems employees present to a comprehensive EAP include the following:

— Alcohol and/or drug problems.
— Marital and family relations.
— Emotional problems.
— Single parenting.
— Financial.
— Legal.
— Work related problems.

RESULTS OF WELLNESS PROGRAMS

How do you measure the effectiveness of the Wellness Program? Mark Wright[173] suggests:

The easy way out on results is head counts.

In 1985, over 100,000 individual uses of the Physical Fitness module are on record—65 percent participation in individual exercise and 35 percent in classes. Unfortunately, all measurements aren't that black and white.

The traditional measurements of productivity, absenteeism and turnover are, needless to say, good measurements. Under close examination, however, questions arise. Say, for example, an employee is absent 25 days in one year, begins coming to the Wellness Center and is absent only 10 days the next year. Great—a 15-day improvement. But, can the Wellness Center take all the credit? Or did the employee change to a better job or take care of a domestic situation or a parenting problem? The measurement of results can be trickly, but as we get more experience, we have better results measurements.

In many areas, too, we can nail down direct costs and direct savings:

☐ $34,440 saved by doing treadmill testing in-house rather than using an outside agent.
☐ $45,000 saved through Cardiac Rehabilitation Program instead of an outside program. More importantly, employees in the Cardiac Rehabilitation Program returned to work 5.2 months sooner than previously experienced. Resulting savings were $430,000!

☐ $232,000 saved through 212 cases handled in the Well Back Program.
Total savings in 1982 amounted to $744,000—exceeding the entire Wellness Center annual budget by about $250,000.

Our positive trends, both usage and cost savings, continued:
☐ Over 100,000 individual uses at the Wellness Center—a solid increase over 1982 usage.
☐ $46,472 saved by doing treadmill testing in-house.
☐ $85,680 saved in Coors Cardiac Rehabilitation Program. Thirty employees, spouses and retirees participated in Cardiac Rehabilitation Program. Return-to-work time for employees was reduced to 3.27 months resulting in savings of $212,220.
☐ $319,305 saved through the Well Back Program that includes:
  — Well Back Clinic
  — Orthopedic Rehabilitation Program
  — Warehouse Back Injury Prevention Program
  Total savings in 1983 amounted to $663,677. Cost savings again exceeded total operating budget for the Wellness Center—this year by more than $107,500.

Perhaps the "unmeasurable" results—healthier, happier, longer-living, more productive, self-responsible and self-managed employees—are, in the long run, the most valid.

Dr. Weisenberger[168] also contributes some figures:

What might a company expect if it chose to start a Wellness Program for its employees. Measuring the cost effectiveness of a wellness program is difficult. When you prevent problems from occurring, counting the number of conditions that did not occur, so to speak, becomes difficult. Using Epidemiological Systems some companies and medical economists have been able to put some dollar figures on savings a company might achieve. Some of these dollar figures are the companies personal experiences, and some are based on state of the art of scientific estimates. Any company considering health promotion activities should try to build a system of evaluation of their programs results into their planning process.

It's estimated that it will cost a company at least $4,000 a year per employee for direct health costs in the year 2000. Other economists feel that we may reach this figure sooner than the year 2000. Certainly this figure will impact upon a corporations' profits. One of the auto manufacturers in Detroit estimated their medical benefits added $800.00 a year to the price of each car that they manufactured.

One department of economics looking at smoking costs per year to a company estimated the short term cost of: absenteeism, which is 25 percent greater in a cigarette smoker, productivity loss, as they describe it, time occupied in the smoking process or with the smoking material; fire losses, Workmens' Compensation costs related to the synergistic effect of smoking with coal, asbestos, etc., to be $346.00 per employee per year. Furthermore, they estimated the long-term costs related to cancer, heart disease, chronic lung disease, vascular disease, and birth defects to be another $255.00 or approximately $600.00 per year per smoking employee and these dollars were 1980 dollars.

At another company that I had personal experience with, we looked at the results in reduced health costs and weekly benefits for the first 200 employees that were successfully rehabilitated from excess alcohol usage. This employee assistance program resulted in $80.00 per month per employee in reduced health costs and weekly benefits or a savings of $192,000 for the first 200 employees in the first two years. Interestingly enough the employee earned $1,100.00 per year more in increased income as they were on the job working rather than being on illness-absence.

An interesting program reported as the employee fitness and lifestyle project (Toronto, 1978) looked at a controlled program of introducing exercise in an insurance company. The results were impressive. A group of employees who adhered to an exercise program which involved exercising two or more times a week shows a 42% average decline in monthly absenteeism. In the same study overall hospital utilization decreased in the test group from .27 days to .09 hospital days per employee per year while the control company, another insurance company, showed an increase of 0.13 to 0.5 hospital days per employee per year.

New York Telephone reports that they feel they have saved $2.00 for every $1.00 invested in Wellness Programs.

Control Data has a program which they call Stay Well, which is similar to the Wellness Program possibilities that I have just described. When they analyzed their health care costs in 1980 to look at what might be their expected savings, they discovered some of the following items.

(1) They discovered in their employees smokers averaged twice as many hospital days as non-smokers.
(2) The hypertensive employees cost $692.00 per year as compared to $326.00 per year for normal or borderline blood pressure employees. There was no significant difference in days hospitalized.

They discovered that sedentary employees averaged $436.00 per year in medical care costs as compared to $321.00 per year for exercisers. Exercisers average 2/3 the number of hospital days as compared to sedentary employees.

It is my opinion that as more and more data comes in we will find even stronger evidence to support a company entering a Wellness Program for its employees. As an employer what is it worth to you to reduce heart attacks per 1,000 male employees per year from 10 to 6? What is it worth to you as an employer to reduce the chances of an employee getting lung cancer from 1 to 200 per year to 1 in 2,000 employees per year?

In recent years there has been an increase in the development of work site programs to improve employee's health. Approximately 40 percent of the Fortune 500 companies now have some form of Wellness Program in place. In many instances the development of these programs has not been based on clear cut cost-effectiveness. When asked why they chose to enter the Wellness arena, a good number of these employers gave the following reasons:

To improve the health of their employees.
To help people help themselves.
To improve morale.
To increase productivity.
To control the rate of inflation of their health insurance premiums.
To educate their employees about how their health habits affect their health.

Although these are all commendable issues, it is my belief that the remainder of these Fortune 500 companies will also turn toward Wellness Programs based on the fact that they will be proven to be cost-effective. It will make good business sense for companies to concern themselves with their investment in the total health of their employee. Additionally, the employee is considered to be the healthiest member of the family unit. When the company hires the employee, in addition to the employee's health cost they also usually assume some part of the cost of the health of that employee's dependents. In many cases these health costs are even greater than the health costs of the employee. I believe companies *should be* and *will be* looking at mechanisms of influencing these health costs as well. How do you get willness into the home? How do you make people more health conscious? How do you teach people that their health is a product of the choices that they themselves make?

Chapter Twenty-Eight

Substance Abuse

The United States economy has suffered a staggering loss because of drug useage.

Bruce Wilkinson

In the first edition of this book we did not discuss substance abuse, not because it wasn't a problem, but the severity of the problem simply wasn't recognized at that time. It probably still is not recognized, but at least it receives much more attention today than ever before.

THE PROBLEM

Bruce Wilkinson[170] describes the enormity of the problem:

> A United States Senate House Committee, assigned the task of studying the impact of drugs on the American economy, reported that narcotics sales in this country have skyrocketed at an annual rate of $10 billion every year since the year 1978. Drug traffickers grossed $110 billion in 1984; which was double the combined profits of all Fortune 500 companies.
>
> This sudden surge has brought the total number of heroin addicts in the United States to more than *one half million* people for the first time since 1973. Deaths attributed to drug overdosing increased by 93% from 1979 to 1983.
>
> A 1977 government sponsored study revealed that the United States economy has suffered a staggering loss because of drug usage. Its 1977 findings, adjusted to 1983 dollars, indicate that the American economy suffered a loss of over $25.8 billion for that year and over $40 billion in 1984.
>
> These losses were outlined as follows:
>
> - Absenteeism, slowdowns, mistakes and sick leave $4.9 billion
> - Drug-related deaths $1.3 billion
> - Imprisonment $2.1 billion
> - Leaving jobs for criminal careers to support habits $8.3 billion
> - Treatment and rehabilitation in hospitals and by doctors; administration of treatment programs, research and training $367 million
> - Federal, state, and local expenditures for courts, police, and prisons $5.2 billion
> - Alarm systems, locks and other preventative steps for businesses and individuals. $113 million

Note: Most of this chapter was authored by Bruce Wilkinson, President, Workplace Consultants, Inc., Gretna, LA.

Some portions of the substance abuse problem have received even more scrutiny than others. Alcohol abuse is one where a number of studies have been made. One such study was made in the railroad industry in recent years:[107]

> Between 1977 and 1979, University Research Corporation studied the drinking practices of workers on seven railroads employing 234,000 workers—47,000 exempt employees (managers and others not working under union contracts), 72,000 operating personnel (contract workers responsible for the operation of trains and engines), and 115,000 non-operating personnel (contract workers who maintain trains, engines, and track). They represent about one-half of the workers on the nation's Class I railroads (railroads with annual revenues of $10 million or more).
>
> The data on which the findings and recommendations are based came from 240 key management, labor, and employee assistance program staff who gave personal interviews and from 5,704 randomly selected employees of the participating railroads who completed mailed questionnaires. The survey enjoyed a 70-percent response rate.
>
> Their conclusions:
> 1. Employee drinking is an important contributing factor to railroad accidents, but the connection between drinking and safety is not being adequately investigated.
> 2. Employee drinking also contributed to increased rates of absenteeism, lost productivity, illness, grievances, and reduced productivity of workers who are problem drinkers.
> 3. The single greatest *known* cost incurred by railroad companies was the reduced productivity of workers who are problem drinkers.
> 4. It costs more to dismiss a problem drinker than it does to rehabilitate him.
> 5. Conservative estimates of the company-incurred costs of employee drinking are:
>
> | | |
> |---|---:|
> | Absenteeism | $3,100,000 |
> | Lost Productivity | 25,000,000 to 100,900,000 |
> | Injuries | 583,000 |
> | Accidents/Damage | 650,000 |
> | Employee Assistance Programs | 1,000,000 |
> | Insurance Premiums | 2,300,000 |
> | Grievance Process | 408,000 |
> | *Total* | *$ 33,941,000* |
> | | to |
> | | *$108,941,000* |

SUBSTANCE ABUSE CONTROL

As big a problem as this is, it is only recently that we have begun to build controls for the problem in industry. The elements of a control program are:

— Preemployment screening.
— Identifying problem employees.
— Employee Assistance Programs.
— Company policies.

Before examining each briefly, following is a list of the more common substances involved:

— Cannabis (Marijuana, Hashish, Hash Oil).
— Amphetamines (Benzedrine, Dexedrine, Neodrine).
— Barbiturates (Seconal, Nembutal, Quaalude).
— Cocaine.
— Opiates (Heroin, Morphine, Codeine, Opium).
— Hallucinogens (LSD, Angel Dust—PCP).

Alcohol is also a drug and is commonly abused in the workplace. It is classified as an intoxicant and its effects are usually identified with those of barbiturates.

PREEMPLOYMENT SCREENING

Business groups of all types are evaluating different methods of screening drug users at the pre-employment level. It stands to reason that drug users will avoid applying for jobs where drug screening is required. Therefore, it simply means that if your company is not screening for drug users, you can expect these potential employees to appear on your job site.

Employers have the right to know of any potential problem that may be encountered when hiring employees. Urine drug screening at the pre-employment level is an excellent loss prevention method used by management. Management has an obligation to provide a safe and healthy work place for all employees and has the right to hire the most qualified applicants for the available job positions.

Urine drug screening is normally included in the pre-employment physical requirements. However, if pre-employment physicals are not required, preemployment drug screening can be required as a separate entity. Local physicians, hospitals, clinics or laboratories may be equipped to provide the urine drug screening for your job applicants.

Employers should fully inform new applicants of the company's drug abuse policy as the first step in the hiring process. In many cases, applicants on drugs will not show up for the pre-employment exam if they are aware of the drug test. Moreover, if the applicant fails the drug test, the subsequent cost of back x-rays and other pre-employment medical tests are avoided.

When it is known that continuous monitoring is required as a condition of employment, the odds of a drug user not showing up for the pre-employment exam will increase.

IDENTIFYING PROBLEM EMPLOYEES

Drug users are sometimes difficult to identify unless some type of urine or blood and plasma screening procedures are incorporated in a company's drug program. However, drug users do establish a pattern of unusual behavioral habits, that, if alert, company representatives can detect.

Some of the symptoms include. . .

— An increase in quality control problems.
— Low production output.
— An increase in automobile liability premiums.
— Increased absenteeism.
— Workers' Compensation rates increase.

— Signs of employee theft surface, through inventory shortages.
— Morale begins to deteriorate.

The Personnel and Training Departments discover problems on a different level. They observe. . .

— Monday morning absenteeism.
— Habitual tardiness.
— An increase in recordable injuries and illnesses.
— Low attention span in training classes.
— An increase in turnover.

The field supervisors are your first line of defense in combating employee drug abuse. Their observations come from working with the employee at the point of operation. Some of the signs they notice are:

— Some employees taking more breaks than others.
— Employees going to the restroom or their lockers in groups.
— A sudden change in individual personalities; short tempers where patience once existed.
— An increase in employee reprimands.
— A noticeable rise in employee complaints.
— Missing tools and equipment.
— Individual loss of work quality or production output.
— Employees sleeping on the job or showing up in areas where they do not belong.

An alert industrial nurse or medic will notice an increase in routine treatments. They'll find that:

— More accidents are occurring, especially unusual ones.
— An increase in minor illnesses, specifically in certain work areas.
— A rise in supervisor visits, complaining about the health conditions of certain employees.

Safety representatives become alarmed. They have a steady well-trained workforce that has been issued the best of personal protective equipment, yet they discover:

— An increase in unsafe acts.
— An increase in the number of first report of injury forms filed.
— A reduction in safety award winners.
— An increase in preventable and chargeable vehicle accidents.
— More forklift and motorized equipment accidents are recorded.
— The unsafe act becomes routine in certain departments.
— More than a few employees fail to use prescribed safety equipment.
— Heavy objects are being dropped and injuries are resulting.
— Tools are being misused and damaged equipment goes unreported.
— Usually cautious employees suddenly become careless around dangerous equipment.

LEGALITY OF DETECTING

Industry currently uses several forms of detection for drugs in the workplace. The ones that have been most effective thus far can be categorized into two areas:

— Contraband searches.
— Urine drug screening.

When drugs are found by either of these two methods, some form of punitive action for the employee should be defined in the company policy.

Contraband searches rely on the use of trained narcotic detection dogs and private sector investigators. The oil and gas industry initially employed this form of detection in an effort to keep its offshore rigs and platforms free from the hazards associated with drug abuse. The popularity of this method increased in the late '70s because of the success in uncovering drugs, alcohol, weapons, ammunition, drug paraphernalia and stolen property.

Contraband searches have been challenged by employees who felt that laws prohibiting unreasonable searches have been violated.

The exposure to litigation is minimized if certain guidelines are followed:

— Searches and inspections must be part of an overall safety effort to provide a safe workplace.
— Employees must be notified in advance in the form of a company policy. They should be informed of the details of the policy as well as what form of punishment or discipline is specified for violators.
— Company policy needs to be posted in conspicuous places such as employee bulletin boards.
— Discipline must be imposed on an equal basis.

In summary, some of the positive aspects of a drug search are:

— Searches uncover items other than drugs, for example, weapons and stolen property;
— It poses a deterrent to those who feel that they can bring drugs on the job without being discovered;
— It results in physical evidence for unemployment and litigation purposes; and,
— Equitably performed searches have been upheld in court.

Some of the negative aspects are:

— They do not always identify the employee working under the influence of a drug at the time of the search;
— Some employees feel uncomfortable with searches and dislike having inspectors looking through their personal belongings;
— Searches cause some temporary down time of operations; and,
— Searches sometimes uncover violations that management does not want to find, for example, policy violations committed by top management or a well liked employee.

Urine drug screening for identification of drug abuse gained prominence in 1981 when the U.S. Navy initiated a program for drug testing among its servicemen. A crash aboard a large aircraft carrier resulted in fourteen fatalities. Marijuana was found in the systems of six of the fourteen fatally injured personnel. A Congressional study during 1983 reported that two-thirds of the Navy personnel aboard another large aircraft regularly used marijuana and amphetamines. The Navy, faced with a growing drug problem, has now increased its anti-drug program to $24 million per year.

Urine testing for drugs has been proven to be a reliable test method and can be performed rather easily. Currently, the type of machine that most examiners use is portable and carried like a piece of luggage. These machines will not normally record a false positive and have the ability to test for the following drugs of abuse in urine:

— Amphetamines.
— Barbiturates.
— Benzodiazephines.
— Cannabinoids.
— Cocaine metabolite.
— Ethyl alcohol.
— Methadone.
— Methaqualone.
— Opiates.
— Phencyclidine.
— Synthetic drugs.

These portable machines are used by consulting firms and trained company representatives who perform their own testing. This particular system produces qualitative results and can detect the presence of drugs in the urine to an accuracy from 95% to 99% depending on the drug in question.

A positive test result, using this system, means that there is a detectable presence of a particular drug in that employee's urine. Further information and confirmation would be obtained in the form of a second methodology in cases where the employer plans to use the urine screening results to determine disciplinary action against an employee.

For example, there are several types of tests that can be performed in a medical laboratory that will give you a confirmation. Many of these tests are compatible with the portable machine results and some will produce a quantitative reading level of the drug in the employee's urine. These methods include a more comprehensive multi-test system that can process numerous samples each hour.

Principles of selected laboratory methods used to detect and/or confirm drugs in the employee's body fluids are:

— Thin Layer Chromotography.
— Gas-Liquid Chromotography.
— Gas Chromotography/Mass Spectrometry.
— Radioimmunoassay.

Urine screening for drugs is commonly used in the areas of:

— Pre-employment.
— Re-entrance employee examinations.
— Random screening.
— Contraband searches.
— Isolated cases.
— Post-accident testing.

Chain of Custody

Legal challenges may also result when there is no procedure for handling and processing the employee's urine sample. The phrase "Chain of Custody," refers to providing that the sample tested positive is actually the sample from the employee who is accused. Some suggestions to avoid these types of problems are as follows:

— Sample should be taken in a dispensable urine cup with a screw top.
— Both the lid and the cup should be labeled with the employee's name and social security number. The sample should also be numbered.
— The cup should be sealed with police evidence tape if it is transported to a lab for confirmation.
— The laboratory should give the company a receipt showing the numbers of the samples received and that the seals were not broken.

EMPLOYEE ASSISTANCE PROGRAMS IN SUBSTANCE ABUSE

EAPs have been proven to be very cost effective. The minimal cost of the program can range from $12 to $30 per employee per year and is more than offset by increased productivity and reduced claims on Workers' Compensation and medical insurance. Counselors assist managers and supervisors in identifying the source of employee performance problems. Counseling may be offered in areas other than drug abuse such as personal finances and legal problems. Counselors who contract directly with the company will tailor a program to match the employee's needs and provide counseling on a 24-hour basis. Moreover, managers have found that incorporating EAPs into their operations have provided supervisors with additional productivity time. It is interesting that companies will spend thousands of dollars to repair expensive machinery but traditionally do not do the same when an employee breaks down.

An EAP is another method management can use to identify employees with drug abuse problems. Employees are referred to EAP counselors by supervisors for either poor job performance, a positive urine screen test, or the employee admitting to a drug or alcohol problem.

Counselors will attempt to assist employees in solving their problems and return a more productive worker to the supervisor. This task may be accomplished through group therapy, one-on-one sessions, or encounters with the employee's immediate family and support groups. Periodic monitoring of an employee's drug habit may be done in conjunction with the company's ongoing urine drug screening program.

All EAPs are not the same. Some EAPs offer only a referral type service. The employee contacts a counselor for assistance. The counselor identifies the nature of the employee's problem and, in turn, refers the employee to an independent specialist. The cost of the specialist is usually additional and is covered by the employee medical policy.

Other, more complex, EAPs offer a total service package in which the employee's problem is handled by a specialist on staff with the EAP contractor. In most cases, these EAP contractors will only refer employees to independent specialists when they determine the employee's problem is more complex than they can handle. Consultation with EAP staff specialists are normally included in the EAP fee schedule.

An EAP should offer features such as:

— 24 hours/7 days a week service
— Confidential services

— Voluntary participation.
— Serves employees and families.
— Free of charge to the employees and their families.
— Covers any problem.
— By appointment.

The EAP contractor should provide services such as:

— Needs assessment.
— Counsel/referral.
— Follow-up.
— Consultation.
— Training.

An EAP program can play an important role in a company's overall drug abuse program. It allows the employer to offer assistance to employees who really need and want the help, while using the other components of the policy to eliminate hard core undesirables from the workplace.

COMPANY POLICIES

The key to a successful drug abuse program is the development of a comprehensive policy. This policy must be detailed and broad enough to address all situations that are likely to be encountered. It is important to define which employees will be covered by the policy.

It would be desirable to consolidate as many elements as possible into one overall policy. When time is taken to develop a comprehensive policy, confusion is avoided when it is necessary to discipline employees. The policy also provides excellent documentation in the defense of a legal challenge.

All elements of a comprehensive drug program from pre-employment urine sampling to EAPs should be included in a company drug policy.

RESULTS

What results can be expected from a substance abuse control program? Here are some figures from a publication of the Accident Compensation Corporation of New Zealand:[2]

> Variations in sample size and group selection make it difficult to compare success rates achieved by different organizations. The information that is available indicates that programmes can achieve a recovery rate of 60% to 80%. 'Recovery' is based on one or two years' sobriety and safety.
>
> Recovery rates experienced include—
>
> **US Postal Service.** Programme for alcohol-impaired employees began in 1968 in several centres. Employees totalled 83,214. By 1972, 1396 had agreed to participate, 1051 were recovering. Only 345 cases were unsuccessful. Recovery rate: 75%.
>
> **Hughes Aircraft Company.** Programme for 34,000 employees in Southern California began in 1967. Recovery rate: 60%. Average age of employee referred: 45. Average length of service: 12 years. Salaried employees made up 66.7% of those on this programme.

Findings concerning cost-effectiveness have included—

US Postal Service. The cost of alcohol impairment was estimated in 1972 at $3000 per employee per year. Incidence of alcoholism estimated at 8% of work-force. Cost impact in total estimated at $168 million a year. Cost benefit of pro-gramme estimated at $170,874 for one accounting period. Benefits of programme reported to exceed costs (including start-up costs) by five to one.

Bell Telephone Company, Illinois. A programme begun in 1951 was extended to cover all employee problems in 1963. Ten-year records on each of 402 employees (five years before and after referral) showed the following related to those 402 employees—

Improvement rate 72%, including 57% achieving sobriety for one year or more.
Incidence of poor job performance reports down from 28% to 12%.
Incidence of high work performance reports up from 22% to 58%.
On-the-job accident rate down 80%.
Off-the-job accident rate down 66%.
Incidence of sick leave exceeding seven days down from 662 cases in the pre-pro-gramme five years to 356 cases in the post-programme five years.

General Motors, United States. An employee-assistance programme there reported an 80% recovery rate and significant savings indicated by sickness and accident benefit payments down 70%, sich leave down 47%, overall man-hour losses down 85%. These figures related to staff taking part in the assistance programme.

Chapter Twenty-Nine

The Future

Massive changes await us.

　　　　　　D.C. Petersen

In the final chapter of the first edition of this book we talked about the "safety program of the future" and what it might look like. While we're not quite there yet (whenever the future is), we might briefly look at what was said to see what has happened:

END OF BUREAUCRACY

Warren Bennis[8] recently predicted that in the next twenty-five to fifty years we will participate in the end of bureaucracy as we know it and the rise of new social systems better suited to twentieth century demands of industrialization. This forecast was based on the evolutionary principle that every age develops an organizational form appropriate to its genius and that the prevailing form of pyramidal-hierarchical organization, known by sociologists as "bureaucracy" and most businessmen as "that damn bureaucracy" was out of joint with contemporary realities.

Our present-day safety programs fall into Bennis's definition of bureaucracy. In fact by any-body's definition the typical safety program of today is a pyramidal-hierarchical organization, or a bureaucracy. Heinrich initially proposed a system of hierarchical responsibility that paralleled the bureaucratic organization, and even our updated axioms cling to traditional concepts of organization.

Bennis suggests that there are at least four relevant threats to bureaucracy. The first is a human, basically psychological one, while the other three spring from extraordinary changes in our environment. These three are (1) rapid and unexpected change, (2) growth in size where the volume of the organization's traditional activities is not enough to sustain growth, and (3) com-plexity of modern technology where integration of activities and persons of very diverse highly specialized competence is required. The bureaucratic safety program faces these same threats.

Rapid and Unexpected Change

Some of these conditions exist right now. For instance, our productivity per man-hour now doubles almost every twenty years rather than every forty years, as before World War II, and the federal government alone spent $16 billion in R&D activities in 1965 and will spend $35 billion by 1980. The time lag between a technical discovery and recognition of its commercial uses was thirty years before World War I, sixteen years between the wars, and only nine years since World War II.

Change is here, and will continue at an ever-expanding pace. With each technological change we face a safety problem. Traditional safety programs are not keeping up. We say traditionally that the responsibility for safety is the first-line supervisor's, and yet he receives no special training nor has he any special technological competence to cope with change. Even safety staff cannot keep up technically.

Growth in Size

Not only have more organizations grown larger, but they have become more complex and more international. Firms like Standard Oil of New Jersey (with fifty-seven foreign affiliates), Socony Mobil, National Cash Register, Singer, Burroughs, and Colgate-Palmolive derive more than half their income or earnings from foreign sales. A long list of others such as Eastman Kodak, Pfizer, Caterpillar Tractor, International Harvester, Corn Products, and Minnesota Mining & Manufacturing make from 30 to 50 percent of their sales abroad. General Motors' sales are not only nine times those of Volkswagen, they are also bigger than the gross national product of The Netherlands and well over those of a hundred other countries.

Safety programs traditionally structured literally have not coped with growth. Programs designed for a single plant have been stretched to multiplant operations. Programs designed for primary metals industries have been superimposed on light assembly operations in a conglomerate. And they have not worked. Some conglomerates have unified safety programs which have tried to cover all subsidiaries. In other cases only the larger or older member companies have a safety program. The newer companies are ignored. Generally foreign subsidiaries are ignored.

Highly Specialized People Are Needed

With the rapid change in technology and growth in most companies, new safety specialties and new knowledge are needed. Our traditional safety program which says the supervisor is responsible with safety staff to help just does not handle the need for new and specialized knowledge. No one person in the role of safety director can know all there is to know in most of our complex operations.

The above three factors are economic in nature.

CHANGE IN MANAGERIAL BEHAVIOR

Bennis[13] mentions a fourth factor which stems from his personal observation that during the past decade there has been a fundamental change in the basic philosophy that underlies managerial behavior. The change in philosophy is reflected most of all in:

- A new concept of man, based on increased knowledge of his complex and shifting needs, which replaces an oversimplified, innocent pushbutton idea of man.
- A new concept of power, based on collaboration and reason, which replaces a model of power based on coercion and threat.

— A new concept of organization values, based on humanistic-democratic ideals, which replaces the depersonalized mechanistic value system of bureaucracy.

These changed views of man, power, and values have gained wide intellectual acceptance by management. They have caused a terrific amount of rethinking on the part of many organizations. They have been used as a basis for policy formulation by many large-scale organizations. This philosophy is clearly not compatible with bureaucratic practices.

The primary cause of this shift in management philosophy stems not from the bookshelf but from the manager. Many of the behavioral scientists, like McGregor[98] or Likert,[88] have clarified and articulated—even legitimized—what managers have only half registered to themselves. In short, not only do the behavioral scientists show that this change in philosophy works but also the modern manager believes in the fundamental concepts of treating mature human beings like mature human beings. According to Argyris,[7] the desire for relationships has little to do with a profit motive per se, though it is often rationalized as doing so. Managers of tomorrow (if not of today) will believe in the worker as an individual, not as a part of the machine. They are more interested in the person, in his or her needs, physical and psychological well-being, and safety. Managers of tomorrow will believe in participation much more than they have in the past. Managers of tomorrow will adhere to organizational goals that go beyond profit. These goals will be more in line with the needs of the community, the consumer, and the worker. The consumerism, ecology, and safety movements of today will be a molding factor for the manager of tomorrow.

As with the others, this fourth factor affects the safety program. Participation must be considered; safety rules developed by management and handed down like stone tablets just will not work. Inherent in the traditional way that safety rules are formulated and administered is that the few at the top of the organization dictate what is proper or improper regarding safety for those at the bottom. The job of the supervisor is to enforce the rules and subordinates are to follow them. Since most people resist being told what they can or cannot do, each rule becomes a restriction of one's personal liberty. Acceptance of restriction is difficult for most people, unless they are convinced of the validity of the restriction. For example, most people stop for red traffic lights because they are aware of the validity of a rule which prevents accidents. Without acceptance, any rule or law will either be ineffective or not enforceable. It is necessary for industry to find an imaginative approach to a system which will initiate and develop a solid set of safe practice specifications which will be accepted and enforced as valid guides to safe behavior by those who must work within their bounds. Acceptance is the key requisite and acceptance hinges on the obvious validity of any rule. One of the basic principles of nondirective counseling is the acceptance of one's own conclusions. A group will tend to accept its own conclusions as well. Positive acceptance of safety rules can result if subordinates can suggest and develop their own safe methods and rules. Future organizations will, I believe, value the safety function much more than it has been valued in the past. It fits in with the trend.

CHANGE IN WORKER BEHAVIOR

Significant changes have occurred in the labor force. The Department of Labor states that the total labor force increased by about 18 percent between 1970 and 1980. People in the labor force between the ages of forty-five and fifty-four actually declined, while those fifty-five to sixty-four increased at below-averaged rates. The major bulge occurred among those twenty-five to thirty-four; there were 49 percent more workers in this age category in 1980 than in 1970.

Moreover, our society will become more heavily dependent on younger people to fill management positions held by older people during former eras.

Who are these people who will make up this important twenty-five to thirty-four year-old market? Some of them are black, some white. Many are Protestant, more are Catholic. Some live in the East, others in the West. But despite their diversity, they all have one thing in common: all are under thirty-six years of age today. In other words, the great work force of today. Numerous studies indicate that about 40 percent of these people have values and attitudes considerably different from the values and attitudes of the past. It does not take a professional psychologist or sociologist to realize that things are changing. It is quite obvious that many people do not think and act like people used to. The question is not whether things are different; they are. That is a fact. Understanding why there is a difference is important because it is a prerequisite to managing people. Also, it is not a question of whether these new values and attitudes and life styles are "right" or "wrong" or whether we "like" them or "dislike" them. It is rather a question of managing them the way they are.

Individuals are not born with values or ideas about what is right and wrong, good and bad, desirable and undesirable. Rather, values are learned as the result of interactions with certain institutions—primarily the family, church, and school—and with lifetime experiences. The reason why the youth of today are different is that they have experienced different institutional influences and lifetime experiences. According to Dr. David Kollatt[78] (Management Horizons), these three institutions (family, church and school) have undergone a series of profound changes that have altered their influence on people's values. He points out, for instance, that children used to begin school at the age of five or six. This has changed. Many children now attend pre-school at ages two to four (37 percent). Thus, children are out of the home at an earlier age and are exposed to people of different social classes. Since about 80 percent of a child's personality is formed before the age of six, this change becomes significant. Also, with the great amount of moving in today's families, children are less exposed to grandparents and the other relatives. More children leave home for college so the influence of the family on them is removed at an earlier age. Also, divorce rates rose at an alarming rate (up 46 percent) with more children involved, and unstable families generally have a more difficult time instilling values in children.

Another factor in the changed family influence is the role of women in the family. In 1970, 41 percent of women were employed compared to 20 percent employed twenty-five years ago. The majority of these women work full-time, and a large percentage of them have children—most commonly between the ages of five and seventeen. Changes have also occurred in the man's job which tends to make changes in his influence in the development of values in his children. Men travel more in business and are away from home more than ever before. For all of these reasons, and others, the family is not as influential in instilling values in its children as it used to be. Therefore, a substantial percentage of youth believe in different things than their parents did.

Kollatt points out that the influence of the church and the school is also different than it used to be. The church has always had much to say about what people should and should not do, and what they should and should not believe in. This influence has declined. Gallup polls have asked the question, "It religion losing its influence on American life?" In 1957, only 14 percent said yes. By 1968 the figure had reached 70 percent, and in 1971, 75 percent felt that religion was becoming less influential.

Schools, like the other institutions mentioned above, have also changed in recent times. Of the changes, the major one has been the emergence of different teaching techniques. It used to be that schools taught by means of memorization: memorize dates, states, the Presidents, the multiplication tables. Students would keep at it until they could get it right. Teaching was authoritarian in nature. Now, the trend has been toward different approaches to teaching which encourages students to think through problems for themselves, not to accept anything that they cannot work out themselves, and not to accept things just because they always have been that way. This teaching encourages analysis and thought and the questioning attitudes that the youth of today have about life, the world, and all else.

The institutional influences that have shaped the young people of today are quite different from the institutional influences that shaped the older generations. The other major area that shapes a person's value system is experience. Here too there are major differences between the young people and the older generation. The older generation's value system was shaped to a great extent by two major events, as described by Kollatt.[78]

> The first critical experience was the Great Depression, particularly the 1929-1933 period. During this unfortunate episode in our history, Gross National Product dropped by 49.6 percent, personal income dropped by 49.8 percent, and unemployment rose from 3.2 to 23.6 percent. Behavioral scientists maintain that this event left an indelible psychological scar on nearly everyone who experienced it. It tended to foster very strong attitudes about risk and speculation and saving money. It also tended to produce a strong sense of company loyalty because an individual was sincerely grateful to a company for employing him while one out of every four of his friends were without work.
>
> The second critical event was the Second World War because it touched the lives of nearly everyone in the country in one way or another. Over 16 million Americans served an average of 33 months. Nearly 292,000 Americans died, and over 671,000 were wounded. Despite the initial controversy about the propriety of the United States involvement in the war, it became a unifying device in our country, tending to weld people together in a common cause and promoting and intensifying a feeling of patriotism.

Kollatt then describes[78] the critical events that have shaped the values of the younger generations:

> The first thing that must be remembered about this generation is that none of them were alive during the Great Depression, nor the Second World War—nor were they old enough to experience even the Korean War in an adult way. Consequently, the critical experiences of the pre-war generation have about as much impact on the values of the post-war generation as the Boston Tea Party had on the pre-war generation. At best, it is something that they listen to politely.
>
> Instead, most studies indicate that the critical experiences of the post-war generation are as follows:
>
> - The nuclear age
> - The civil rights movement
> - Poverty
> - Space exploration
> - The Viet Nam War
> - Ecology
> - University experiences
> - The communications revolution

As a result of the critical experiences of the post-war generation, certain key perspectives have emerged. They are:

- Belief that national priorities are inverted
- Distaste for social rigidity
- Concern about institutional rigidity
- Lack of political influence and power
- Loss of self to technology and institutions
- Intolerance of hypocrisy
- Absence of meaningful relationships

It is true the older generation has experienced these latter events also, but as Dr. Kollatt points out they have also experienced a larger and broader variety of events and, consequently, have a more realistic and mature perspective on these developments.

As a result of the different critical experiences that have shaped the values of these different generations, they look at things differently. For instance, older people probably see our country's economy as a powerful productive system, taking pride in the advantages of capitalism and competition. Youth are likely to agree, but also see that our system tends to pit company against company, individual against individual; and they see that it brings out some of the worst attributes of people. Older people tend to point with pride to our industrial power and success, while young people tend to point rather to certain other achievements that came with urban decay, water pollution, air pollution, and traffic congestion. Older people speak about our affluence, while youth prefers to look at the fact that in 1969, 33 percent of all black people were "poor" according to the definition of poverty used by the Department of Health, Education, and Welfare. Older people tend to talk about our successes, while young people tend to look at how far we must yet go. Young people also tend to remember that with all of our progress, we have not succeeded in attaining happiness for them or for ourselves. In fact, they may well be remembering the deterioration of a family relationship caused by a working mother or a traveling father.

The values and attitudes of our youth are the result of changes that have occurred in the family, the church, and the school and of their lifetime experiences. The new ideas and attitudes are different as a result. While these people will grow more conservative, their values and attitudes will no doubt remain qualitatively different from those of previous generations. These changed values have resulted in changed life styles. Consider some of these changes:

1. *Instant Everything.* It used to be that a person in debt was viewed as a social outcast; debt was considered something to be avoided. We used to save before buying. Our attitude about credit has completely changed. This postponed gratification value has been replaced by an emphasis on instant satisfaction.
2. *Pleasure.* There is a trend toward living in the present rather than the future, toward having fun now rather than later. The call for "doing our own thing" has enjoyed remarkable success.
3. *The Changed Work Ethic.* The Puritan ethic states that work is good. It should be maximized and fun minimized. People used to feel guilty about having fun. Executives justify taking vacations on the grounds that it will allow them to achieve more after getting away. According to Margaret Mead, "People used to live to work, now they work in order to live." The younger people often reject the old work ethic.
4. *Simplification.* Another change is the trend toward simplification. People are interested in products and services that take the work out of life, that allow them to do things quicker and easier, leaving them more time and energy to do what they want.

5. *Safety and Health.* People are becoming increasingly interested in safety and health products and in their appearance. Interest in safety is at an all time high in this country.
6. *Naturalism.* The young tend to reject artificial behavior. They don't dig "phonies." They like to be themselves and expect others to also be natural.
7. *Personal Creativity.* This trend is apparent on the job (being recognized with job enrichment, participation, etc.) and off the job. Everyone wants to "do his own thing."
8. *Reliance on Others.* Problems seem of such a magnitude (pollution, energy, and traffic) that people feel they personally can do little. We must rely on "them," on government, on business, and others, to solve our problems. The trend is away from self-reliance.
9. *Loss of Confidence in Government, Business, Management, etc.* Kollatt[78] states that the mood of our country has changed. During the Depression people were pessimistic about their own personal futures, but were optimistic about the future of the country as a whole. In recent times, people are optimistic about their individual futures, but are very pessimistic about the future of society. There is an important linkage between people's confidence in society and their attitudes toward business. The loss of confidence in institutions becomes, in reality, the loss of confidence in business. This is one of the underlying reasons for the growth of the consumerism movement.
10. *Consumerism.* As we know, this major change is upon us in total in the safety profession.

The younger people of today (tomorrow's work force) are different. How will they react to a bureaucratic safety program? It seems obvious that they will react rather strongly and negatively to most programs. Where does this leave us? With the need to change.

As we look at the above comments, and at the situation as we put together this second edition, we find that most of what was said by Bennis, Kollat, and others have occurred and are continuing to occur at an ever increasing rate. We have changed and we need to continue to change. As we look at the above comments and other things that have occurred that we did not expect, it's obvious that some things have been a real surprise (some pleasant, some not so pleasant).

SOME NOT SO PLEASANT SURPRISES

1. *More Federal Intervention.* In the first edition we had just recovered from the initial impact of OSHA. Some of us hadn't foreseen that OSHA was just a starting point for what was to come in terms of federal legislation. While OSHA became livable in the mid- and late 70s (both from our getting used to it and because of OSHA's own gyrations—from safety to health in emphasis, from adversarial to cooperative in approach, from one administration to the next and back), we found that being able to live with OSHA was not enough as the Congress passed more and more additional safety related laws (MSHA, Toxic Substances, Hazard Communications, Asbestos). Much time has been spent on all of these—understanding the laws, reacting to the laws, doing what the law says. While these changes were doubling the work load of managers and safety professionals, we found at the same time (in the early 80s) an economic slowdown that resulted in slimmed safety staffs.
2. *The breakdown of the Worker's Compensation System.* When Worker's Compensation came into being early in this century, it guaranteed that injured employees would be compensated without any consideration of fault. It also protected management, as it was the sole recourse of the injured employee, who could not sue. That system is systematically being dismantled in this country. In many states employees can sue in lieu of or in addition to Worker's Compensation benefits. They can sue the company, the boss, the executive, the safety professional, or all of the above, and it is happening regularly.

3. *The advent of criminal liability.* For the first time we find that management has more than financial and morale responsibility for the safety of the worker; in certain situations they now could also end up behind bars. Company executives have been indicted for manslaughter for knowing of hazardous conditions and ignoring them.

4. *The spectacular accidents.* While we've always had major incidents occurring, it seems in the last few years some have been truly spectacular:
   - Bophal. The toxic leak that killed thousands.
   - Chernobly. The nuclear melt down that killed more thousands.
   - The shuttle disaster which killed seven astronauts before our eyes on television.

5. *The growth of stress claims.* In the last edition, for all practical purposes, and stress claims didn't exist. Today they are the number one occupational disease. And we've just started.

SOME PLEASANT SURPRISES

1. *More attention being paid to safety.* Due to the general theme of improving the quality of work life from management books like *In Search of Excellence* there are strong efforts towards a people orientation in many (maybe most) organizations. This typically (after time) grows to include safety and health. Other reasons for this interest are numbers 1, 2, 3, 4, and 5 above.

2. *Acceptance of accountability concepts.* Most major companies have changed their systems to include clearer role definition, more valid measures for safety performance, and more meaningful reward structures. Most of this change has been within the last dozen years.

3. *Acceptance of SBO.* As indicated in Chapter 24, the concepts of MBO, so foreign to our thinking a dozen years ago, has become an accepted way of managing safety today.

4. *Acceptance of positive reinforcement concepts.* While positiveness is still foreign to safety thinking, at least we have conceptual acceptance to listen and to experiment with it as indicated in Chapter 21.

5. *Acceptance of employee involvement in safety programs.* Most companies are recognizing the importance of employee involvement. Employee-run safety programs are no longer uncommon.

THE FUTURE

Donald Wallace[166] (University of Southern California) discusses management and safety management in the 1980s decade, and looks to the future with these thoughts:

> Many of those who attempt to foresee and evaluate the future have described the decade of the 1980s as a period of serious difficulty and fundamental change for the United States. This transformation is one in which we are moving towards a post-industrial society comparable to the 14th century move from a medieval to a renaissance society, and as a period in which we will experience:
>
> 1. The end of an era of unlimited resources and growth.
> 2. An increased acceptance of our international interdependence.
> 3. A re-examination of our basic attitudes and beliefs about:
>
>   a. our institutions, e.g. government, business, education.
>   b. people.

Yankelovich and Lefkowitz summarize the findings of their own survey work and the surveys of others—Harris, Gallup and Michigan Survey Research Center—as follows:

1. Public attitudes towards government have shifted sharply in the past two decades:

> In 1964, a 59 percent majority of the American public had faith in the competence of government officials ("They know what they're doing"); by 1976, the number of Americans holding this view dropped to 44 percent; by 1978, it had dropped further to 40 percent.
>
> At the end of the 1950s, a 56 percent majority of the public expressed the view that "you can trust the government in Washington to do what is right most of the time." Two decades later, their level of trust had been cut virtually in half to 29 percent.

2. More Americans, for the first time, now believe that the future will not be as good as the present and the present, however bad, is better than the past.
3. The public's attitude toward business has shifted as indicated by response changes to these questions:

    a. People who agree with the statement: "Business strikes a fair balance between profits and the public interest"
       (Yankelovich)   1968 – 70%
                                1977 – 15%

    b. "People who feel a great deal of confidence in major companies."
       (Harris)   1966 – 55%
                                1979 – 18%

The last, but by far not the least, social trend worth our attention is the movement towards an increase in the level of participation in the government of those factors affecting the quality of life. The increasing search for alternative models of living may be a manifestation of our dissatisfaction and disillusion about government, business, and our future and may underlie the increasing demand for a greater voice in shaping the future.

IMPLICATIONS FOR THE FUTURE

Is the drive for economic well being in the 80s incompatible with our changing values and the demand for a better quality of worklife? Can we more widely adapt the models being developed here and in other countries and survive the coming decade as an economy and as a people? Scholars in response to these questions seem to be saying with increased frequency that:

1. Productivity and the quality of worklife are inextricably linked.
2. The methods developed by the Japanese and others demonstrate that efforts to improve productivity can also improve the quality of worklife.
3. Our economic well-being and, perhaps, our survival, may depend on the extent to which we can adapt to our culture what has been learned in other countries and in parts of the United States.

The common threads which appear to be present in these workplace innovations are:

1. A primary emphasis on human growth and dignity at work.
2. Innovative and more effective designs of organizations and jobs.
3. An increased level of employee responsibility, involvement, and participation in the decision making process.
4. The creation of a climate of cooperation between employees and management.

IMPLICATIONS FOR THE SAFETY MANAGEMENT PROFESSIONAL

1. *A broader scope.* Many have suggested that the theoretical and conceptual base upon which the safety profession has sought solutions to the safety problem in the past 50 years has been too simplistic. An over reliance on a strictly technical or engineering approach is another often heard theme from critical safety thinkers suggesting that our profession must now broaden it's view into a *multi-* or *interdisciplinary* approach. The empirically obtained data derived from the research obtained in all fields bearing on the safety question must now be brought to bear on the answers (e.g., the behavioral sciences, medicine, engineering, ergonomics, management science).

2. *The Professional Safety Manager's Role.* This wider view of our profession may then allow us to speculate with some degree of accuracy upon a new perspective of the safety manager's role in 1990:

   A. As a manager, the safety professional will have knowledge of *all* of those disciplines that can make a contribution and will be able to exploit their expertise in the solution to safety problems.
   B. As an influence for change, the safety professional will be expert in developing innovative organizational forms, especially those that have been shown to be effective in maximizing the contributions of *all* of the members of the organization in the accomplishment of its objectives.
   C. As an educator, the safety professional will be a consultant and advisor to all levels of management in order to facilitate a systematic and on-going examination of their assumptions, attitudes and beliefs about people, organizations, work, and safety.
   D. As a trainer, the safety professional will understand and be expert in applying those skills and methods which exploit the energies and talents of *all* of the members of the work group in the solution of safety issues.
   E. As a student, the safety professional will systematically study how to apply an increasing body of knowledge based on sound theory and valid research from all disciplines to the safety problems of 1990.

FUTURE PROGRAMMING

Future safety programs should then consider the relevant factors brought out by Bennis, Argyris, Kollatt, and others. In structuring the program for the future company these things should be considered:

1. The role of management would seem to be first to allow safety to happen; to allow staff safety to tap technical competency where it is needed; and to allow employees to make

decisions on rules and programs. Their second area will be to demonstrate conclusively that their value system includes safety, both product safety and employee safety, and that it includes it at a rather high priority level. In most organizations that is not yet accomplished. Employees of the future will not be as subservient to profit-only goals; they will demand that safety be an equally important organizational goal.

2. The safety program will be better equipped for rapid change. Safety program components will be more flexible. What works in one department, or plant, will not necessarily be right for the next. What works this year, similarly, probably will not work next year. Also, the safety professional must cope better with technological change; will be more aware of coming changes; will have the ability to plan for change; and will have access to personnel with the technical ability to cope with change. This probably means losing some of the present responsibility and authority to people in engineering, research, etc.

3. The safety program will have to cope with size and diversity. Safety programs must be better packaged, more flexible, and more adaptable to individual location use. Modules, such as the No Strain Campaign discussed in *Techniques of Safety Management*[127] might serve this need.

4. Locations with diverse operations will be allowed to structure their own programs. This does not mean they will be ignored. Accountability systems can tightly control results while location responsibility allows flexibility. We will let them "do their own thing" but make sure they do something.

5. Employees will be allowed to structure their own programs. After all, who knows more about the work and its hazards than the people actually doing the work? Motivation to follow the safe practice, then, would come from within each person as a part of the group, and would not be the result of outside pressure. This approach to rule-making gives a number of people, perhaps for the first time in their lives, a degree of participation in directing the activities of the department. This can create a sense of involvement and commitment to the goals and the objectives of the department—goals and objectives that they have a part in establishing.

   Major decisions on safety rules, safety program components and motivational approaches will be decided by employees. Technical and system controls will be conceived by those technically competent and coordinated by safety staff. Management's primary role will be that of a catalyst.

6. Safety programs, structured to demonstrate that management's value system includes a high priority on customer and employee safety, will be more technical, complicated, and unprogrammed. They will rely on intellect instead of muscle. They will be too complicated for one person to comprehend, to say nothing of control. Essentially, they will call for the collaboration of specialists in a project or a team form of organization.

7. The structure of organizations of the future in general, and of safety programs, will have some unique characteristics. The key word will be *temporary;* there will be adaptive, rapidly changing temporary systems. These will be *task forces* organized around problems to be solved by groups of relative strangers who represent a diverse set of professional skills. The groups will be arranged on an organic rather than mechanical model; they will evolve in response to a problem rather than according to programmed role expectations. The safety director thus becomes a coordinator or "linking pin" between various task forces, someone who can speak the diverse languages of research, with skills to relay information and to mediate between groups. Organizational charts will consist of project groups rather than functional groups. This trend is alrecdy visible

today in the aerospace and construction industries, as well as in many professional and consulting firms.

Adaptive, problem-solving, temporary systems of diverse specialists, linked together by coordinating and task-evaluating specialists in an organic flux—this is the organizational form that will gradually replace bureaucracy as we know it.

8. The role of the safety professional will change. There will be fewer safety professionals, but they will function as key advisors to top executives (The Type I Professional), or as a purely technical resource for both safety and health (The Type II Professional).

9. The first line supervisor (our old key man) will disappear from many (maybe most) organizations. Between the increasing span of control and the high performance concepts (no supervisors whatsoever), the traditional supervisory function will almost disappear. Safety programs in these situations will have to be run by employees.

10. Safety meetings as we know them will disappear. The meetings (if they happen at all) will simply be team discussions.

11. Safety committees as we know them will also disappear; or be replaced by Safety Circles or by the team (participative management). With either committees are irrelevant.

12. Safety training of the future will concentrate on allowing workers to find safety problems rather than telling them the problems and the solutions.

13. Safety programs will concentrate on group involvement.

14. Safety will be used as a means of enriching individual jobs.

15. We will finally solve the measurement problem. The solution will stem from behavior sampling concepts, Statistical Process Control concepts, and climate survey concepts.

16. The Worker's Compensation system will be changed even more markedly. It might disappear and be replaced by something quite different, or be federalized; probably first the latter. What will a federal Worker's Compensation system mean? Look at the F.E.L.A. system for railroads, the longshoremen's system, etc. It means a doubling, trebling, or worse, of management's cost. It also means the liberalization of what is covered and of benefits.

17. The criminal liability of managers and safety professionals will grow at an ever increasing rate. At a recent accident in an Eastern manufacturing plant, the first outsider on the scene was the state police wanting to pinpoint blame for criminal arrest. This environment will not get better; it will probably get worse.

18. Cost containment will become critical. Medical case management and claim administration will be crucial aspects of control and getting much more management attention.

19. Academia will react to needs (eventually) by increasing occupational physician training (for medical case management needs), by maintaining industrial hygiene training, and by reducing (and eventually phasing out) safety management training. The Type I Safety Professional will be selected from top line management or top junior executive ranks rather than from graduates of safety schools. Safety technology and industrial hygiene technology might well end up combined in an academic curriculum for the Type II Safety Professional.

20. Accident frequencies and costs will continue to spiral upward.

NOTE: Numbers 1, 2, 3, 4, 5, 6, 7, 8, 9, 10, 11, 12, 13, 14, and 16 above pertain only to private industry and not to government or the insurance industry. Government will probably remain the same due to a number of constraints. Insurance companies also could well remain the same, unless there is a marked redefinition of their role in safety and they employ the Type II Safety Professional described above as a service to their policyholders.

Part VIII

Appendixes

Appendix A
Safety Sampling

The following material is adapted from my **Techniques of Safety Management.**

SAFETY SAMPLING[127]

One of the newer methods of fixing accountability, using statistical methods, is safety sampling. Safety sampling measures the effectiveness of the line manager's safety activities, but not in terms of accidents. It measures effectiveness before the fact of the accident by taking a periodic reading of how safely the employees are working.

Like all good accountability systems or measurement tools, safety sampling is also an excellent motivational tool, for each line supervisor wants to be sure the employees are working as safely as possible when the sample is taken. To accomplish this, he or she must carry out some safety activities, such as training, supervising, inspecting, and disciplining.

In those organizations that have utilized safety sampling, many report rather a good improvement in their safety record as a result of the improved interest in safety on the part of line supervisors.

Safety sampling is based on the quality control principle of random sampling inspection, which is widely used by inspection departments to determine quality of production output without making 100 percent inspections. For many years industry has used this inspection technique in which a random sampling of a number of objects is carefully inspected to determine the probable quality of the entire production. The degree of accuracy desired dictates the number of random items selected which must be carefully inspected. The greater the number inspected, the greatest the accuracy.

Procedure

I. *Prepare a Code.* The element code list of unsafe practices is the key to safety sampling and supervisor training. This list contains specific unsafe acts which occur in your plant. These are the "accidents about to happen."

    The element code list is developed from the accident record of each plant. In addition, possible causes are also listed. The code is then placed on an observation form (see Exhibit A.1).

II. *Take the Sample.* With the code attached to a clipboard and with a theater counter, the time has come to start sampling.

    The inspector identifies the department and the supervisor responsible. Proceeding through the area and observing every employee who is engaged in some form of activity, the inspector instantaneously records a safe or an unsafe observation of the employee.

DEPARTMENT

SAMPLING WORKSHEET Page 1 of 1 Safe observations Unsafe acts	DC & Service	Maint. power	Tool room	Foundry & pattern	Stock & shipping	Rotor	Shaft	Punch press	Body & frame	Bracket	Small winding	Large winding	Small assembly	Lg. assem. & pck.
(1) Improper lifting														
(2) Carrying heavy load														
(3) Incorrect gripping														
(4) Lifting w/o protective wear														
(5) Reaching to lift														
(6) Lifting and turning														
(7) Lifting and bending														
(8) Improper grinding														
(9) Improper pouring														
(10) Swinging tool toward body														
(11) Improper eye protection														
(12) Improper foot wear														
(13) Loose clothing — moving parts														
(14) No hair net or cap														
(15) Wearing rings														
(16) Finger/hands under dies														
(17) Operating equip. at unsafe speeds														
(18) Foot pedal unguarded														
(19) Failure to use guard														
(20) Guard adjusted improperly														
(21) Climbing on machines														
(22) Reaching into machine														
(23) Standing in front of machine														
(24) Leaning on running machine														
(25) Not using push stick (jigs)														
(26) Failure to use hand tools														
(27) Walking under load														
(28) Leaning — suspended load														
(29) Improper use of compressed air														
(30) Carrying by lead wires														
(31) Table too crowded														
(32) Hands and fingers between metal boxes														
(33) Underground power tools														
(34) Grinding on tool rest														
(35) Careless alum. splash														
(36) One bracket in shaft piling														
(37) Feet under carts or loads														
(38) Pushing carts improperly														
(39) Pulling carts improperly														
(40) Hands or feet outside lift truck														
(41) Loose material under foot														
(42) Improper piling of material														
(43) Unsafe loading of trucks														
(44) Unsafe loading of skids														
(45) Unsafe loading of racks														
(46) Unsafe loading of conveyors														
(47) Using defective equipment														
(48) Using defective tools														
(49) Evidence of horseplay														
(50) Running in area														
(51) Repair moving machines														
(52) No lock-out on machine														
Total unsafe acts Additional unsafe acts														
(53)														
(54)														
(55)														
(56)														
(57)														
(58)														
(59)														
(60)														
(61)														
(62)														

Date_____ Time _____ Sampler _____

Exhibit A.1 Safety sampling worksheet.[127]

Each employee is observed only long enough to make a determination, and once the observation is recorded, it should not be changed. If the observation of the employee indicates safe performance of the job, it is counted on the theater counter. If the employee is observed performing an unsafe practice, a check is made in the column which indicates the type of unsafe practice by the element code number.

III. *Validate the Sample.* The number of observations required to validate is based on a preliminary survey and the degree of accuracy desired. The following data must be recorded on the preliminary survey: total observations and unsafe observations. The percentage of unsafe observations is then calculated. Using this percentage *P* and the desired accuracy, which we will determine as plus or minus 10 percent, we can calculate the number of observations *N* required by using the following formula:

$$N = \frac{4\,(1 - P)}{Y^2\,(P)}$$

where N = total number of observations required
P = percentage of unsafe observations
Y = desired accuracy

(See Exhibit A.2 for a table based on this formula.)

Percentage of unsafe observations	Observations needed	Percentage of unsafe observations	Observations needed
10	3,600	30	935
11	3,240	31	890
12	2,930	32	850
13	2,680	33	810
14	2,460	34	775
15	2,270	35	745
16	2,100	36	710
17	1,950	37	680
18	1,830	38	655
19	1,710	39	625
20	1,600	40	600
21	1,510	41	575
22	1,420	42	550
23	1,340	43	530
24	1,270	44	510
25	1,200	45	490
26	1,140	46	470
27	1,080	47	450
28	1,030	48	425
29	980	50	400

Exhibit A.2 Number of observations needed for 90 percent degree of accuracy.[127]

SAFE PRACTICE SAMPLING REPORT

Plant ___1___   Period covered ___October___

Department Supervisor	Unsafe practice code number													Observations		Per-centage unsafe
	1	2	7	9	11	17	26	34	36	59				Total	Unsafe	
E. Jones - supt.														1,094	39	3.4
Smith-gen. for.														246	9	3.5
Jolas			1				1	1						90	3	3.2
Johnson		3			1		1	1						156.	6	3.8
G. McArthur														226	11	4.6
Mantle				1	1		1	1						101	4	3.8
Williams				1	1									53	2	3.6
Nedstrom					1		1	1	2					72	5	6.5
Mack														284	13	4.4
Peters		1			3	1								96	5	5.0
Sadelri							3	1	1					73	5	6.4
Altert	1		1											64	2	3.0
Anderson	1													51	1	1.9

Exhibit A.3 Safety sampling report.[127]

For example, if the preliminary survey produced the following results—(1) total observations 126 and (2) unsafe operations 32—the percentage of unsafe to total observations would be 32 divided by 126, which is 0.254 or 25 percent. Thus

$$N = \frac{4(1-P)}{Y^2(P)}$$

$$= \frac{4(1-0.25)}{(0.10)^2(0.25)}$$

$$= \frac{3}{0.0025}$$

$$= 1{,}200 \text{ (no. of observations required)}$$

Thus this study must have a minimum of 1,200 observations to give effective results.

IV. *Report to Management.* The results can be presented in many different forms; however, the report should include the following:

1. Total percentage of unsafe activity by department and by shift.
2. Percentage of unsafe activity by supervisor, general foreman, or superintendent.
3. Number and type of unsafe practices observed.
4. Breakdown of types and number of unsafe observations by supervisory responsibility.

For examples of these see Exhibits A.3 to A.6.

Correlation of Results

During 1967, the author collaborated with Paul Mueller, corporate safety director of the Green Giant Company, and Jim Young, safety consultant, Employers Insurance of Wausau, on an experiment using this tool of safety sampling.

This experiment was cut short and did not accomplish its stated objectives, but it did give us an opportunity to study the tool thoroughly and to correlate our findings with other normal indicators of safety performance.

SAFETY SAMPLING REPORT

Plant _____1_____   Month of _____October_____

Department	Total observations	Unsafe observations	Percentage of unsafe activity	
			This month	Previous month
Manufacturing-Prod	442	77	17.4	12.1
Press	1,815	244	15.3	19.7
Assembly	1,699	59	4.0	4.0
Welding	322	70	21.0	11.2
Subtotals	4,278	450	14.4	14.2
Production Eng.	339	55	16.2	21.5
Plant Engineering	341	51	14.9	26.7
Subtotals	680	106	15.6	23.6
Plant totals				

Exhibit A.4 Safety sampling report.[127]

Sufficient samples were taken to ensure validity, and the results were then compared with cost, all-accident rate, number of accidents, and cost per man-hour.

Here are the results:

Correlation of:

Accident claim costs with the safety sample taken . . . . +0.353
Accident claim costs with the all-accident rate . . . . +0.364
Number of accidents with the sample taken . . . . +0.446
Accident cost per man-hour with the sample taken . . . . +0.401
Accident cost per man-hour with the all-accident rate. . . . . . . +0.245

The results are interesting and seemingly significant. Sampling seems to show the same trends as claim costs, number of accidents, and accident cost per man-hour, although it correlates better with the all-accident rate (all reported accidents per 1,000 man-hours).

This seems to mean that sampling provides an excellent indicator of accident problem areas— before the accidents occur. Of course, by far the best value of sampling is motivational. We found that sampling aroused extreme interest in safety where there was little interest before.

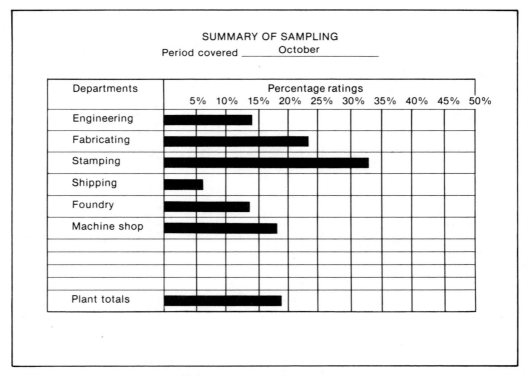

Exhibit A.5 Safety sampling report.[127]

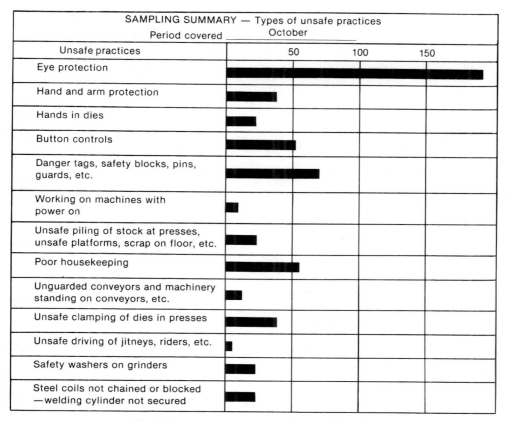

Exhibit A.6 Safety sampling report.[127]

Sampling also seems to create a great deal of interest among supervision, for management knows that sampling, a measurement of worker activity, is perhaps its best indicator yet of what the supervisor is actually doing about safety.

Appendix B
Performance Measurement Systems

The following material is adapted from my Techniques of Safety Management.

A PERFORMANCE MEASUREMENT SYSTEM[127]

It is conceivable that a measurement system could be devised to measure a line manager's safety performance in terms of ability, role perception, and effort. Perhaps through such a system management could decide what achievement level on safety should be expected for a given manager. Such a system of measurement would have to be based on self-rating and/or rating by management. A system of rating the three elements that go into safety performance (in terms of points, percentage points, or whatever seems logical) would help to identify supervisory performance in safety before the fact, i.e., before the accident occurs. Perhaps some equation could be devised:

Effort x ability x role percention = safety performance

It is possible that rating systems could be devised for each of the three factors, resulting in a final number which would be an indicator of a manager's safety performance. Such a rating, to my knowledge, has not been tried; it might be worth experimenting with. For example, we might construct the following rating system.

Effort

Using a rate of 0 to 100 (the actual value would be determined first by the manager), the manager would rate efforts on safety compared with efforts in other supervisory functions. (Quantity of production might rate high, 95, for example, while safety would rate lower, say 40). The boss might make a similar rating, and the two ratings could then be compared.

It is also possible that the value of management's rewards could be rated on a scale from 0 to 10 units. The "effort-rewards probability" could also be rated on a scale ranging from 0 to 10. The product of these two would then become the percentage effort rating. Ratings by both the supervisor and the manager would be valuable.

"Reward" here means much more than just financial reward. It includes all those things that motivate people: recognition, chance for advancement, ability to achieve, and increased pay. In fact, most research today into supervisory motivation indicates that the rewards of advancement and responsibility are the two greatest motivators.

355

Ability

A rating ranging from no ability (0 unit) to maximum ability (1.0 unit) could be made by the supervisor and the manager. This rating might initially be set by a test score; however, it probably would eventually become a high constant.

Role Perception

A rating from no perception or inaccurate perception (0 unit) to completely accurate (1.0 unit) could be assigned by both the supervisor and the manager. This point rating could be based on the supervisor's job description and safety performance standards knowledge.

Supervisor A might rate as follows:
Value of rewards . . . .   5 units
Effort-rewards probability. . . . .   3 units
   Effort rating. . . . .   15 units
   Ability rating . . . .   1.0 unit
   Role perception . . . .   0.8 unit
      Safety performance . . . .   12 units
Supervisor B might rate like this:
Value of rewards. . . . .   7 units
Effort-rewards probability. . . . .   6 units
   Effort rating. . . . .   42 units
   Ability rating . . . .   0.9 unit
   Role perception . . . .   1.0 unit
      Safety performance . . . . .37.8 units

This measurement system has never been applied in practice, and it may or may not be feasible. The purpose in presenting it is not to promote this or any other particular system, but rather to sell the idea that we do need a system of rating supervisory safety performance. We need a system, or several systems, of fixing accountability for safety through measurement of line performance.

SCRAPE[127]

SCRAPE is a systematic method of measuring accident prevention effort. As we have said, most companies are set up to measure accountability through analysis of results. Monthly accident reports at most plants suggest that the supervisor should be judged by the number and cost of accidents that occur under his or her jurisdiction. This is good, but we should also judge line supervisors by what they do to control losses. SCRAPE is a simple way of doing this. It is as simple as deciding what we want supervisors to do and then measuring to see that they do it.

The SCRAPE rate indicates the amount of work done by a supervisor and by the company to prevent accidents in a given period. Its purpose is to provide a tool for management which shows before the accident whether or not positive means are being used regularly to control losses.

The first step in SCRAPE is to determine what specifically we wish the line manager to do in safety. Normally this falls into the categories of (1) making physical inspections of the department, (2) training or coaching the people, (3) investigating accidents, (4) attending management meetings, (5) establishing safety contacts with the workers, and (6) orienting new people.

With SCRAPE, management selects which of these are things it wants supervisors to do and then determines their relative importance by assigning values to each. Let us suppose that management believes the six items above are the things it wants supervisors to do and believes that (1) and (2), inspections and training, are the most immediately important, followed by accident investigations and individual employee contacts, and that attending meetings and orienting new people are relatively less important at this time. Management might then assign these values:

Item	Points
Departmental inspections	25
Training or coaching (e.g., 5-minute safety talks)	25
Accident investigations	20
Individual contacts	20
Meetings	5
Orientation	5
Total	100

Depending on management's desires, the point values can be increased or decreased for each item. They should, however, total 100 points.

Every week each supervisor will fill out a small form (see Exhibit B.1), indicating weekly activity. Management, on the basis of this form, spot checks the quality of the work done in all six areas, and rates the accident prevention effort by assigning points between 0 and the maximum.

For example, in department A the supervisor makes an inspection and makes six corrections. The safety director later inspects, finding good physical conditions. Supervisor A rates the maximum of 25 points.

Department _____ Week of _____ Points _____

(1) Inspection made on _____ # corrections _____ _____
(2) 5-minute safety talk on _____ # present _____ _____
(3) # accidents _____ # investigated _____
          Corrections _____
          _____ _____

(4) Individual contacts:
          Names _____
          _____
          _____
          _____ _____

(5) Management meeting attended on _____ _____
(6) New men (names):   Oriented on (dates):
    _____   _____
    _____   _____
    _____   _____ _____

Exhibit B.1 SCRAPE activity report form.[127]

In department **B** the supervisor made an inspection but no corrections. The plantwide inspection, however, indicates that much improvement is needed. Supervisor **B** might get only 5 points for having done a sloppy inspection.

In department C there were five accidents. Only one individual lost time. Supervisor C turned in only one investigation, and thus gets only 5 points.

In department **D** there are 43 employees, but only 3 were individually contacted during the week. This might also be worth only 5 points.

Management must decide relative values by setting maximum points and must also set the ground rules about how maximum points can be obtained.

Each week a report is issued (see Exhibits B.2 and B.3).

SCRAPE can provide management with information on how the company is performing in accident prevention. It measures safety activity, not a lack of safety. It measures before the accident, not after. Most important, it makes management define what it wants in safety from supervisors and then measures to see that it is achieving what it wants. SCRAPE is a system of activity accountability.

Week of							
Department	Activity						
	Inspect (25)	5-min talks (25)	Acc inv (20)	Ind cont (20)	Meet atten (5)	Orient (5)	Total rate (100)
Average							

Exhibit B.2 SCRAPE weekly report.[127]

Week of							
Department	Activity						
	Inspect (25)	5-min talks (25)	Acc inv (20)	Ind cont (5)	Meet atten (5)	Orient (100)	Total rate
A	25	15	20	15	5	5	85
B	5	10	20	5	5	5	50
C	25	10	5	5	5	5	55
D	15	25	20	20	—	5	85
E	10	5	—	—	5	—	20
F	20	20	15	5	—	—	60
Average	17	14	13	8	3	3	58

Exhibit B.3 SCRAPE weekly report.[127]

SBO

Safety by Objectives is perhaps the best of the current accountability systems. If properly done it provides a valid measure of performance, adequate reward for performance, and great flexibility in determining performance details. See Chapter 24.

Appendix C
Biorhythms

The following article from the Summer, 1972 issue of Family Safety *is a good summary of the concept of Biorhythms. The article is by Paul Dreiske of the National Safety Council.*

ARE THERE STRANGE FORCES IN OUR LIVES?

Startled at seeing the ghost of Hamlet's father, Horatio was given this matter-of-fact explanation by the Prince of Denmark:

"There are more things in heaven and earth, Horatio, than are dreamt of in your philosophy."

Hamlet's classic understatement could well apply today—in view of some recent theories that just might shake up a few notions in the field of accident prevention.

Some curious investigators fiddling with their computers now say that accident-prone conditions can be created by the relative positions of the earth, moon and sun. So what else is new, Grandpa might snort, recalling that no one with a grain of sense planted potatoes unless the phase of the moon was right.

Other explorers of the unknown point the finger of accident fate at a host of influential phenomena such as sunspots, electromagnetism, barometric pressure and even winds.

Startling? To say the least. For no matter how true such claims may be, they still come off sounding like "Safety in Wonderland." And perhaps none of those phenomena are more phenomenal than the theory of *biorhythms.*

If you were born on April 10, 1924, and today happens to be July 5, 1988, take care. Biorhythmic charts indicate you are a reluctant candidate for an accident!

Biorhythm claims we have various biological rhythms or cycles, all of which begin at birth, and all of which can be computed with mathematical accuracy throughout our lives.

Those click-like cycles, biorhythmists say, not only explain our "ups and downs," our "good" days and our "bad" days—but can actually be used to pinpoint precise times when we are most likely to have an accident.

Astrology? Occultism? Metaphysical con game? Not according to the Ohmi Railway Company of Japan.

Ohmi operates more than 700 buses and taxis in heavily populated Kyoto and Osaka. Faced with a wave of costly traffic accidents years ago, the company began storing the biorhythms of its drivers in a computer. At the beginning of each shift, drivers scheduled to have "bad" days were given cards reminding them to be extra careful.

Ohmi safety officials report that in the first biorhythmic year, 1969, accidents dropped 50 percent. The downward trend has continued despite an increase in traffic.

On the other side of the world, similar success is reported by the transportation system of Zurich, Switzerland, which has used biorhythm for years. (The latter's recognition of the human being's internal "clocks" isn't surprising—in a country famous for its man-made time-pieces.)

Other countries, too, are either using—or getting into—biorhythm. According to its proponents, several European airlines credit their outstanding safety records to biorhythm. And in the United States biorhythm is catching on with both government and private industry.

Both have submitted lists of names and birth dates to Biorhythm Computers, Inc., a New York firm long active in the field. Purpose? To chart the "critical" days of personnel.

The firm's president, George S. Thommen, has written a book entitled *Is This Your Day?* (Crown Publishers, Award Books). In it, Thommen cites the following studies:

R. K. Anderson Associates, a safety-consulting company in Rutherford, New Jersey, examined a total of 300 accidents in four factories. Its findings: 70 percent of them happened on biorhythmically critical days.

Thommen has charted accidents and produced computerized forecasts for the Guggenheim Aviation Safety Center at Cornell University, our Armed Forces and many American corporations.

Some safety engineers remain skeptical, however. Accepting the fact that accident rates have been reduced, they ask whether biorhythm had anything to do with it. As one safety man puts it: "If I told a group of employees that I had a way of computing their accident-prone days and reminded them to be extra careful on those days, I would, in fact, be conducting a safety program. And I'd expect an improvement in accident rates even if biorhythm were a lot of baloney."

Whether biorhythms of a fixed duration in days are fact or fiction, there is no question that biological cycles of some sort do exist.

In the introduction to a new report on *Biological Rhythms in Psychiatry and Medicine,* Dr. Bertram S. Brown, director of the U.S. National Institute of Mental Health, writes: "Most of us are dimly aware that we fluctuate in energy, mood, well-being and performance each day, and that there are longer, more subtle behavior alterations each week, each month, season and year. Through studies of biological rhythms, many aspects of human variability—in symptoms of illness, in response to medical treatment, in learning, and in job performance—are being illuminated. . . "

What, exactly, is biorhythm—and how does it work?

In the early 1900's, biorhythmic cycles were discovered independently by two European physicians, Dr. Wilhelm Fliess in Berlin and Dr. Hermann Swoboda in Vienna. Their detailed studies were later expanded by others.

Those doctors revealed the existence of three different life rhythms, each of which begins at birth. The first is a 23-day physical cycle, the second a 28-day sensitivity cycle and the third a 33-day intellectual cycle.

The 23-day physical cycle affects our physical strength, energy, endurance and resistance.

The 28-day sensitivity cycle governs our emotions and the efficiency of our nervous systems.

The 33-day intellectual cycle indicates our changes in intelligence, memory and mental alertness. (See Exhibit C.1.)

During the first half of each cycle, our respective powers are in high gear; during the second or lower half of each curve, our powers are recuperating and recharging like batteries—either physically, emotionally or intellectually, or any combination of the three.

So far so good. But now comes the most important aspect for accident prevention—the so-called "critical" day.

Biorhythmically critical days occur whenever one of our three cycles changes either from high to low or from low to high. Expressed by a curve on a chart, the critical day appears at the point where the curve crosses the zero or median line, either upward or downward (see illustra-

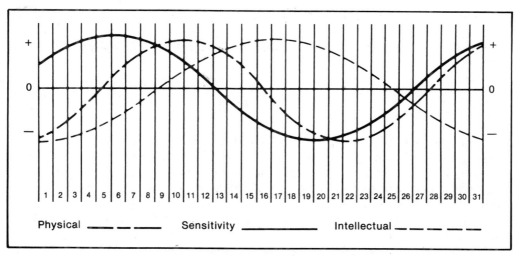

Exhibit C.1

tion). On those days, our systems are in a state of change or flux and, according to biorhythm, we should be extra careful.

In terms of accident potential, biorhythmists say the 33-day intellectual curve isn't too important—not nearly as important as the 23-day physical and 28-day sensitivity cycles.

A critical day can change from bad to worse when two curves cross the median line at the same point. When that happens, we have a "double critical" day and should be doubly cautious. But, perhaps fortunately for us all, those days are relatively rare.

Although biorhythmic statistics show we are 7 to 8 times more accident prone on critical days, there is no cause for undue alarm. Biorhythm wasn't intended to predict the future or spoil our fun, Thommen says. He only regards it as a valuable tool in our everyday lives—an early warning system that can help us avoid accidents by making us more aware of our "inner space."

Do you want to figure out your own biorhythms? If you're mathematically inclined, follow the instructions in the accompanying explanation. If arithmetic isn't your bag, a chart kit and pocket calculator are on the market.

How to Chart Your Biorhythms

All you need are your birth date and some minor mathematical moxie! Let's say you were born on April 10, 1924.

First, figure the number of days you've lived from your birth date to your last birthday, April 10, 1972, by multiplying 365 times your age—48 in this case. Then add a day for each leap year—12 in this hypothetical example.

$$
\begin{array}{rcl}
365 \times 48 & = & 17{,}520 \\
\text{Leap years} & = & \underline{12} \\
& & 17{,}532 \text{ days}
\end{array}
$$

Now let's assume today is July 6, 1972, which means you've lived 48 years plus 87 days, or a total of 17,619 days.

Now to compute your 23-day physical cycle, you simply divide the total days you've lived by 23:

$$\overline{23\,\big|\,17,\!619}\;\;\; 766 \text{ "and I left over"}$$

Or, as of July 6, 1972, you've lived 766 23-day physical cycles plus one day.

Now get yourself some graph paper like that used in the accompanying illustration (or improvise your own). At the bottom, write in the 31 days of July, from left to right.

Again supposing that today is July 6, 1972, mark the number 6. On that day, you've lived through 766 complete 23-day physical cycles *plus one day*.

Now simply backtrack or subtract one day and, on July 5, therefore, you start a new 23-day physical cycle or rhythm.

So on July 5, you draw a 23-day curve, rising then falling above the median line, then—halfway or 11½ days later on July 16—it crosses the median line and zooms down, then up, once again intersecting the median or zero line on a new upward swing on July 28.

All of this means you'll experience "critical" days—physically—on July 5, July 16 and July 28.

For your 28-day sensitivity cycle and 33-day intellectual cycle, repeat for each by dividing your total number of days by 28 and 33, respectively. Then check your arithmetic with the curves below.

Human "periodicity" isn't confined to the 23, 28 and 33-day cycles of biorhythm. Scientific evidence now indicates we are also subject to a daily or "circadian" cycle. Snychronized with the 24-hour rotation of the earth, we sleep and wake and experience clock-like alterations in our strengths, weaknesses and numerous body processes.

Those daily cycles of ours are now being applied to safety procedures in industry, medicine— and modern jet travel.

Traveling across time zones in a jet airliner can seriously upset our daily rhythms and throw our internal clocks out of whack. Airlines have known this for some time and have scheduled their pilots accordingly.

Sun Spots and Moon Phases

Following the Adventures of Safety in Wonderland, one encounters even more arcane arts.

The scientists at Sandia Laboratories in Albuquerque, NM, ordinarily engaged in sophisticated studies for the Atomic Energy Commission, published a report in 1970 entitled: *Intriguing Accident Patterns Plotted Against a Background of Natural Environment Features.* The key word in the title is "Intriguing"—and the study is certainly that. However, the laboratory cautions that the data base in inadequate to draw significant conclusions.

In an investigation of accidents to Sandia employees, the researchers discovered the "intriguing" possibility that accidents might be related to phases of the moon, solar cycles and other natural phenomena.

There seemed to be some evidence "of a heightened accident susceptibility for people during the (moon) phase similar to that in which they were born, and for the lunar phase which is 180 degrees away from that in which they were born."

The scientists then plotted accidents against the 27-day cycle of disturbances in the earth's magnetic field, which is associated with the sun's rotation. They found another possible correlation, and while they were at it, they also detected a link between accidents and sunspot activity.

The Sandia investigation even delved into barometric pressure and concluded that most accidents seem to happen when the barometer is either rising or falling sharply.

Wind Can Do You In

Winds, too, have come under the scrutiny of today's scientists. Throughout history a number of ill winds have blown no good on several continents: Italy's sirocco, France's mistral, California's Santa Ana, the Rocky Mountains' chinook. And the Middle East's khamsin wind is so ill it apparently causes increases in automobile accidents and crime rates.

To help solve the problem, a nine-year study has just been completed by an Israeli group led by Professor Felix Gad Sulman of the Hebrew University's Department of Applied Pharmacology in Jerusalem.

The oppressive khamsin creates a hot, dry atomsphere that builds up an excess of positive ions. The victim's hair becomes electrified and literally stands on end. Accompanying this symptom are migraine, nausea, tension and irritability.

Sulman's staff has sucessfully countered those and other khamsin side effects with drugs and a nifty little machine called the Ionotron. About the size of a small radio, the device produces negative ionization indoors and is bringing relief to countless "overcharged" sufferers.

Hamlet may have been right after all. Perhaps there *are* "more things in heaven and earth" than are dreamed of in our philosophies!

References

1. Aberle, E.: *SBO at Salt River Project,* unpublished paper, 1985.
2. Accident Compensation Corporation: *Employee Alcohol Impairment,* Accident Compensation Board, New Zealand, 1981.
3. Aldis, O.: "Of Pidgeons and Men," *Harvard Business Review,* pp. 59-63, July-August, 1961.
4. American Society of Safety Engineers, "Scope and Functions of the Professional Safety Position," Chicago, 1966.
5. Argyris, C.: *Personality and Organization,* Harper, New York, 1957.
6. Argyris, C.: "The Individual and Organization: Some Problems of Mutual Adjustment," *Administrative Science Quarterly,* pp. 1-24, June 1957.
7. Argyris, C.: *Interpersonal Competence and Organizational Effectiveness,* Irwin-Dorsey, Homewood, IL, 1962.
8. Barenklau, K.: "A Formula for Motivating Top Management," *Environmental Control and Safety Management,* pp. 19-32, November 1971.
9. Barich, B.: "Occupational Health Programs," *Occupational Health and Safety,* September 1985.
10. Barr, A.: *Strategies for Health-Care Cost Containment,* unpublished paper, 1985.
11. Bass, B. M., and V. A. Vaughn: *Training in Industry: The Management of Learning,* Wadsworth, Belmont, CA, 1966.
12. Bennis, W.: "Organization of the Future," *Personnel Administration,* vol. 30, pp. 17-29, 1967.
13. Bennis, W.: in *Visualizing Change,* by Lippitt, G., University Associates, LaJolla, CA, 1973.
14. Benson, H.: *The Relaxation Response,* Morrow, NY, 1975.
15. Berne, E.: *Games People Play,* Grove Press, New York, 1964.
16. Bird, F., and L. Schlesinger: "Safe-Behavior Reinforcement," *Journal of the ASSE,* June 1970.
17. Blake, R., and J. Mouton: *The Managerial Grid,* Gulf, Houston, Texas, 1964.
18. Blanchard, K. and S. Johnson: *The One Minute Manager,* The Berkley Publishing Group, New York, 1981.
19. Bracey, H.: *An Investigation into the Effectiveness of Safety Directors and Influenced by Selected Varaibles,* Louisiana State University, Baton Rouge, 1969.
20. Brody, L.: "Auto Accidents—Physicians Problems?" *Journal of the American Medical Association,* April 1953.
21. Brown, H. V.: "History of Industrial Hygiene: A Review with Special Reference to Silicosis," *American Industrial Hygiene Association Journal,* May-June 1965.
22. Brown, R.: *An Introduction to Applied Behavioral Analysis in Occupational Safety,* unpublished paper, 1976.

23. Brown, R.: *A Systems Approach to Safety Motivation,* unpublished paper, 1976.

24. Bureau of Labor Statistics: *Occupational Injuries and Illnesses in the United States by Industry, 1978,* U.S. Department of Labor, Washington, 1980.

25. Burns, R.: "Participative Safety: A Motivating Factor," *ASTME Vectors,* vol. 5, pp. 19–20, 1969.

26. Carey, H.: "Consultative Supervision," *Nation's Business,* April 1937.

27. Carlson, S.: *Executive Behavior,* Strombergs, Stockholm, 1951.

28. Cohen, H.: "Employee Involvement," *Professional Safety,* June 1983.

29. Cohen, H., R. Smith and S. Cohen: "Characteristics of Successful Safety Programs," *Journal of Safety Research,* 1978.

30. Currie, R.: "Optimizing Safety and System Effectiveness," *National Safety News,* August 1968.

31. Dalton, G.: *Influence and Organizational Change,* unpublished paper, Howard University, 1969.

32. Dalton, M.: *Men Who Manage,* Wiley, New York, 1959.

33. Danielson, L., and N. Maier: "An Evaluation of Two Approaches to Discipline in Industry," *Journal of Applied Psychology,* vol. 50, pp. 319–329, 1956.

34. Drucker, P.: *The Practice of Management,* Harper, 1954.

35. English, W.: "What Does it Take to be an Effective Manager?," *National Safety News,* September 1980.

36. Ferster, C., and M. Perrott: *Behavior Principles,* Appleton-Century-Crofts, New York, 1968.

37. Festinger, L.: *A Theory of Cognitive Dissonance,* Harper and Row, New York, 1957.

38. Forsgren, R.: "A Model of Supportive Work Conditions through Safety Management," *Journal of the ASSE,* February 1970.

39. Frank, K.: "The Plant Manager," *Professional Safety,* 1987.

40. Gausch, J.: "Loss Prevention and Proper Motivation," *Financial Executive,* pp. 28–32, January 1970.

41. Gausch, J.: *Workbook on Stress,* unpublished, 1985.

42. Gausch, J. P.: "Balanced Involvement," Monograph 3, American Society of Safety Engineers, Chicago, 1973.

43. Gellerman, S.: *Motivation and Productivity,* American Management Association, New York, 1963.

44. Georgopoulos, B.: "The Normative Structure of Social Systems: A Study of Organizational Effectiveness," unpublished dissertation, University of Michigan, 1957.

45. Ghiselli, E., and C. Brown: *Personnel and Industrial Psychology,* McGraw-Hill, New York, 1948.

46. Goldiamond, I.: "Stuttering and Fluency as Manipulatable Operant Response Classes," in L. Krasnas and L. Ullman (eds.), *Research in Behavior Modification,* Holt, New York, 1969.

47. Guion, R.: *Personnel Testing,* McGraw-Hill, New York, 1965.

48. Hammer, C.: "Behavior Modification and the Bottom Line," *Organizational Dynamics,* Spring, 1976.

49. Haner, C.: "Motor Vehicle Reports: Who Needs Them," *National Underwriter,* Oct. 6, 1972.

50. Hannaford, E.: *Supervisors Guide to Human Relations,* National Safety Council, 1976.

51. Harris, T.: *I'm OK–You're OK,* Harper, New York, 1971.

52. Haskins, V.: "Effects of Safety Communication Campaigns: Review of Research Evidence," *Journal of Safety Research,* vol. 2, pp. 58–88, 1969.

53. Heath, E.: "Creative Problem Solving: Its Application to Accident Prevention," *Journal of the ASSE,* August 1964.

54. Heinrich, H. W.: *Industrial Accident Prevention,* McGraw-Hill, New York, 1st ed. 1931, 2nd ed. 1959.

55. Herzberg, F.: "Management of Hostility," *Industry Week,* Aug. 24, 1970.

56. Herzberg, F., B. Mausner, and B. Snyderman: *The Motivation to Work,* Wiley, New York, 1959.

57. Herzberg, F., B. Mausner, R. Peterson, and D. Capwell: *Job Attitudes: Review of Research and Opinion,* Psychological Service of Pittsburgh, Pittsburgh, 1957.

58. Herzberg, F.: *Work and the Nature of Man,* World, Cleveland, 1966.

59. Herzberg, F.: "One More Time: How Do You Motivate Employees?" *Harvard Business Review,* January-February 1968.

60. Herzberg, F.: "Managers or Animal Trainers," *Management Review,* pp. 2-15, July 1971.

61. Hovland, C., L. Janis, and H. Kelley: *Communication and Persuasion,* Yale University Press, New Haven, 1953.

62. Hueter, J.: "Creativity—Choice or Chance?" *The Journal of Industrial Engineering, October 1966.*

63. Hughes, C.: "Making Safety a Meaningful Part of Work," *Journal of the ASSE,* pp. 20-25, June 1967.

64. Humble, J.: *Management by Objectives in Action,* McGraw-Hill, New York, 1970.

65. Insurance Institute for Highway Safety, "Safety Belt Ads Have No Effect," *Status Report,* vol. 7, no. 11, 1972.

66. Janus, I., and B. King: "The Influence of Role Playing on Opinion Change," *Journal of Abnormal Social Psychology,* vol. 7, pp. 340-358, 1972.

67. Johnson, J.: "The Safety Organization—Line or Staff?" *Journal of the ASSE,* June 1963.

68. Kaestner, N., E. Warmoth, and W. Syring: "Oregon Study of Advisory Letters: The Effectiveness of Warning Letters in Driving Improvement," *Traffic Safety Research Review,* vol. 11, no. 3, pp. 77-79, 1967.

69. Kahn, R.: "The Prediction of Productivity," *Journal of Social Issues,* vol. 12, no. 2, pp. 41-49, 1956.

70. Katz, D., N. Maccoby, and N. Morse: *Productivity Supervision and Morale in an Office Situation,* Institute for Social Research, Ann Arbor, 1950.

71. Katz, D., N. Maccoby, G. Gurin, and L. Floor: *Productivity, Supervision and Morale Among Railroad Workers,* Institute for Social Research, Ann Arbor, 1951.

72. Keenan, V., W. Kerr, and W. Sherman: "Psychological Climate and Accidents in an Automotive Plant," *Journal of Applied Psychology,* vol. 35, pp. 108-111, 1951.

73. Kelly, J.: *Organizational Behavior,* Irwin-Dorsey, Homewood, IL, 1969.

74. Kelvin, P.: *The Bases of Social Behavior,* Holt, Rinehart and Winston, New York, 1970.

75. Kerr, W.: "Accident Proneness of Factory Departments," *Journal of Applied Psychology,* vol. 34, pp. 167-170, 1950.

76. Ketchum, L.: "Improving the Quality of Working Life—A Case Study," *Proceedings of the Professional Development Conference,* American Society of Safety Engineers, Chicago, 1978.

77. Kibler, R. J., L. L. Backer, and D. T. Miles: *Behavioral Objectives and Instructions,* Allyn and Bacon, Boston, 1970.

78. Kollatt, D., and R. Blackwell: *Direction 1980, Changing Life Styles,* Columbus, Management Horizons, 1972.

79. Konz, J.: unpublished paper.

80. Laner, S., and R. Sell: "An Experimental Study of the Effect of Specially Designed Safety Posters," *Occupational Psychology,* vol. 37, pp. 153–169, 1960.

81. Larson, J.: *The Human Element in Industrial Accident Prevention,* New York University, New York, 1955.

82. Lawrence, P.: "How to Deal with Resistance to Change," *Harvard Business Review,* vol. 32, no. 3, pp. 49–57, 1954.

83. Lennon, R., and G. P. Hollenbeck: "The Impact of Behavioral Science on Business and Industry," *Organizational Dynamics,* Spring 1972.

84. Leventhal, H., and J. Watts: "Sources of Resistance to Fear–Arousing Communications on Smoking and Lung Cancer," *Journal of Persuasion,* vol. 34, pp. 155–175, 1966.

85. Levinson, H.: *The Great Jackass Fallacy,* Harvard Businss, 1973.

86. Levy, S., and S. Greene: *The Effectiveness of Safety Education Materials,* Social Research Inc., Chicago, 1962.

87. Lewin, K.: *Field Theory in Social Science,* Harper, New York, 1951.

88. Likert, R.: *New Patterns of Management,* McGraw-Hill, New York, 1961.

89. Likert, R.: *The Human Organization,* McGraw-Hill, New York, 1967.

90. Lindzay, G., and E. Aronsen: *The Handbook of Social Psychology,* Addison-Wesley, Cambridge, MA, 1968.

91. Luthans, F., D. Lyman, and R. Otteman: *Organizational Behavior Modification (OB Mod),* unpublished, 1972.

92. Luthans, F., and D. White: "Behavior Modification: Application to Manpower Management," *Personnel Administration,* pp. 41–47, July-August 1971.

93. McClelland, D., J. Atkinson, R. Clark, and R. Lowell: *The Achievement Motive,* Appleton-Century-Crofts, New York, 1953.

94. McDonell, M.: *Current Content Issues on Organizational Change-Safety,* unpublished paper, 1982.

95. McFarland, R.: "Human Factors Engineering," *Journal of the ASSE,* pp. 9–20, February 1964.

96. McGehee, W., and P. Thayer: *Training in Business and Industry,* Wiley, New York, 1961.

97. McGregor, D.: "The Human Side of Enterprise," *Management Review,* vol. 46, no. 11, pp. 22–28, 1957.

98. McGregor, D.: *The Human Side of Enterprise,* McGraw-Hill, New York, 1960.

99. McGuire, F.: "A Typology of Accident Proneness," *Behavioral Research on Highway Safety,* p. 32, January 1970.

100. McKay, Q.: "Group Nerves Affect Our Safety Attitudes," *Industrial Supervisor,* February 1970.

101. McKelvey, R., J. Ingen, and M. Peck: "Performance Efficiency and Injury Avoidance as a Function of Positive and Negative Incentives," *Journal of Safety Research,* June 1973.

102. Mager, R.: *Preparing Instructional Objectives,* Fearon, Belmont, CA, 1962.

103. Mager, R., and P. Pipe: *Analyzing Performance Problems or You Really Oughta Wanna,* Fearon, Belmont, CA, 1970.

104. Mager, R.: *Developing an Attitude Toward Learning,* Fearon, Belmont, CA, 1968.

105. Maier, N.: *Psychology in Industry,* Houghton Mifflin, Boston, 1965.

106. Mann, F., and J. Dent: "Appraisals of Supervisors and Attitudes of Their Employees in an Electric Power Company," Institute for Social Research, Inc., Ann Arbor, 1954.

107. Mannello, T.: *Problem Drinking Among Railroad Workers,* University Research Corporation, 1979.

108. Marks, L.: "Guidelines to a Fitness Program," *Occupational Health and Safety,* June 1984.

109. Marten, J. A.: "Large Plant Safety Program Management," *Journal of the ASSE,* May 1963.
110. Matteson, M., and J. Ivancevich: *Controlling Work Stress,* Jossey Bass, San Francisco, 1987.
111. Mayfield, E.: "The Selection Interview: A Reevaluation of Published Research," *Personnel Psychology,* pp. 239-260, 1964.
112. Mintz, A., and M. Blum: "A Re-Examination of the Accident Proneness Concept," *Journal of Applied Psychology,* vol. 33, pp. 195-211, 1949.
113. Murphy, J.: "Is It Skinner or Nothing?" *Training and Development Journal,* pp. 2-8, February 1972.
114. Myers, M.: "Who Are Your Motivated Workers?," *Harvard Business Review,* pp. 73-88, January-February 1964.
115. Myers, M.: "The Managers Role in Motivation," presented at Western Printing & Lithographing Company, 1966.
116. National Industrial Conference Board: *Behavioral Science,* 1969.
117. National Safety Council, "Accident Facts," Chicago, 1973.
118. Newport, D.: "A Review of Training Fundamentals," *Training and Development Journal,* vol. 22, no. 10, 1968.
119. National Institute for Occupational Safety and Health: *Stress Study,* unpublished, 1977.
120. Noble, L.: unpublished paper, 1987.
121. Norem, D.: *Safety Training Practices in Companies with High Versus Low Accident Experience,* unpublished paper, 1976.
122. Odiorne, G.: *Management by Objectives,* Pitman, New York, 1965.
123. Peters, T., and N. Austin: *A Passion for Excellence,* Random House, New York, 1985.
124. Peters, T., and R. Waterman: *In Search of Excellence,* Harper & Row, 1982.
125. Petersen, D.: *Human Error Reduction and Safety Management,* Aloray, Goshen, NY, 1984.
126. Petersen, D.: *SBO–Safety by Objectives,* Aloray, Goshen, NY, 1977.
127. Petersen, D. C.: *Techniques of Safety Management,* McGraw-Hill, New York, 1971.
128. Petersen, D. C.: *A Participative Safety Program at Lozier Corporation,* unpublished, 1971.
129. Pickering, W. D., and G. W. Rogers: "The Anatomy of Teamwork," *The Management Psychologist,* September 1965.
130. Pigors, P., and F. Pigors: "Let's Talk Policy," *Personnel,* July 1950.
131. Piotrowski, Z., and M. Rock: *The Perceptanalytic Executive Scale: A Tool for the Selection of Top Managers,* Grune & Stratton, New York, 1963.
132. Planek, T., G. Driessen, and F. Vilardo: "Industrial Safety Study," *National Safety News,* August 1967.
133. Pollina, V.: "Safety Sampling," *Journal of the ASSE,* August 1962.
134. Pope, W. C., and T. J. Cresswell: "Safety Programs Management," *Journal of the ASSE,* August 1965.
135. Porter, L., and E. Lawler: *Managerial Attitudes and Performance,* Irwin-Dorsey, Homewood, IL, 1968.
136. Powell, R., and J. Schlarter: "Participation Management—A Panacea?," *Academy of Management Journal,* June 1971.
137. Rahe, R.: "Social Stress and Illness Onset," *Journal of Psychosomatic Research,* 1981.
138. Reddin, W. J.: "Making Sense of Management Style Theory," Social Science Systems, 1966.

139. Robertson, L.: *An Epidemiological Investigation of Injuries in a Metal Working Industry,* unpublished paper, 1982.
140. Rubinsky, S., and N. Smith: "Safety Training by Accident Simulation," *Readings in Industrial Accident Prevention,* Petersen, D., Ed., McGraw-Hill, New York, 1980.
141. Safety Council of Nebraska: *An Evaluation of the Relationship Between a Safety Plan and Accident Rates,* unpublished paper, 1983.
142. Sanders, M.: *The Effect of Organizational Climate and Policy on Coal Mine Safety,* Naval Weapons Support Center, 1976.
143. Sayles, L.: *Managerial Behavior,* McGraw-Hill, New York, 1964.
144. Scanlon, B.: "Philosophy and Climate in the Organization," *ASTME Vectors,* vol. 5, pp. 30-35, 1969.
145. Schulzinger, M. S.: *Accident Syndrome,* C. C. Thomas, Springfield, IL, 1956.
146. Schwab, D.: "Why Interview? A Critique," *Personnel Journal,* pp. 126-129, 1969.
147. Shafai-Sahrai, G.: *Determinants of Occupational Industry Experience,* Michigan State University, 1973.
148. Sherif, M.: "Experiments in Group Conflict," *Scientific American,* pp. 54-58, 1956.
149. Sinaiko, H., and E. Buckley: "Human Factors in the Design of Systems," Naval Research Laboratory, Washington, 1957.
150. Sinaiko, H.: *Selected Papers on Human Factors in the Design and Use of Control Systems.* Dover, New York, 1961.
151. Skinner, B.: *Science and Human Behavior,* Macmillan, New York, 1963.
152. Strauss, G.: "Management by Objectives: A Critical View," *Training and Development Journal,* vol. 26, pp. 10-15, 1972.
153. Surry, J.: *Industrial Accident Research,* University of Toronto, 1968.
154. Tarrants, W.: "The Role of Human Factors Engineering in the Control of Industrial Accidents," *Journal of the ASSE,* pp. 9-16, February 1963.
155. *Training Magazine,* "Who, What and Where," October, 1984.
156. Trommler, K.: "Supervisory Safety Training," *Professional Safety,* August 1984.
157. Tweeddale, R. E.: "3-D Executive," *Executive,* January 1967.
158. Ulrich, L., and D. Trumbo: "The Selection Interview, Since 1949," *Psychological Bulletin,* pp. 100-116, 1965.
159. *U.S. News and World Report,* "New Breed of Workers, Sept. 3, 1979.
160. Uslan, S., and H. Adelman: *Research on Accident Reduction by Reinforcement of Safe Behavior,* Human Potential Development Corporation, Canoga Park, CA, 1983.
161. Vernon, H.: *Accidents and Their Prevention,* Macmillan, New York, 1937.
162. Vilardo, F.: "Human Factors Engineering: What Research Found for Safety Men in Pandora's Box," *National Safety News,* August 1968.
163. Vogel, A.: *Why Don't Employees Speak Up,* unpublished paper, 1967.
164. Wagner, R.: "The Employment Interview: A Critical Summary," *Personnel Psychology,* pp. 17-46, 1949.
165. Wald, R., and R. Doty: "The Top Executive—A First-Hand Profile, Skills that Build Executive Success," *Harvard Business Review,* pp. 31-40, September 1965.
166. Wallace, R.: *Business, Industry and the Safety Professional in 1990,* unpublished paper, 1981.
167. Weaver, D. A.: *Strengthening Supervising Skills,* Employers Insurance of Wausau.
168. Weisenberger, B.: *Decreasing Costs through Medical and Wellness Programs,* unpublished paper, 1985.

169. Wesolowski, Z.: "The Future of Bureaucracy," *Personnel Administration,* pp. 32–36, January-February 1971.

170. Wilkinson, B.: *Drug Abuse and the Workplace,* Associated Builders and Contractors, Washington, 1985.

171. Wilson, S.: "The Incentive Approach to Executive Development," *Business Horizons,* April 1972.

172. Woodson, W., and D. Conover: *Human Engineering Guide for Equipment Designers,* University of California Press, Berkeley, 2nd rev. ed., 1965.

173. Wright, M.: *Coors Wellness Program,* A. Coors Company, Golden, CO, 1988.

174. Wright, O.: "Summary of Research on the Selection Interview Since 1964," *Personnel Psychology,* pp. 391–413, 1969.

175. Zenz, C.: "An Ergonomic Checklist," *National Safety News,* August, 1968.

176. Zohar, D.: "Safety Climate in Industrial Organizations: Theoretical and Applied Implications," *Journal of Applied Psychology,* 1980.

Index